D1259439

Human Evolution

Human Evolution

AN INTRODUCTION TO MAN'S ADAPTATIONS

Second Edition

by Bernard G. Campbell

University of California, Los Angeles

ALDINE PUBLISHING COMPANY / *Chicago*

Bernard Campbell has achieved a position of eminence in anthropological circles in a relatively brief period. He is Professor of Anthropology at the University of California, Los Angeles. After taking an honors degree in Zoology, Botany, and Chemistry at Cambridge and a Diploma in Anthropology, Dr. Campbell received a Ph.D. at Cambridge in Primate Taxonomy. He has been a lecturer in Anthropology at Cambridge and Harvard Universities and has taken part in research and teaching in East and South Africa. Among his many contributions to the field of anthropology are Sexual Selection and the Descent of Man *(Aldine, 1972).*

First Edition published 1966
Second Edition published 1974 by

Aldine Publishing Company
529 South Wabash Avenue
Chicago, Illinois 60605

ISBN 0–202–02012–6 cloth
ISBN 0–202–02013–4 paper
Library of Congress Catalog Number 72–140006
Printed in the United States of America

Contents

Preface

IN WRITING THIS BOOK I have attempted to gather under the title of *Human Evolution* a range of matter not normally included in texts of this kind. My object has been to try to expand our view of evolving man from skeletal and dental transformations to something more complete. While I have clearly fallen far short of describing the whole man, I have made a move to stretch the subject into one which requires knowledge of many different fields of biology. Though there are many gaps, I have attempted to define what I consider the proper study of human evolution.

My aim has been to make the book intelligible to readers with no previous knowledge of biology, but they will need a real interest to carry them through a text as dense as this. For those who are more advanced I have added references as well as reading lists so that they may pursue topics that may interest them. If the references appear thin in places, it is a reflection of the small number of papers that follow my own approach and interest.

My main object in quoting references has been to indicate sources of richer data than I have given; I have not attempted to justify my statements by this means or to acknowledge the sources of my ideas. The references do not therefore adequately indicate how much I owe to the work of my colleagues and predecessors: I owe them so much that to acknowledge them all would expand the book considerably. What I have written I do not claim to be original, but rather the freshly digested work of others.

Many of my colleagues have read parts of the text and to these I record my grateful thanks: Theodosius Dobzhansky, Edmund Leach, Ernst Mayr, Kenneth Oakley, Colwyn Trevarthen, and Emile Zuckerkandl. The final typescript has been read and carefully criticized by C. R. Carpenter, F. Clark Howell, Stanley M. Garn, Gabriel W. Lasker, John Pfeiffer, David Pilbeam, Anne Roe, Elwyn Simons, and Sherwood L. Washburn. I have also received an immense amount of help with both text and figures from my publisher, Alexander J. Morin, who has made many invaluable suggestions. That is not to say, of course, that those who have helped me necessarily agree with either my intentions or my methods, but they have not hesitated to give me their own ideas and the benefit of their advice, and I am most grateful to them. Sherwood Washburn started this particular ball rolling, and I thank him, for I have enjoyed the game that ensued.

Preface to the Second Edition

IN THIS NEW EDITION, I have retained and strengthened the original plan of the book. I have brought it up to date, and made substantial additions to a number of chapters, especially 3, 10 and 11.

Eight years ago, I predicted that new fossil evidence of man's evolution would make the first edition out of date within ten years. Discoveries at Olduvai Gorge, and more recently in Northern Kenya and Ethiopia, have borne this out. In particular, the contribution made to our understanding of man's evolution by the Leakey family and their staff in East Africa has been magnificent. The new fossils have enormously enlarged our knowledge of early man and helped us to understand more clearly the variability and varieties of Hominidae. As a result I have been able to revise the classification used in the previous edition. The most striking change will be seen in the interpretation that I have placed on the South African *Australopitheci*, which are now all classified as races of *Australopithecus africanus*. This represents a considerable simplification of the taxonomic framework, which is set out in chapter 3.

Developments in the study of living primates and especially in our knowledge of the chimpanzee, which we owe mainly to the remarkable observations of Jane van Lawick-Goodall, have thrown much light on the evolution of human behavior and especially communication. This new work is frequently cited and has necessitated the revision and extension of certain sections.

Updating the text has caused me to recognize the speed with which this evolutionary approach to understanding mankind has developed. This represents some real progress in knowledge, since I believe that no full comprehension of man and his works is possible without an understanding of the biology of man, set in its evolutionary perspective. Neither our anatomy nor our behavior can be fully understood in terms of the present. Mankind is the product of successive adaptations to a number of different environments, none of which is anything like that of Western Man. An understanding of these past adaptations is therefore an essential prerequisite to an understanding of—let alone a solution to—man's present predicaments.

In both small ways and great I have acted on the constructive suggestions of my critics, and for their help I wish to thank them. It would be impossible here to list all those colleagues and friends who have corresponded with me over the years or discussed matters touched on in the first edition. I am extremely grateful to them all for their generosity.

Acknowledgments

FOR THE PRIVILEGE of including quotations long enough to require special permission, grateful acknowledgment is made to the following publishers: Aldine Publishing Company for permission to quote from A. H. Schultz, in *Social Life of Early Man,* edited by S. L. Washburn (Viking Fund Publications in Anthropology, No. 37, Chicago, 1961); Harper and Brothers for permission to quote from an article by C. J. Herrick in *Science* (1946); Longmans, Green and Company for permission to quote from W. R. Inge, *Outspoken Essays* (London, 1922); Princeton University Press for permission to quote from W. Penfield and L. Roberts, *Speech and Brain Mechanisms* (Princeton, N.J., 1959); the University of Chicago Press for permission to quote from A. I. Hallowell, in *Anthropology Today*, edited by A. L. Kroeber (Chicago, 1953); Yale University Press for permission to quote from L. Z. Freedman and A. Roe, in *Behavior and Evolution*, edited by A. Roe and G. G. Simpson (New Haven, Conn., 1958), and G. G. Simpson, *The Meaning of Evolution* (New Haven, Conn., 1949).

Figures 6.15 and 6.16 have been redrawn from "The Evolution of the Hand," by John Napier, with special permission from W. H. Freeman and Company. Copyright © 1962 by Scientific American, Inc. All rights reserved. Figures 7.20 and 7.21 are here reprinted from *Olduvai Gorge,* Vol. III, by M. D. Leakey, by permission of Cambridge University Press; © Cambridge University Press 1971. Figure 4.11 is reprinted from *Early Hominid Posture and Locomotion,* by J. T. Robinson, by permission of the author and University of Chicago Press; © 1972 University of Chicago Press. Figure 9.8 is reprinted from *Primate Behavior: Field Studies of Monkeys and Apes,* by Irven DeVore, by permission of the author and Holt, Rinehart and Winston; © 1965 Holt, Rinehart and Winston. Figure 11.3 is reprinted from *Sexual Selection and the Descent of Man*, edited by Bernard Campbell, by permission of Aldine Publishing Company; © 1972 Aldine Publishing Cohpany.

For original drawings and for redrawings that involved original work, I have pleasure in thanking Diana Darke, Mrs. E. K. Frankl, Rosemary Powers, and Jo Reed. The frontispiece, "Adam and Eve," by Lucas Cranach the Elder, is used with the kind permission of the Trustees of the Courtauld Institute Galleries of London. The other figures and photographs are cross-referenced to the list of references, and I wish to express my appreciation to these authors and their publishers for allowing me to use their material.

Human Evolution

Introduction: Method and Plan

SCIENCE DOES NOT claim to discover the final truth but only to put forward hypotheses based on the evidence that is available at the time of their presentation. Well-corroborated hypotheses are often treated as facts, and such a fact is that of organic evolution. If a hypothesis is fairly general in its presentation, it is difficult to test, but a detailed hypothesis like that of organic evolution is readily susceptible to disproof. The evidence for evolution is overwhelming, and there is no known fact that either weakens the hypothesis or disproves it.

As part of organic evolution, the phenomenon of human evolution also amounts to a fact, but as yet its detailed path is not known with great certainty. We shall not aim merely at showing that human evolution has occurred, for this has already been well demonstrated (see, for example, Le Gros Clark 1971 and Pilbeam 1972). My intention here is to examine what evidence we have available for the detailed path of human evolution, in an attempt to discover the origin of man. Such a detailed hypothesis as I have here presented is likely to prove a fallible achievement, but its fallibility will not altogether detract from its value. Not only does the presentation of such a hypothesis have a high heuristic value, but it is by the erection and testing of hypotheses that science progresses. They may indeed be tested and found wanting; science demands only that they should be consistent with the available evidence and as far as possible self-consistent.

With the heuristic as well as the scientific value of the exercise in mind, I have attempted to synthesize into a coherent account the evidence now available of the course of human evolution. This evidence is derived from two sources, from fossils and from living animals. When we examine the evidence carefully, however, we realize that neither source supplies directly pertinent facts, but inferences from both sources must be made as to what happened in time past. The evidence is always indirect; inference after inference must be made as to the course of the evolution of man. The whole truth will perhaps never be known, but that does not negate the value of making a hypothetical interpretation of what evidence we can lay our hands on.

Our interpretation of the evidence of human evolution rests on three kinds of inference:

1. A first inference, which lies nearest to the facts, may be obtained from a study of fossil bones and teeth. The evidence that these ancient fragments furnish is not *directly* relevant because we cannot tell whether or not any fossil actually belonged to a human ancestor. Indeed, such a coincidence would be unlikely for any particular fossil individual, but whether it be the case or not can never be known for sure. However, even if such fragments do not lie on the main stem of human evolution but are side branches, they can tell us something of the main stem from which they themselves evolved. Their relevance can of course only be understood in the light of the other evidence that we have at our disposal.

Studies of fossil bones and teeth tell us directly of the skeleton and dentition of the animals of which they were a part. We can, however, make a second inference from their structure, which will tell us more about the animal. We can make deductions about the size and form of the nerves and muscles with which they formed a single functional unit. Muscles leave marks where they are attached to bones, and from such marks we can assess the size of the muscles. At the same time, such parts of the skeleton as the cranium give us considerable evidence of the size and form of the brain and spinal cord.

2. Having built up a picture of at least some part of our fossilized creature, we can now make a further inference as to its whole biology and its way of life. In the first place, knowledge of its body will tell us something of its mode of locomotion (swimming, running, jumping, climbing, burrowing, etc.), from which it is not a long step to infer its environment (marine or fresh water, terrestrial or arboreal). At the same time, supporting evidence derived from a study of the geological context of

the fossil may confirm the presence in the prehistoric times when it lived of seas, lakes, plains, or forests. Another line of inference also begins with the teeth. These may suggest the diet on which the animal fed (herbage, roots, flesh, etc.), and here again we find some indication of how the animal lived. In this way, by rather extensive deductions, we can build up a picture of the whole biology of the extinct animals.

3. But these interpretations are based on one further and essential line of evidence, that of living animals. This third kind of inferred evidence, though rather remote, will prove of inestimable value. If we assume from the fact of evolution that all animals are related, it is reasonable to deduce that those most similar are most recently descended from a common stock. We must therefore compare the anatomy and physiology of living animals—and especially the monkeys and apes—with that of living man. This method can also be used to assess the closeness of relationship of different fossils. We can then assume that fossil bones most similar to any living animal are the remains of an ancestor of that animal, or a near relative of that ancestor. In turn, geological data that indicate the age of the fossil deposit will give us some idea of the succession in which the fossils lived. And we can make a further inference from the study of the comparative anatomy of living animals, for it is possible to predict within certain limits what a common ancestor was like, even in the total absence of fossils. Indeed, we can in that way infer the existence of ancient forms and from them infer their way of life.

In view of the methods of studying human evolution outlined above, it is not surprising that our concern will be mainly with bones and teeth, for they are the only parts preserved as fossils. But that is not as limiting as might be supposed, since the skeleton is the most useful single structure in the body as an indicator of general body form and function. The teeth, in turn, are very valuable in assessing the relationships of animals because their basic form is not affected by the environment during growth. It is, however, a feature of the present study to infer from the bones and teeth—by consideration of their function—the maximum possible amount about the body as a whole and the way of life of the animal. In that way we shall attempt to trace the evolution of the whole human being as a social animal. We shall move from a study of the evolving human body to consider evolving human behavior and human society.

In synthesizing evidence to discover the course of human evolution, we therefore draw on three kinds of data; we study living animals, fossil animals, and their geological age. At one point, we shall be comparing the living primates—man's nearest relatives—with man himself, to gain insight into their differences; at another we shall compare their structure with that of fossil remains of their ancestral relatives. The structure of this approach is shown in the accompanying diagram.

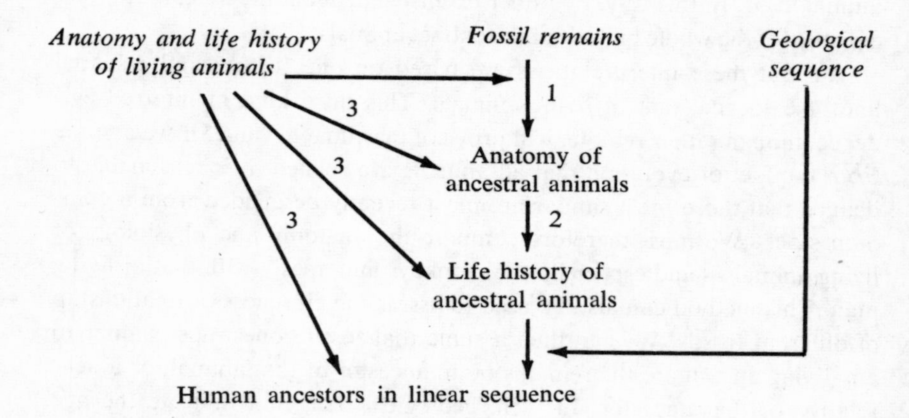

The underlying argument should be clear at any stage. As we shall see in chapter 3, the living primates form a series, from the most primitive to the most advanced, which suggests a general trend of evolutionary development. We shall investigate that trend to see what light it can throw on the origin of man, who is the most advanced of all primates. On the basis of our varied evidence, we shall finally postulate how man may have evolved from his primate ancestor. Such a final inference of the origin and evolution of *Homo sapiens* will, as we have said, be valid yet hypothetical: it will always be subject to revision as new evidence becomes available.

The only way to infer the whole life history of an animal from a few fragments of bone is to investigate the function of those fragments. For example, the form of fossil shoulder and arm bones will be informative in that respect only if we try to find how the muscles were attached to the bones and compare them with those of living animals. We may thus be able to determine the difference in function of our fossil bones from that of the bones of living animals. Was this fossil shoulder joint the type associated with animals that walk quadrupedally upon the ground or with animals that hang by their forelimbs from trees? An answer to that question gives us immense insight into the whole life history of the animal, and such insight is gained from our knowledge of living animals.

This *functional* approach involves rather detailed study of the working of each part of the body—in particular the structural parts—and for that reason the method adopted in this book is to study the evolution of the different parts in turn. Such an approach presents considerable problems, however, since the body is a single complex mechanism, not a collection of discrete mechanisms. Any number of different "functional complexes" can be recognized, but every stage in the subdivision of the animal for descriptive purposes means a loss of truth. We must therefore examine a functional complex in its broadest possible interpretation.

The idea of a functional complex of characters is not new, and it is certainly the most informative and valuable way of analyzing the biology of an organism. A classic study of man's origin along these lines was published in 1916 under the title *Arboreal Man* by F. Wood Jones. According to that approach, therefore, the body is divided not into skull, dentition, trunk, arms, and legs, but into the different parts involved in the different functions, which means, of course, that we must face a considerable overlap of subject matter in different parts of the book. It must also be stressed that the meaning of each great functional complex itself cannot be understood alone; on the contrary, even consideration of the whole physical body as a single unit is ultimately meaningless without consideration of its psychological and social correlates.

Function, in fact, which describes the structure and operations of an organism, is only half the study of biology. We must also be concerned with *behavior,* which describes how the organism interacts with the environment (which is everything other than the organism): how the organism actually receives its sensory input, and how it delivers its motor output. The social and cultural behavior of man, which is the correlate of his complex functions, is in essence no more than an extension and development of the simple behavior of the smallest and simplest protozoa: it is the means of interaction between the organism and the environment that makes possible the function of the individual and the survival of the species.

Our study of human evolution will therefore be set forth as follows: We begin with an introductory chapter on the nature of evolution, followed by a survey of the background to human evolution—that is, mammals and man's nearest relatives, the primates, together with a short review of some fossil evidence (chapters 2 and 3). We then trace in detail the evolution of certain broad functional and behavioral complexes: posture, locomotion, and manipulation (chapters 4, 5, and 6); sense reception, feeding, ecology, and the head (chapters 7 and 8); reproduction and society, communication and culture (chapters 9 and 10). Finally, chapter 11 will review those different evolving complexes in a time sequence, and the last stages of the story of human evolution will there be told.

It should be said that the text does not include a complete account of primate anatomy or taxonomy at even the simplest level; such accounts should be consulted elsewhere (Le Gros Clark 1971, Napier and Napier 1967). Our consideration of anatomy will be limited to features that have evolved in such a way as to differentiate man from other primates and in so doing made man the creature he is. Our aim is at all times to build a picture that shows how primate biology was modified in the course of time into human biology. This study will show how the primate pattern of life associated with the forest was changed into a totally different pattern of life adapted to a different environment, our own.

In such a task, and in an attempt to achieve consistency with recent research, an effort has been made to make this account of human evolution up to date by incorporating new ideas and data (often untested) in their appropriate places. Yet we have attempted to maintain a properly balanced picture; new research has not been included simply because it is new, and the reader may be disappointed to find scant reference to topical and controversial issues. What is of central importance is not the controversial problem of classification of fossil man but the anatomy and ecology that we can learn from fossil bones. If famous sites like Sangiran, Choukoutien, Sterkfontein, and Olduvai, from which our fossil evidence has come, receive only occasional reference, it is because we are not attempting a historical review of the science of paleoanthropology but a study of the paleobiology of man. We must not, however, lose sight of our debt to the men who have discovered the relics of our ancestry, for they make possible the preparation of a book of this kind.

In spite of the contribution of archeologists, geologists, and paleontologists, our knowledge is still small. Crucial gaps in our account appear in every chapter, and we have not overlooked them. What we do not know is as exciting as what we do know, especially at a time like the present, when the subject is developing very fast. Let us hope that in ten years this account of human evolution will be thoroughly superseded.

1

Evolution and Environment

1.1 Organic evolution

When, in 1858, Charles Darwin and Alfred Russel Wallace published their theory of evolution by natural selection, they provided a rational and convincing explanation of the causes as well as the fact of evolution in plants and animals.

These two naturalists both traveled widely and had observed minutely the variation that clearly existed within each species. Members of species, they observed, are not identical but show variation in size, strength, health, fertility, longevity, behavior, and countless other characters. Darwin, in particular, realized that natural variation was used by man in the selective breeding of plants and animals, for, by selection, man would breed only from the particular individuals possessing the qualities desired by the breeder.

In due course, the key to how a similar kind of selection operated in nature to transform wild species of organisms came to both men, and it arose from the same source. The first edition of a book entitled *An Essay on the Principle of Population,* by an English clergyman, T. R. Malthus, appeared as early as 1798, and in that book the author showed how the reproductive potential of mankind was far in excess of the natural resources available to nourish an expanding population. Malthus showed that in practice the size of populations was limited by such lethal factors as disease, famine, and war, and that such factors alone appeared to check what would otherwise be an expanding population.

Both Darwin and Wallace read Malthus' essay independently, and, remarkably enough, both men record in their diaries how they realized (in 1838 and 1858, respectively) that in that book lay the key to understanding the cause of the evolutionary process. It was clear to both that what Malthus had discovered for human populations was true for populations of plants and animals: their reproductive potential was vastly in excess of that necessary to maintain a constant population size. They realized that the individuals that in fact survived must for that reason be in some way better equipped to live in their environment than those which did not survive. Thus it followed that in a natural interbreeding population any variation would most likely be preserved that increased the organism's ability to leave fertile offspring, while the variations that decreased that ability would most likely be eliminated.

The theory that Darwin and Wallace formulated on that basis (at first, independently of each other) may be stated as four propositions and three deductions. Both propositions (P) and deductions (D) have since been well corroborated by careful observation.

P.1. Organisms produce a far greater number of reproductive cells, and, indeed, young individuals, than ever give rise to mature individuals.

P.2. The number of individuals in populations and species remains more or less constant over long periods of time.

D.1. Therefore there must be a high rate of mortality both among reproductive cells and among immature individuals.

P.3. The individuals in a population are not all identical but show variation in all characters, and the individuals that survive by reason of their particular sets of characters will become the parents of the next generation.

D.2. Therefore the characters of those surviving organisms will in some way have made them better adapted to survive in the conditions of their environment.

P.4. Offspring resemble parents closely but not exactly.

D.3. Therefore subsequent generations will maintain and improve on the degree of adaptation realized, by gradual changes in every generation.

As a result of the weight of evidence presented by Darwin in his famous book of 1859, *On The Origin of Species by Means of Natural Se-*

lection, or the Preservation of Favoured Races in the Struggle for Life, biologists became convinced of the value and truth of the theory of evolution that he and Wallace proposed. Since that date, scientists have closely investigated the processes involved in the different propositions and deductions, and geneticists have come to understand the mechanism that accounts for the origin of variation and the transmission of characters (see, for example, Dobzhansky 1962). Of direct interest to students of human evolution are the concepts involved in deductions 2 and 3, for, while the genetic processes of man are no different from those of the rest of the animal kingdom, the selective factors that caused human evolution were unique.

The directing force in evolution is natural selection, and we must satisfy ourselves as to exactly how selective action came to effect the evolution of mankind.

1.2 Natural selection and fitness

IT IS CLEAR that only a proportion of individuals in a population survive long enough to reach maturity and in their turn bear offspring. The environment itself determines the fate of each and, in destroying a proportion, selects the remainder. Through its effect upon each individual the environment controls to a decisive extent the direction and rate of evolution, and for that reason it may be considered to be one creative factor in the process of evolutionary change.

Although natural selection acts on individuals, it is the population that evolves, since the genetic plan of an individual is unalterable and remains constant throughout its life. A novel genetic plan arises only in the production of germ cells (*gametes*) and in the fusion of male and female germ cells in sexual reproduction. Not only are successive generations therefore necessary for the introduction of new gene combinations, but they are in fact the source of variation on which natural selection acts. This is not to deny the existence of evolution among animals and plants that reproduce asexually, but the sources of variation are more limited in them. In this book we shall consider only sexually reproducing animals.

A series of successive generations reproducing sexually relates individuals not only through the dimension of time but also in the dimension of space. Animals must find a mate among their contemporaries, and if they mate more than once in a lifetime (as most of them do), sexual relationships will be spread widely. Thus, the unit of evolution, the breeding population (or Mendelian population), includes all the individuals able to mate with each other. The size of the population may vary, but it is the breeding unit, with its network of sexual relations, that evolves in the course of time.

The *fitness* of such a population requires not only ability to cope with the existing environment and to reproduce but also the potentiality to evolve in the future in response to environmental change. This potentiality requires not only genetic stability (see 1.7), which reflects the broad stability of the environment, but also genetic variability (consequent upon sexual reproduction), which reflects the instability of the environment. That is to say, a population cannot afford to vary greatly in a stable, competitive, and hostile environment, for random variation may be lethal; the population must remain well adapted. At the same time, the population must be able to change, evolve, in adaptation to environmental change. This necessary genetic stability, accompanied by flexibility in the form of adaptability, is the basis of Darwinian fitness, and the balance struck between these two factors determines how fit a population is.

The dynamic stability of all the genetic components of a population (called the *gene pool*) makes possible adaptation to the environment without losing the possibility of modifying such adaptation in the presence of environmental change in the future. The absence of such modification can result in extinction; a proper balance must be found between stability and flexibility, and it is an alteration in the form of this balance that, among other things, characterizes the evolution of man.

It is clear that every gene, every character of the individual, its anatomy, physiology, and psychology, contributes to the biological fitness of the population, and it is in this sense and this sense only that a particular character is of evolutionary significance. There is no reason to suppose that any character can be neutral in this respect. Whatever characters evolve in a population, it is the contribution that they make to the population's fitness that results in their selection, in their survival. It is the population that evolves, not the individual.

At the same time, since all parts of an organism require energy for their maintenance, any part that ceases to have a function will be rapidly lost in the process of evolution. Not only any part but any process will also be lost. As an example, color vision is believed to have been evolved by the reptiles and then lost in the very early period of mammalian evolution; it was evolved a second time in the evolution of the primates, but other mammals cannot see color because it has not been selected during their evolution. Thus we do not often find characters without functions, a fact that may well apply to so-called vestigial characters; it seems probable that they have at least a reduced function.

The function of a character can therefore be understood fully only as an activity that is necessary and contributes to the overall reproductive advantage of the population in which it has evolved. The function of any character that cannot be interpreted in that light cannot be said to be properly understood. It follows that in order to understand the evolution of man it is desirable to consider the function of each new character that was evolved and to discover how it bestowed upon the population in which it became established a greater probability of survival in a changing environment.

How populations have survived by changing their nature is the story of evolution. The concept of fitness involves both adaptedness and adaptability, but, like evolution itself, though it may be elucidated in the past, it can only be surmised in the present.

1.3 Genotype, phenotype, and the environment

THE GENOTYPE of an individual organism is its heredity—the factors that determine the path that development may follow in different environments. The determinants are called *genes,* and they take the form of very complex chemical substances—in particular, that known as *DNA, deoxyribonucleic acid.* This substance, which forms immensely long chain-like molecules, is found in the nucleus (the controlling center) of every living cell. The DNA of the male and female gametes (the sperm and egg), together with some other genetic factors outside the nucleus, when combined in the fertilized egg cell or *zygote,* determine the form and structure of the new individual into which the zygote will develop. From that time onward, as the zygote divides many times, the DNA molecules are copied exactly, and similar copies lie within every cell of the growing individual's body. Once the genotype is determined at fertilization, therefore, it is fixed throughout the life of the individual organism. It is a highly condensed encoded bank of information, which during the development of the zygote determines the kind of organism that the zygote will grow into in a given environment.

The *phenotype* is the whole individual, the manifest characters of an organism, the discrete biological unit, the man himself or the worm. It is formed as a result of the living interaction of the genotype and the environment, an interaction called assimilation or growth. The phenotype is neither constant in form and structure nor permanent. Through its interaction with the environment, the genotype determines not only the phenotype's mature form (its *morphology*) but also its total growth pattern, that is, its form at all ages, together with its growth rate, its capacities, and its life-span. While the genotype is constantly copied in cell division, its form does not change, but the phenotype itself does change as it grows from a fertilized egg into old age.

Variation in the genotype occurs between generations as a result of sexual reproduction. In the process of fertilization, a single set of paternal genes is combined with a single set of maternal genes to produce a novel genotype. With the exception of identical twins, no two individuals of *Homo sapiens* are ever likely to be similar in spite of their vast numbers, so efficient is the shuffling of the pack of tens of thousands of genes that occurs in sexual reproduction. By combining different genotypes in fertilization, sexual reproduction is an essential source of genetic variation in evolution (see Dobzhansky 1962), and the process itself has been selected as a reproductive mechanism for that reason. It is through selective breeding that plant breeders can create such remarkable phenotypes as the hybrid roses we enjoy in our gardens. Created by sexual reproduction, they are in turn preserved by asexual vegetative propagation in which no further variation of the genotype can occur.

But, although sexual shuffling is of great importance, it is not the fundamental source of genetic variation. There is also the phenomenon of so-called "spontaneous" change in the genotype (*mutation*), which results from "imperfections" in its natural and ordered reduplication during cell division. This valuable and inherent "flaw" in genetic replication is most common in the production of the paternal and maternal gametes, when the paired gene sets of the normal body cells are separated. In this process of division, the DNA chains undergo stress, which very often results in chemical modifications. This kind of mutation is random and therefore is usually lethal, for no highly complex mechanism is likely to benefit from random interference.

Thus the mutations that allow survival of the gametes (let alone of the individual) and are subject to natural selection are themselves already selected by the need for internal coordination in the cell. This internal selective process therefore limits the kinds of mutations available to be selected by the environment (Whyte 1965). The nature of organic mechanisms is such that slight changes that might improve adaptedness are possible. Organic characteristics (for example, growth rate, body temperature, size, and so forth) vary continuously (not discontinuously, in jumps), so small adjustments in the mean or average form of these characters may be advantageous at any time in a changing environment. It therefore follows that spontaneous mutation, if of very slight effect on the phenotype, can be of value to an individual, and if it is, it may spread by the aid of sexual reproduction to a whole population over a period of time. But the value of such mutations can be realized for the population only by natural selection, the selective interference of the environment upon the gene pool.

In practice, the environment can interfere with the genotype in three ways, which blend into each other:

1. Random interference from the environment is perhaps the least common. The most obvious example is the radiation that enters the earth's atmosphere from outer space and penetrates organic substance. It may be *mutagenic*; that is to say, it may cause mutations in the genotype of the cell that it penetrates. Should this cell be a gamete or newly fertilized egg cell, the radiation either may kill the cell or, by changing the chemical structure of one or more individual genes that do not lethally affect cell function, may cause an increase in variation in the gene pool. Atomic radiation is powerfully active in this way, and although man-made and localized, its effects on the genotype are random.

2. Most of the other ways that the environment acts upon individuals are more or less selective, and they may be either lethal or sublethal. Lethal interference of a particular kind (for example, famine) may kill some kinds of individuals (for example, thin ones) more readily than others, and for this reason it will change the nature of the gene pool of the population (individuals with reserves of fat are selected). Sublethal interference will also affect the nature of the gene pool if it affects the reproductive capacity of any individuals as a result of a particular character they possess (for example, disease or displacement of the uterus may result in abortion). Such selective interference is enough to result in evolution, even if all the individuals survive; it is only necessary for them to vary in the rate at which they reproduce themselves.

3. Finally, the environment might interfere with a population in a non-selective manner, though it would be hard to prove that any feature of the environment had absolutely no effect on the reproductive capacity of an individual. The environment affects strongly the growth and devel-

opment of individual phenotypes because they show a certain amount of developmental *plasticity* (see 1.5), yet this plasticity is itself sometimes adaptive and as such helps to insure a constant and high rate of reproduction in the face of environmental variation. The effect of the environment upon the gene pool will ultimately depend on the intensity of the interference (or "selection pressure," as it is usually called) and the degree of adaptation developed by the species in response to this pressure.

It should be noted that the individual (and, of course, the population) will alter the environment to some small extent merely by its existence, but, while the genotype cannot change in the face of selective interference, the gene pool can and does. Such change is evolution; failure to change may result in the final extinction of the population.

It must be recognized that as a correlate of the theory of evolution we recognize an element of competition that is invariably associated with organic life. Because organisms reproduce at a rate much higher than that necessary to maintain a stable population, there is always competition within a single population for the available food and space. In this way, every individual changes the environment. Similarly, there is competition among populations of the same species, as well as among species. It is noteworthy, however, that members of a single species are in more immediate competition because they require the same food and habitat, and therefore they alter their own environment more than members of different species that are adapted to different ecological niches (see Birch 1957).

Natural selection resulting from competition within species (intraspecific competition) is therefore as much a cause of evolutionary change as selection resulting from competition between species (interspecific). Intraspecific competition is a permanent feature of organic life, as is natural selection, and a noncompetitive situation cannot exist for more than an extremely short time because the reproductive potential of species is always greater than that which is required to maintain even an expanding population in times of beneficial environmental change. This intraspecific competition means in effect that the evolution of further adaptation may occur even when the overall external environment is fairly constant. And evolution within one population will of course come to change the environment of its neighbors.

It is clear that the total interaction among individuals, populations, and their environment is intricate, and the totality of living plants and animals upon the earth's surface forms a dynamic system of great complexity. This complexity is seen to be immense when we consider further that among animals the connecting link between the environment and the phenotype is the medium of behavior, since what an animal *does,* as well as what it *is,* determines its ability to survive.

1.4 Variability and Speciation

THE SOURCES OF variation which operate in animal populations and which are described above, result in a great variety of forms in any particular population. The fact that every individual has a different genotype (unless a monozygotic twin), and has been subject to slightly different environmental stress, means that every individual, when mature, will be different in a large number of features. Beyond this we can expect variability due to age: babies and mature adults vary not only in size but in bodily proportion, and in the presence and absence of certain features. The two sexes also may vary considerably, not only in their generative organs but in other features as well—the secondary sexual characters, which may affect stature, weight, and even shape to a surprising extent. For example, male gorillas are about twice as large as females, and the proportions of parts of the skull differ between them. In the past, paleontologists have made the understandable mistake of classifying rare fossils of different age and sex as different species or even genera.

Beyond the variation that we can expect within a population, there will also be variation between different populations of a single species. Populations living in different geographical areas will experience slightly different environments, for no two places on the earth's surface are exactly the same in every feature. These different *geographical races,* as they are called, vary in adaptation to the unique features of their differing environments: this is geographical or racial variation. Local races at the opposite extremes of the range of a species may well show very considerable morphological differences. In this case we can only be sure that they belong to a single species by tracing the continuity of populations over the area of land between them. And if a single species is represented, then the continuity of individuals will be associated with a continuity of morphology, which implies that genes are being exchanged between neighboring populations. This exchange of genes between populations is termed *gene flow* and it maintains the integrity of the species.

Should some barrier become interposed between parts of the species' range, then gene flow will be interrupted, and different populations will become geographically isolated. This *geographical isolation* will allow the separated populations to vary independently, perhaps in different ways. If the isolation is complete and remains for a long time, the independent adaptations of the two groups of populations may well result in their morphological divergence to the point where they are very different. If the difference is so extensive that on coming into contact again at a later date, no further gene flow occurs, then the populations will have *speciated*: that is, one species will have split into two. The necessary conditions for speciation are the variability which is present in all species, the operation of natural selection which is equally a feature of all organic life, and geographical isolation. This last feature is the immediate cause of speciation.

The appearance of geographical barriers is not an everyday occurrence. Changes in climate and particularly rainfall may result in the increased size of rivers, lakes, and seas to form impenetrable barriers to certain animals. Tectonic movements may result in new mountain ranges or deep rift valleys with associated changes in climate. Far less extreme changes in climate and environment may result in what amount in effect to geographical barriers, depending on the mobility and behavioral plasticity of the species. But the formation of the barrier is the means by which one species splits into two, which is the way a single evolutionary lineage also splits into two, and so may bring about a whole new evolutionary radiation.

This discussion leads us to the definition of species as *groups of interbreeding natural populations that are reproductively isolated from other such groups* (Mayr 1963). The critical feature of successful speciation is failure in interbreeding—the absence of gene flow and the build-up of morphological divergence. Biological species, or *biospecies* for short, are the groups which fulfill this definition, the species of plants and animals which surround us. The nature and recognition of fossil species, sometimes called *chronospecies*, present some special problems that are discussed in chapter 3.

We have seen in this section that a species is a natural unit that can be defined and recognized in nature generally without much difficulty. Species are often divided into smaller units called *races*. The term "race" is usually used to refer to a group of populations with a certain number of features in common which distinguish them from other such groups. However, since they are not isolated by genetic discontinuity, their boundaries are not easily recognized and the definition of particular races is therefore always a matter of discussion and often of disagreement. Important and large racial groups are sometimes called *sub-species*, while minor ones are called *local races*. Almost without exception, animal and plant species can be divided into these lesser units according to their geographical range and variability. The presence of different races is a typical character of living species and represents their potential for further evolution and speciation.

Finally, species themselves are grouped into units called *genera* (singular *genus*), which are groups of species with major adaptive features in common. Like races, these units are subjective and do not represent easily defined natural categories, but if properly used do represent realities about the species they contain.

Two Latin names or *nomina* are used to label each species; the first is the generic name, the second the species name. Thus *Homo sapiens* is the name given by Linnaeus in 1758 to mankind: it indicates that he belongs to the genus *Homo* and the species *sapiens*. A third name may sometimes be added to label a particular sub-species, as in *Homo sapiens afer*: the name given by Linnaeus to the African subspecies of modern man. A summary of taxonomic practice is given in Table 1.1.

TABLE 1.1. THE USE OF TAXA

TAXONOMIC CATEGORY	*TAXON*	
	Example 1	*Example 2*
Kingdom	Animalia	Animalia
Phylum	Chordata	Chordata
Class	Mammalia	Mammalia
Order	Primates	Primates
Super-Family	Hominoidea	Hominoidea
Family	Hominidae	Pongidae
Genus	*Homo*	*Pan*
Species	*H. sapiens*	*P. gorilla*
Sub-species (geographical race)	*H. sapiens afer*	*P. gorilla berengei*
Race (local race)	Nilo-Hamite	Virunga race

Names of categories above the species begin with a capital letter. Generic, specific and sub-specific names are italicized. Note that Man and Gorilla belong to the same superfamily. For a classification of Primates, see Tables 3.1–3.3.

1.5 Homeostasis of the individual

BEFORE WE EXAMINE FURTHER the interaction of the breeding population and the environment, it is necessary to consider the interaction of the individual and the environment, in an attempt to understand how one is related to the other. An individual organism is a very delicate and complex living system and is clearly very unstable, since it can stand only limited outside interference; a momentary electric shock or shortage of oxygen may destroy the living system entirely. Its survival depends upon its property of self-regulation.

A living organism is a self-regulating dynamic system that maintains a more or less steady state or equilibrium, both within itself and between itself and its environment. Cannon (1932) proposed the term *homeostasis* for the self-regulating property of organic systems. The particular character of an organic homeostatic system arises from the fact that it is an open system that depends for its continued existence on an energy supply obtained from the environment by chemical interchange, yet at the same time must maintain itself as an integrated and discrete mechanism separate from the environment. Furthermore, the system is maintained throughout the growth and development of each individual, and during this period the whole cellular structure is being continually broken down and replaced as it matures. The same body in infancy and old age has little in common beyond its genotype and a dynamic system of a particular kind. Human personality, the continuing identifiable nature of an individual man, is an aspect of this system.

Cannon has described in his book, *The Wisdom of the Body* (1932), how the homeostatic physiological mechanisms of man's body maintain what has been termed his "internal environment" at a constant level. He has described how the body maintains constant (within narrow limits) its water content, salt concentration, level of sugar, fat and protein in the blood, oxygen supply, temperature, and many other features of its organization. The efficiency of these systems is remarkable; for example, man can survive dry heat temperatures up to 128° C. without an increase in body temperature above normal (37° C.). Arctic mammals can similarly survive 35° C. below zero without a drop in body temperature. This particular homeostatic mechanism is, as we shall see, a property of mammals (and birds), and it is of the utmost importance in the story of human evolution. It enables an animal to survive changes in the external environment that otherwise would destroy its delicate chemical systems, changes it could not otherwise survive. Animals with a limited range of homeostatic mechanisms are limited to more or less constant environments; animals with a wide range of homeostatic mechanisms can occupy unstable environments and survive external conditions that in no way approximate their internal environments.

Thus marine organisms live in a relatively constant environment of water, containing salt, food, and oxygen. For deep-sea animals, changes in temperature associated with the seasons are negligible. In the course of the evolution of simple marine creatures into the terrestrial vertebrates, homeostatic mechanisms have been developed that maintain their internal environment so that it will approximate that of their marine ancestors. Every living cell of the body of mammals is bathed in a fluid called lymph. The composition of this watery matrix is kept constant by the diffusion of salt, proteins, sugars, oxygen, and other substances from the blood vessels. Thus we can survive dry heat and cold because the living cells of our bodies are preserved in a saline environment of constant temperature and unvarying composition.

The homeostatic mechanisms of mammals are numerous and complex; they may operate in such varied ways as, for example, sweating or sun tanning (the former lowers the body temperature; the latter lowers the penetration of ultraviolet light). Both these adjustments to hot, sunny weather are methods of coping with environmental change and, as such, increase the chances of survival of their possessors. Such internal adjustments, however, are not the only kind of homeostatic adjustment that is made, for the maintenance of the organism depends on chemical interchange with the environment. Thus a low water content will cause the sensation of thirst and lead to the drinking of water; a low oxygen content will cause an acceleration in the respiratory rate to increase oxygen intake from the atmosphere. Food is required to maintain sugar, fat, and protein levels. The sensations of thirst, breathlessness, and hunger are therefore aspects of the homeostatic mechanisms involved, and so in turn are the behavioral responses necessary to satisfy these needs. If there is not food, water, and oxygen in the environment surrounding the organism (as is the case for a flourishing colony of bacteria), the organism must go to them. Behavior, then, is part of the homeostatic mechanism because it is the process whereby animals satisfy their need to maintain their internal equilibria. Indeed, the maintenance of this steady state is survival, and animal behavior exists to that end.

Physiological homeostatic mechanisms are not the only means by which an individual maintains a steady state in its relationship with the environment. Another kind of homeostatic adjustment is discernible during individual growth and development; it is called, appropriately, *developmental homeostasis* and is a mechanism, as yet not fully understood, whereby the development of the phenotype seems to involve some self-regulating mechanism that maintains its integrity in the face of environmental variation (Mayr 1963, p. 220). Like any self-regulating mechanism, developmental homeostasis is based on a factor of stability (due to the genotype) and a factor of variability, which is the *plasticity* of the phenotype during its development. Examples of such plasticity are found among lower animals and plants more easily than among vertebrates, which have other methods of dealing with environmental change (that is, versatility of behavior). Yet the plasticity of individual vertebrate development is clear. For example, man's physique is affected by the kind of life he leads; muscles enlarge with use or atrophy as a result of disuse; fat deposition depends upon activity and climate as well as upon genes. The digestive processes can become adjusted to different diets, and jaw strength and size respond to the demands made upon the masticatory apparatus by the food. This kind of plasticity is a valuable character, which has evolved like any other and is under genetic control. The mechanisms of developmental homeostasis control the dynamic interaction between the genotype and the environment, so that the phenotype is enabled to survive changes in the environment during its growth.

1.6 Behavior

THE STUDY OF BEHAVIOR refers to what animals do in their interaction with the environment, and it is concerned to a great extent with the activity of locomotion. Broadly, it is concerned with how animals come into contact with their environment—how they breathe, how they touch and move upon the ground, how they eat the portion of the environment that constitutes food, how they escape from the portion that constitutes predators, and how they communicate with and copulate with the portion that constitutes their own species. Behavioral scientists describe the way that an animal is related to the environment, and much of the interaction between the phenotype and the environment occurs through the medium of behavior. Therefore, an understanding of the behavior of an animal is necessary for a full understanding of the interaction of a population and its environment; such interaction must be studied in an analysis of the process of evolution.

How is animal behavior determined? That is a difficult subject, but some broad generalizations may be made as a basis for our discussion of the evolution of human behavior. In the first place, behavior, like morphology, arises as a result of the interaction between the genotype and the environment. The potentialities of all behavior patterns are genetically determined, but in the absence of a suitable environment a behavior pattern may not mature, just as an individual may not. Behavior that is derived from information coded in the genotype, with little contribution from the environment, is often called *innate* and is well exemplified among lower animals such as insects. A most remarkable example of such behavior is found in the honeybee. Since it concerns a visual "language," it is of some interest to students of human evolution and will be referred to again in chapter 10. By a remarkably skillful piece of research conducted over many years, von Frisch (1950) has shown conclusively that a worker bee that has discovered a new source of food will, on returning to the hive, perform a dance on the face of the honeycomb. This dance has been shown to transmit information to other worker bees about the direction, distance, and nature of the food source and is a rare example of a descriptive language among animals. It is significant that it is found among one of the insects with a highly organized society. This descriptive information is in condensed, coded form and accurately enables other bees to find the flowers described. The bees' capacity for this very complex behavior pattern is innate, but, since it records variable information, it is determined in its details by the conditions of the external environment.

There is a great deal of variation among different species of animals as to the proportion of information input (which determines behavior) that comes from the genotype and from the environment. There is an almost equally great variation in this factor within each individual species. The chart in Fig. 1.1 shows in simple form the situation in a species such as man. Although some behaviors are described as innate, none can appear in the total absence of appropriate environmental input. For example, the nipple-searching and sucking reflex of a newborn baby appears as a response to contact with the mother's skin (or the nipple of a bottle). At the other end of the scale, all behaviors are ultimately based on potentials that are coded in the genotype. The ability to play a musical instrument, while learned, is based on a potential not evenly distributed among the population.

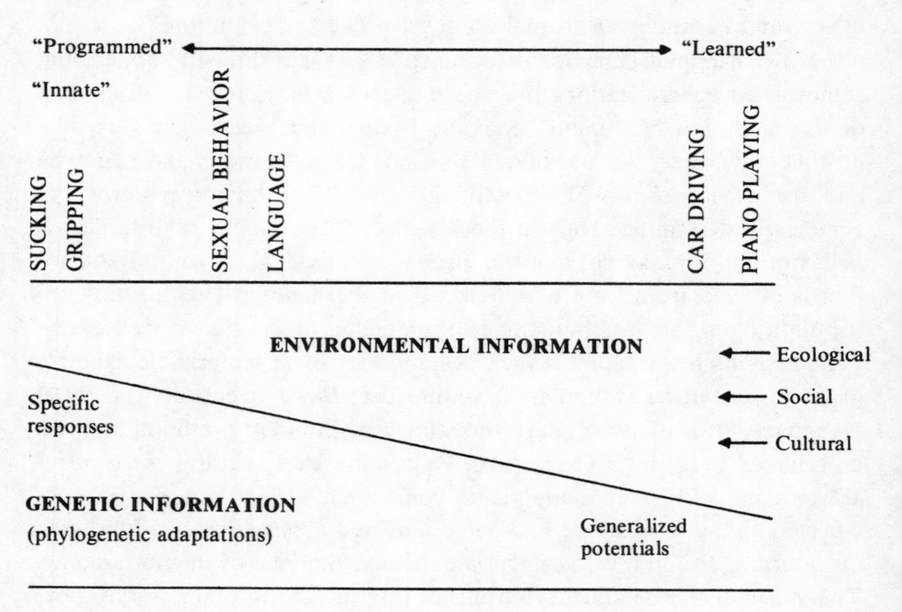

FIG. 1.1. Diagrammatic representation of the contribution of genetic and environmental information to the development of behavior.

Thus if we consider the genetic or innate component only, we see that it can vary from a detailed, programmed motor response (such as sucking) to a very generalized potential such as musical ability. An instance of even more generalized potential is car driving. Here the innate component lies in the physical capability of the individual and the possibility of rapid coordination between eye, hand and foot; as we shall see, all these characters are associated with arboreal primates.

The environmental component also varies. At the left of Fig. 1.1 it merely consists of minimal conditions for life and an appropriate stimulus: in our example, the mother's skin. At the right hand side of the chart, the information input may be not only from the general environment but from other members of the social group, and in this case it may take on the quality of *learning*. However, only a small proportion of the input from the environment can be called learning.

In the case of dancing bees, the environment influences each behavioral activity anew every time it occurs, obviously a necessity in the case of the bee dance in order that a correct report be given of each fresh discovery (see Fig. 1.2, *left*). But among many other animals it is clear that the effect of the environment upon each action may be recorded in some way in the central nervous system so as to influence future action. In the latter case some degree of learning has occurred (see Fig. 1.2, *right*).

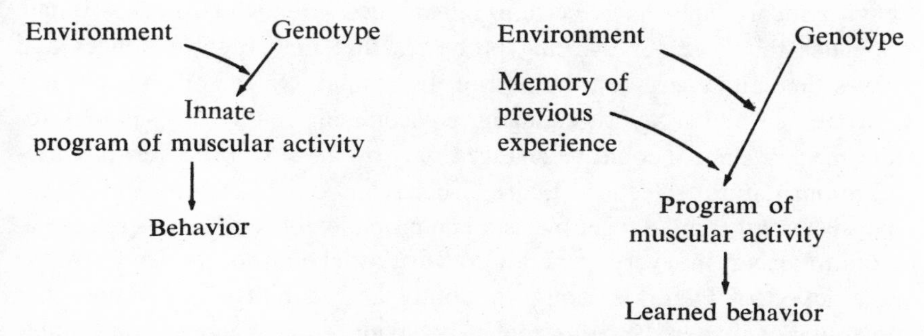

FIG. 1.2. Behavior is always a product of both genotype and environment. The memory of previous experience is not a component in the production of innate behavior (left) but an essential factor in learning (right).

In practice it is impossible to draw a line between innate and learned behavior. Even the lowly flatworms, with a much simpler internal organization than insects, have been shown to be capable of benefiting from experience; but the more complex the animal is, the greater the part that learning appears to play in determining behavior. Among birds, for example, there is wide variation in the relative contribution of learning and inheritance in the development of the mature song repertory. Some birds (the cuckoo is a classic example) develop their songs quite normally when reared away from their own species, while others develop abnormal songs when reared in this way. But, most commonly, broad characteristics are inherited and only the details are learned.

Such data have been gathered by separating birds from their kind, an experiment that cannot be performed easily with mammals, because so much behavior is learned that growth is deeply disturbed in the absence of a mother and it is impossible to isolate the effect of deprivation upon any particular behavior pattern. But, even in the absence of relevant experimental data, it is clear that in nearly all the behavior of the higher mammals there is an element of learning, and it may be very great. Genes supply the capacity for behavior of different kinds, together with some basic components that help form the full behavior pattern; learning makes possible this full pattern of activity. For example, studies of bird song suggest that in many species genes make possible the capacity to sing and, especially, to sing certain notes and trills; the actual pattern of the song is, however, usually learned.

Clearly, behaviors that are primarily determined by information from the genotype and those primarily determined by information from the environment each have certain advantages. Behavior that is innate obviates the need for learning processes that may involve danger and saves time and energy on the part of the animal. On the other hand, it is not flexible and in a fast-changing environment could be disastrous for the species, since it could be changed only by the slow processes of variation and natural selection. Learned behavior, on the other hand, is expensive in time and danger but is much more flexible and allows readaptation to occur in every generation. Just as evolution results from the interaction of factors causing variability and stability, so behavior has this dual character. Heredity makes behavior patterns more or less stable (though subject to evolution), while learning introduces a flexibility that may be of great survival value.

There are three different ways of learning: by trial and error, by observation and imitation, and by instruction. The trial and error method is found in all animals that can learn and is the sole means of learning in invertebrate animals. Compared with other methods, it is lengthy and dangerous. It is the method usually referred to in discussions of learning and in experiments testing learning ability. For example, rats can learn their way through a maze by trial and error (just as humans can) so long as some reward is offered. The rate at which learning can take place by this method has been used as a rough indication of the degree of complexity evolved in the central nervous system (Rensch 1956). Learning by trial and error reaches its highest development in mammals. As a method, it is highly flexible and adaptable to environmental change, and therefore it has been evolved to a remarkable extent among some groups of mammals—particularly among the primates.

Learning by observation and imitation is, however, a speedier and more sophisticated means of building up behavior patterns. Imitation depends in the first place on the ability of an animal to recognize and copy another member of its species, usually its mother. Imitation is a short cut to learned behavior, but it introduces a certain inflexibility in the behavior pattern that is not found in trial and error learning and is therefore more valuable when behavior is concerned with the more constant features of the environment. Rats can learn to pass through mazes more quickly than by trial and error alone, if they are allowed into the maze, during a learning period, with other rats already trained in the art. Many birds learn their song repertories to a great extent by imitation. Monkeys and apes also learn a great deal in that way, and children learn still more (and faster). During the early stages of human development, imitation is the most important factor in developing behavior patterns.

Instruction is, of course, a uniquely human way of creating behavior patterns, for it involves conscious thought and intent. It is the means of learning that is essential to the development of culture yet is dependent upon culture for its existence (see chapter 10).

There is another and very different determinant of behavior that has not yet been mentioned; it may be termed "intelligent thought." It is in all probability a uniquely human characteristic and so is a novelty of great importance in man's evolution. Action determined by intelligent thought may draw on learning but is itself not the result of a learned behavior pattern; it is original. It is discussed in more detail in section 10.4.

With its three determinants—heredity, learning, and intelligence—the study of behavior is highly relevant to the study of human evolution, for the evolution of behavior itself has played a vital role in the evolution of man. In the following sections we shall see more clearly the importance of the less easily detected changes in learning and behavior; we must study them as well as the more obvious changes in anatomy and physiology that characterize our evolution.

The part played by emotion in the generation of behavioral responses is discussed in section 8.6.

1.7 Homeostasis of the population

JUST AS THE physiological and behavioral mechanisms that we have discussed maintain a steady state in each individual organism during its lifetime, so there are mechanisms that maintain the steady state of the whole population over long periods of time, keeping an equilibrium between it and the environment.

The environment is always changing, and, at the same time, natural variation in individuals from generation to generation will tend to cause random changes in the average characteristics of a population. Living populations are homeostatic in the face of both these kinds of variation insofar as they remain in a relatively steady state in spite of constant minor environmental and genetic change. This phenomenon has been termed *genetic homeostasis* (Lerner 1954). Such a mechanism will maintain an equilibrium between the population and the environment and will thus tend to insure the survival of the population. Gross and long-term environmental change will result in a homeostatic readjustment on the part of successive generations of the population.

For example, the fittest mean weight and size of a species in a more or less constant environment will be maintained in the face of individual variation. Natural selection will cut down the reproductive rate of the more extreme variants and will thus operate to maintain this character constant. On the other hand, a gross change in external conditions (such as increased food supply resulting from climatic change) may affect what is the fittest weight, and so a homeostatic adjustment will be made to a new mean weight and size (probably larger) that is homeostatic, in this case not for individuals but for the population. Such an adjustment constitutes evolution and fulfills Dobzhansky's definition (1940) of evolution as a systematic change in the genetic structure of the population (Fig. 1.3). Such a change is termed an *adaptation*—an adaptation to changed environmental conditions. This, then, is evolution by natural selection: shifts in the mean of existing variable characters, called adaptations, yet entirely passive as far as the population is concerned. For example, even one unusual season (an extreme drought or a very hard frost) can, by killing animals and plants, almost instantly cause shifts in the mean of the physiological characters of the populations to which they belonged. New mutations are not necessary for such changes, but variability is essential both for change and for survival.

We see, then, that at the level of both the individual and the population, homeostatic mechanisms are operating to maintain a steady state. This steady state is the means of survival, and the maintenance by evolution of a means of survival is the property of living organisms. From this point of view, the phenotype is relevant only insofar as it secures the survival of the genotype.

Many different kinds of phenotype have in fact survived, and it is of interest to consider whether any one form of survival mechanism is in any way better than another.

BROADLY STABLE ENVIRONMENT — SELECTION PRESSURE CONSTANT

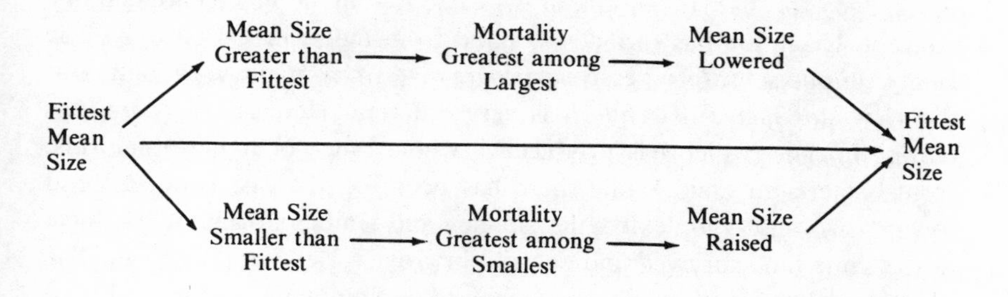

LONG TERM ENVIRONMENTAL CHANGE — SELECTION PRESSURE ALTERS

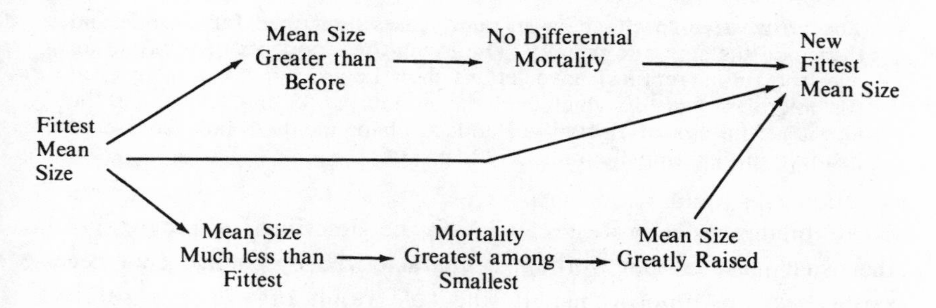

FIG. 1.3. Homeostatic action of natural selection.

1.8 Evolution and progress

THE NATURE OF EVOLUTION has been indicated by the foregoing discussion. *Evolution* is an adjustment in the mean characters of a population as a result of natural selection which results in an increased likelihood of its survival in its particular environment. There is no force operating within individuals that causes evolution, nor are systematic changes in the genetic structure of a population necessarily "advances" on the *status quo*; they are changes which maintain that status. Some changes may prove retrospectively to result in more success than others, depending entirely on the definition of success in a particular context.

The idea of progress and the idea of evolution are, however, not unrelated. The idea of progress was hardly in existence three hundred years ago, for like evolution, it was, above all, a nineteenth-century concept. Evolution has formed mankind from the primeval ooze, so it is clearly in one sense progressive—progressive from man's viewpoint. But the view *sub specie aeternitatis* is very different. Here is a process, operating in time, which has produced a whole range of living organisms, from bacteria to men. While there has been, on the one hand, a trend toward size and complexity, the smaller and simpler forms of life have at the same time survived and evolved in great variety and numbers. The plant kingdom as a whole has achieved a dominant position without the evolution of a nervous system or musculature. Man's greatest competitors in the world today are not other mammals but minute and relatively simple viruses, bacteria and protozoa. As Inge has written:

> The progressive species have in many cases flourished for a while and then paid the supreme penalty. The living dreadnoughts of the Saurian age [the ruling reptiles] have left us their bones, but no progeny. But the microbes, one of which had the honour of killing Alexander the Great at the age of thirty-two, and so changing the whole course of history, survive and flourish [1922, p. 166].

Evolution as a process can hardly be described as progressive in the usual sense of that word; it is characterized by change (not necessarily involving improvement), the sole result of which is survival. It is in that case reasonable and much more valuable to examine the process from man's viewpoint in order to discover whether man is in any sense progressive compared with the rest of the animal world. In what sense is man an improvement upon other animals?

The success of a species can be measured in terms of its numbers, or the gross weight of its *biomass* (the total mass of all of its individual members); it can be judged in terms of its stability in time, its variety of adaptations, or its geographical range. Man is in fact actually increasing his total biomass, at present faster than any other organism, yet that can hardly be called a sign of real progress, since we know it to be both undesirable and dangerous. Other species have had population explosions; so even if man excels in such characters, they do not explain his special evolutionary position. If we look at man himself, we see, for example, that he is very complex (especially his nervous system) and self-conscious (which is probably unique in the organic world). He has also intentionally altered his environment. But again we cannot be sure that it is necessarily progressive to be complex, or to be self-conscious, or to alter the environment. Any of those characters could prove disastrous to their possessor, and they may do so to mankind. But their combination does reflect one single evolutionary trend in which man appears to be progressive: man has an improved homeostatic response to environmental change, or, as Herrick (1946, p. 469) puts it, progressive evolution is "change in the direction of increase in the range and variety of adjustments of the organism to its environment." (See Simpson [1949] for further discussion of evolutionary progress.)

Let us examine the mechanism of homeostasis further in order to discover how such an improvement may have come about. A homeostatic mechanism consists basically of three parts: (1) the receptor, which measures the status of the environmental character that the mechanism controls (for example, the hypothalamus at the base of the brain senses the temperature of the blood, as described in chapter 2); (2) the controller (regulator or mediating mechanism), which acts when the signal from the receptor indicates a need for an adjustment (again, in our example, it is the hypothalamus); and (3) the effector, which the controller causes to correct deficiency or excess (in our example, the effectors are various and cause shivering, sweating, etc., as described in chapter 2). Their interaction can be summarized as shown in Figure 1.4.

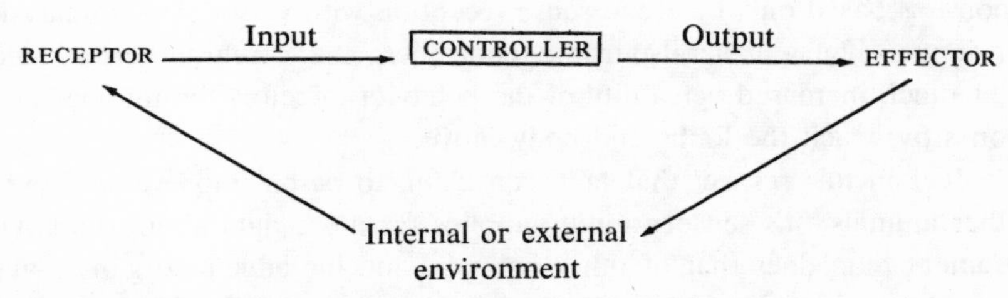

FIG. 1.4. Diagrammatic illustration of a homeostatic circuit.

Receptors that sense the status of the internal environment are of course internal, and they usually are to be found in the brain itself. The survival value of response to changes in the external environment (for example, absence of food) has resulted in the evolution of the sense organs or sense receptors (which in our example can detect food chemically or visually). The evolution of such receptors makes possible response to the external environment, but by themselves the sense organs are not sufficiently accurate as generators of input to allow very precise homeostatic adjustment. For example, the very simple eyespot of certain single-celled animals is able to detect the direction of light sources, but it required not just the eye but the vertebrate visual system as a whole to recognize distinct shapes within the overall visual pattern of the environment. Thus the evolution of the sense organs must be accompanied by the evolution of analytic mechanisms in the brain that can differentiate between different intensities and patterns of stimulus. The evolution of such a refined input mechanism has been accompanied by the evolution of appropriate controllers and effectors.

With the early mammals we find not only the full range of sense organs that we ourselves have, which respond to changes in the external environment, but also the full range of internal homeostatic mechanisms already mentioned (see 1.5) and described by Cannon (1932). But during the more recent stages of human evolution, during the primate stage, we find the development of a more or less new kind of general-purpose controller—the cerebral cortex of the brain—and a rather general-purpose effector—the versatile trunk and limbs, which make possible a wide range of behavior. In this study of human evolution we shall place great emphasis on this new kind of homeostatic mechanism, which is not specific but is generalized and versatile. It takes the outward form of learned and intelligent behavior.

Thus we see in human evolution that increased homeostatic response is based on (1) broad sense reception with very detailed analysis of input, (2) a general-purpose controller, the cerebral cortex, and (3) much increased versatility of the behavior effector, the muscles and bones by which the limbs and body move.

It is in this respect that man can claim to be an improvement over other animals: his sensory input supplies far more data about the environment than does that of other animals, and his effector organs make possible a greater range and greater versatility of behavioral activity. The evolution of the large brain is the necessary concomitant of this extended interaction with the environment.

How do we reconcile the recognition of improvement in human evolution with the definition given at the beginning of this section? The answer is that evolution involves two kinds of adaptation. The first and simplest is an adjustment by existing homeostatic mechanisms (different mean weight, different salt concentration, different coloration). The second and the more complex is an alteration in the functional homeostatic systems themselves. For example, in the evolution of many internal parasites, such as the tapeworm, various functional homeostatic mechanisms have been lost because the present external environment (the body of another animal) is more constant than that in which the tapeworm originally evolved as a free-living organism. On the other hand, the evolution of the temperature-regulating mechanism in primitive mammals was something altogether new and certainly progressive, since it gave mammals improved homeostasis and so a greater independence from environmental change. Since the appearance of mammals, further improvement in homeostasis has arisen from behavioral rather than physiological mechanisms. Man's body in its structure and function is very similar to that of other mammals; his difference from them lies in his behavior and the mechanisms determining it. Such differences are far-reaching because the environment in which man evolved has itself changed drastically.

1.9 Environmental change and evolution rate

ENVIRONMENTAL CHANGE is the prime determinant of homeostatic adjustment in the gene pool of organic populations. Environmental change may take a great variety of different forms, but its ultimate cause is change in the earth's climate. Not only will a long-term climatic change in a geographical region directly affect every single living organism within that region, but every organism will be indirectly affected by the change in every other organism that the change in climate causes. For example, a lowering of rainfall in a particular region will have the effect of changing the vegetation from plants adapted to a relatively high rainfall to plants adapted to a relatively low rainfall. Such a change will immediately affect all herbivorous animals, which may be unable to find enough of their usual food plants. If they survive, such animals will have slowly adapted their behavior and digestive processes to the new food plants that become dominant. In turn, their carnivorous predators will be affected. Change in rainfall may also, for example, affect the transmission and occurrence of parasites, for they may depend on damp conditions to survive the period of transmission between their host animals or plants. Drinking water will perhaps become scarce, and animals may survive only by an adaptation in their water-storage capacity requiring only one visit to a water hole per day. Ultimately, the climate may become subject to occasional extremes of drought—a situation that may be responsible for the extinction of certain species.

It is clear that a single change in one feature of climate may have very complex and far-reaching effects on the whole nature of the environment. Homeostatic adjustment among the different populations of plants and animals may be a lengthy process, for every change in one species will affect the environment and in turn effect change in every other. In stating therefore that the ultimate cause of environmental change is a change in climate, we refer to the whole causal chain of adjustments that result from a change in the prime features of the earth's environment; the level of solar radiation, the nature of the earth's crust and atmosphere, and the earth's rotation, for these features determine climatic change.

Evolution, which is the resultant of genetic variation and environmental change, can therefore occur in the following ways.

1. In a more or less constant environment, a population may evolve better adaptation to it. Such evolution may, however, be limited in extent and will presumably follow a change in the environment.

2. From such a more or less constant environment, a population (usually part of an existing population) may move into a neighboring and different environment to which it may prove to be already to some extent adapted. New adaptations may then follow.

3. The environment may change, and the population may adapt to the change. That is of course the most common situation; it results in situation 1 and very often causes the changes in distribution that result in situation 2. It is, however, worth noting that environmental change may not necessarily result in evolution, since, if the change takes the form of movement of climatic belts on the earth's surface, the fauna may move with the belts, so no new adaptation is necessary. Such a situation would be particularly likely in animals like monkeys or the herbivores of the plains, which are highly mobile and would move within shifting climatic belts without any modification in their way of life.*

* It has been claimed that a change in gene frequency resulting not from selection by the environment but from random genetic variation in small populations (known as *genetic drift*) can result in significant evolutionary change. Evidence for that claim is, however, slight, and the matter does not require further discussion here (Mayr 1963, p. 214). This does not mean, however, that we understand the selective factors that have caused the evolution of every character of living species.

The rate of evolution must in turn be related to its two causative factors, genetic variation and environmental change. Environmental change is the prime determinant of the evolution rate, but in times of fast change the maximum rate will depend upon the amount of genetic variation available from mutation and the sexual shuffling of genes. How is the mutation rate itself determined? It varies considerably among different animals and plants, and it seems most probable that the rate is selected in evolution like any other character. A fast rate will allow fast evolution yet will give a higher ratio of ill-adapted members in a stable population. A low rate will allow well-adapted conformity in a population, yet in times of environmental change a slow rate might inhibit sufficiently fast adaptation to permit survival. Like every other character, the mutation rate of a population is adapted to the environment, and in particular to its overall rate of change. Genetic stability and variability, as we have seen, are important features of biological fitness (1.2).

It is possible that both climatic change and genetic variability may be simultaneously affected by one and the same factor, the level of cosmic radiation. An increase in such radiation could possibly speed up the rate both of climatic change and of genetic mutation. The significance of this suggestion is hard to assess at present, but its importance may well become clear when we understand better the causes of the remarkable climatic changes that have occurred during the earth's history, and in particular those which rapidly changed the climate during the most recent geological period, the Pleistocene, when *Homo sapiens* finally evolved from his more primitive ancestors.

The extent to which a population adapted to a particular environment is termed its degree of *specialization.* As we shall see, some primates are more highly adapted to arboreal life, are more specialized in this respect, than others. Their degree of specialization can be seen as their degree of commitment to their environment: highly specialized species may not survive environmental changes that less specialized species might survive. Specialization in fact seems to cut down the possibility of further evolution, and, in the long view, highly specialized groups can be seen to have been all too often doomed to ultimate extinction when gross changes in the earth's environment have occurred. From the point of view of survival, therefore, a species must remain fairly *generalized* in its overall character; specialization may ultimately involve extinction—and indeed it has usually done so.

There is, however, one remarkable way out of the dilemma, and from our position of hindsight we can see that it has been followed on a number of occasions. It is a process that effects despecialization and is called *neoteny* or *pedomorphosis*. The process operates somewhat as follows: we can suppose that the adult form of an individual is determined by two sets of genes; the first determines the adult characters, and the second determines the rate of growth and time of onset of sexual maturity. Since specializations must be present in the adult stage, they usually appear during growth and are determined by the first set of genes. The second set of genes establishes the permanent adult form by its power to time the onset of sexual maturity. By speeding up the onset of sexual maturity, the adult form may become established before all the specializations can be fully developed. The animal becomes sexually mature when in the young stage; the result is a very remarkable evolutionary change, which is due to a very simple alteration in growth pattern. The phenomenon, discussed by De Beer (1954), has been an important factor in evolution, and perhaps the most famous instance is the evolution of the chordates (the ancestral group of the vertebrates) from the echinoderms (the sea urchins). We have evidence that the primitive chordate evolved from a creature like the larva of a sea urchin that had become sexually mature. There is no doubt that the adult echinoderms are a very unusual and specialized group, which could not possibly evolve into chordates; yet by the process of neoteny that specialization is believed to have been cast off, with the result that there appeared what amounted to an altogether different kind of animal of unspecialized form—the sexually mature echinoderm larva.

This neoteny gives evolving groups a way out of the dead end of overspecialization, and, with limited mutations affecting mainly growth rate and time of maturity, it can result in a saving generalization of form. As might be expected, there is evidence that such a development is a frequent evolutionary adaptation and that it has played some part in the evolution of man himself. While the process in no sense accounts for the evolution of man, nor does it account for all his characters, there is no doubt that many characters considered typically human are present in the early stages of the development of primates and that they have been retained in man by retardation in the growth rate of certain parts of the body. Two examples will suffice: the limited body hair of *Homo sapiens* and the fair skin of certain races may be the result of the arrested development of the hair follicles and the melanin cells, respectively. A list of man's neotenous characters is given by Montagu (1962), who discusses this interesting phenomenon.

If we do not again refer to this phenomenon, it is because it does not relate to human adaptations themselves but only to the developmental and genetic means by which certain changes have occurred. All evolution is due to developmental and genetic changes, and the character of neoteny is of particular interest only because it is a shortcut to drastic change through change in growth rate, without change in form-determining genes. It is also likely that mutations affecting growth rate and the timing of the onset of sexual maturity are less likely to encounter the effects of internal selection (see 1.3; Whyte 1965) than are those affecting structure, because the former threaten less fundamentally the harmony of the whole organism. Though the phenomenon of neoteny gives evolving species a way out of the trap of overspecialization, it does not require us to change our ideas about the action of natural selection and the process of evolution.

1.10 Organism and evolution—a summary

THE ORGANISM can be described as a self-regulating, self-reproducing open system that passes vital information (coded in DNA molecules) to copies of itself. It can be considered to be of two parts: the genotype, which consists of the informational nuclear molecules of DNA within each cell of the individual, and the phenotype, which is the whole individual formed by the genotype through its interaction with the environment.

The genotype interacts with the environment through a causal chain of which the phenotype and its behavior are the connecting links. This interaction maintains the genetic homeostatic equilibrium of the individual and population that is necessary for its survival. The individual phenotype system maintains the physiological and developmental equilibrium of the internal environment for the preservation of the genotype.

Reproduction by the fusion of gametes from different individuals creates in time a network of sexual relations between individuals that constitutes the interbreeding population. In the formation of the gametes, variation arises in the information store of the DNA, and thus gives variability to the population and makes possible changes in it. Both the extent of variation at a given moment (which is ultimately related to the mutation rate) and the long-term alterations in the gene pool (which constitute evolution) are controlled by natural selection, which is the process whereby the reproductive capacity of individuals not well adapted to the environment is depleted by environmental interference. The change in the population brought about by changes in the environment (which constitute selection pressure) can be described as adjustments made in the population that serve to maintain a homeostatic equilibrium between it and the environment. Because living processes are so unstable and delicate, such equilibria must be maintained at all times for the survival of life.

Variability between populations in different geographical regions may lead to speciation if gene flow is interrupted by geographical barriers for a certain period of time. Isolation allows the evolution of morphological and behavioral divergence which may eventually replace the geographical barriers as an effective isolating mechanism so that genetically discrete species are evolved.

Progress in human evolution is seen as the appearance of new homeostatic mechanisms that make possible survival in the face of ever greater environmental variation. The mechanisms themselves increase in turn the complexity of the genotype and phenotype. Progressive evolution in any but an anthropocentric sense cannot be recognized. Evolution is merely a homeostatic adjustment in the mean or average characters of a living system in response to environmental change and can be the outcome only of such adjustment and nothing more. By a study of the anatomy, physiology, behavior, and environment of living organisms related to man, it is possible to learn how human nature was formed and to attempt to understand how each character of our species serves its survival.

SUGGESTIONS FOR FURTHER READING

There is a vast literature on all aspects of evolution, although the concept of homeostasis has not been greatly stressed until recently. For general reading see G. G. Simpson, *The Major Features of Evolution* (New York: Columbia University Press, 1953). For a more technical treatment see E. Mayr, *Animal Species and Evolution* (Cambridge, Mass.: Harvard University Press; London: Oxford University Press, 1963). For a treatment of genetics and the general biology of human evolution see T.

Dobzhansky, *Evolution, Genetics and Man* (New York: John Wiley & Sons; London: Chapman & Hall, 1955), and *Mankind Evolving* (New Haven, Conn., and London: Yale University Press, 1962). For a treatment of classification and nomenclature see G. G. Simpson, *Principles of Animal Taxonomy* (New York: Columbia University Press, 1961). For studies of the relationship of behavior and evolution see A. Roe and G. G. Simpson, *Behavior and Evolution* (New Haven, Conn., and London: Yale University Press, 1958).

2

Progress in Homeostasis

2.1 *The origin and evolution of mammals*

IN TRACING THE origin of man it is not our concern here to describe in detail the part of his evolutionary history that he shares with other mammals. Our interest lies rather in the most recent stage of his evolution, in which his own peculiar nature was evolved. It would, however, be a serious omission to neglect to mention the features that, although shared with other mammals, are the absolutely necessary basis for the evolution of an organism as complex as a human being. The evolution of man cannot be fully understood without some consideration of the main features of his class, the Mammalia.

Mammals are separated by zoologists into three subclasses, the monotremes (such as the platypus), the marsupials (such as the kangaroo and opossum), and the placentals. The first two groups are limited today both in species and in range, and, although they may have had a period of temporary dominance in the past, they were very soon superseded by the *placental mammals*—the group to which man belongs (see Table 1.1). Although what follows applies in part to the two primitive subclasses, we shall concern ourselves from now on with the great *radiation*—the evolutionary divergence and dispersal—of the placental mammals.

While the most important characters of mammals are poorly recorded in the fossil record, consisting as they do for the most part of soft tissues, the fact that certain characters are common to such a large and varied group of organisms suggests that they were evolved by a common ancestral lineage. Paleontologists see the origin of the mammals from reptiles of the Triassic period, and, by Cretaceous times, the main orders of the placental mammals were established (Table 2.1 and Fig. 2.1).

Mammals are characterized by evolutionary novelties of the utmost importance to the story of human evolution. The characters that relate mammals to each other and separate them from other groups, such as birds and reptiles, are very numerous indeed, yet they arise from the evolution of only four great complexes, all of a homeostatic nature. These four complexes can be seen in turn to contribute to a single adaptive development, which made it possible for this most remarkable class of animals to succeed the mighty reptiles that had previously dominated the earth, yet which shared with the mammals a common origin. This all-important adaptation was the ability for constant and lively activity: the maintenance of a steady level of activity in the face of basic changes in the external environment. From being dependent, like reptiles, upon the external temperature with its daily and seasonal rhythms, mammals were enabled by the evolution of new homeostatic mechanisms to remain active at any time.

The four adaptive complexes that made this constant activity possible will be considered in turn: homoiothermy, mastication, improved reproductive economy, and new ways of determining behavior.

2.2 *Homoiothermy*

THE CHEMICAL REACTIONS that underlie homeostasis and constitute the process of living, like any other chemical reactions, can generally occur more quickly at a higher than a lower temperature, but such a temperature should not be so high as to cause breakdown in the relatively unstable organic molecules. At the same time, a constant body temperature makes possible the regularity of the highly complex chemical reactions that underlie the activity of higher organisms, together with the highly elaborate patterns of activity in the brain, which make possible the very rapid and precise control of motor activity and the storage of an extensive memory. The evolution, perhaps dating from 150 million years ago, of *homoiothermy*—the phenomenon of constant and appropriate body temperature—enabled the early mammals to maintain a high level of activity at times when reptiles and amphibians became sluggish (at night and in cold weather). During the Cretaceous period an animal had for the first time gained a tactical advantage over the great ruling reptiles of that age such as dinosaurs. At the same time, this animal could move north from the tropical regions to which terrestial life was then more or less limited and could exploit the vast forests of the temperate zones.

The maintenance of body temperature, at first approximate but later exact, involved a complex regulating mechanism under control of the *hypothalamus*, a small structure at the base of the brain (see Fig. 8.11).

TABLE 2.1 EVENTS IN HUMAN EVOLUTION APPROXIMATELY DATED

Stratigraphical Phases of the Earth's History		*MILLIONS OF YEARS* Extent	*Beginning Before Present (B.P.)*	*Event*
Quaternary	Pleistocene	2	2	Rise of man
Tertiary	Pliocene	10	12	Rise of hominids
	Miocene	13	25	
	Oligocene	10	35	Rise of monkeys and apes
	Eocene	23	58	
	Paleocene	5	63	Rise of placental mammals including primates
Mesozoic	Cretaceous	72	135	First placental mammals; extinction of ruling reptiles
	Jurassic	46	181	First mammals, the pantotheres
	Triassic	49	230	
Paleozoic	Permian	50	280	
	Carboniferous	65	345	First reptiles
	Devonian	60	405	First amphibians
	Silurian	20	425	
	Ordovician	75	500	
	Cambrian	?100	?600	First fossilized animals

As has been explained (see 1.5), the hypothalamus is the thermostat of the mammalian body, and it brings into action the effectors of temperature regulation when the blood departs from that set point to which the species has become adapted. Temperature regulation involves physiological and behavioral adjustments of many different kinds, some of which are listed below.

To lower the temperature of the body during hot weather or great exertion, the usual rate of heat loss is increased as follows:

1. An increase in effective surface area is effected by spreading the limbs (and the ears in some mammals); increasing the ratio of surface area to volume increases heat loss.

2. Fur or hair, which has been evolved to form an insulating layer, is laid flat upon the skin, to reduce its effective thickness.

3. Respiration rate increases to allow the loss of heat by the vaporization of water in the lungs, throat, and mouth. Panting is most commonly seen in dogs, though it is a human character as well in times of great overheating.

4. The *capillaries* (small blood vessels) under the skin dilate and so increase the circulation of the blood near the surface; blood supply to the internal organs is reduced. Heat loss through the skin is thus increased.

5. Sweat glands produce water on the skin, which vaporizes and so lowers skin temperature. Blood temperature is in turn lowered.

To raise the body temperature, the usual level of heat loss is decreased as follows:

1. A reduction is made in effective surface area by curling up into a ball; a sphere is the shape with the minimum ratio of area to volume.

2. The fur is raised to form a thick insulating layer.

3. The capillaries in the skin are constricted and reduce circulation near the surface of the body.

There are two positive reactions to a lowered body temperature: the generation of body heat is increased by an increased rate of metabolic oxidation of foodstuffs and involuntary muscular activity in the form of shivering.

The maintenance of constant body temperature can thus be seen to be made possible by the evolution of certain new organs, common to mammals. The most important are the fur or hair and its *follicles,* with their erector muscles, which raise and lower the animal's coat, together

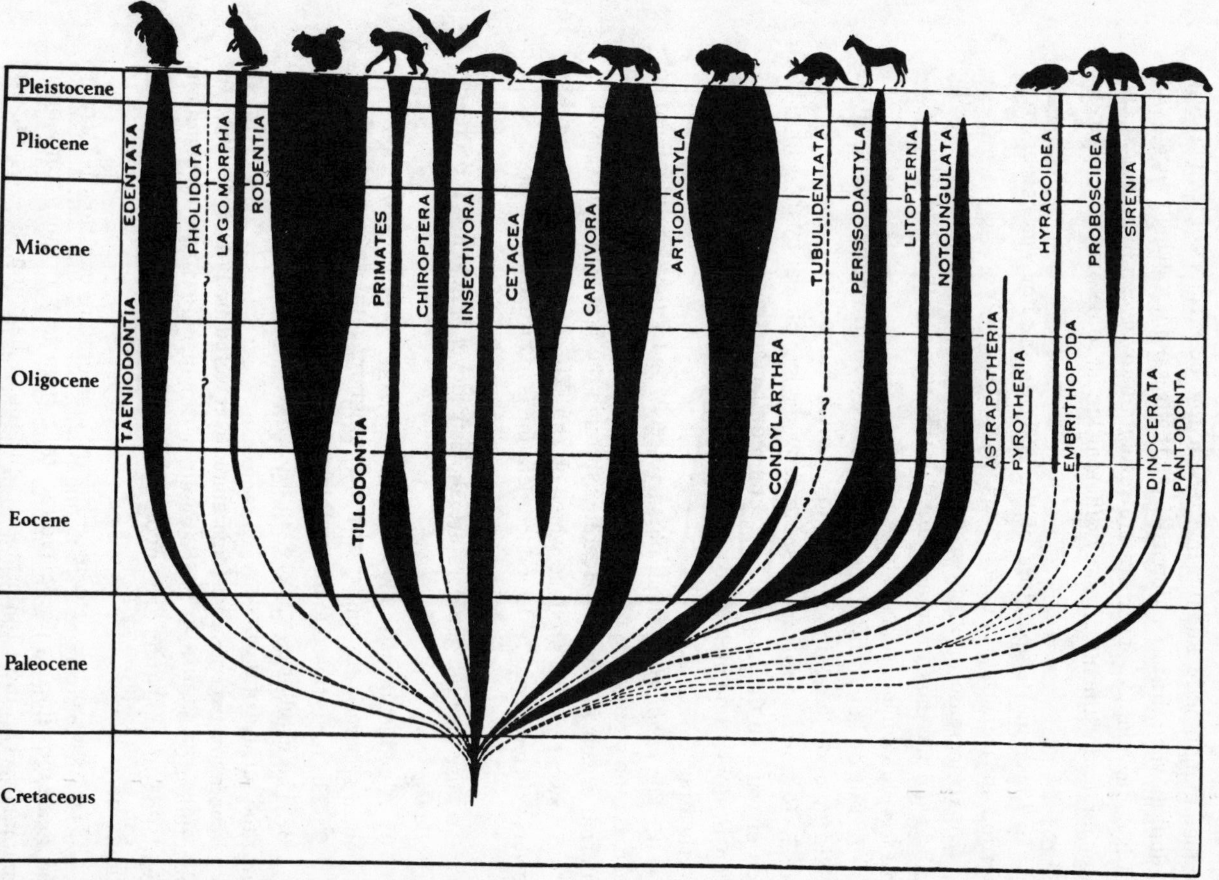

FIG. 2.1 Diagrammatic representation of the evolution of the placental mammals. (After Romer 1945.) See Table

with the sweat glands in the skin. Mammals are also the bearers of a layer of subcutaneous fat below their outer layer of skin, which helps to insulate them from the external environment and acts as a food store. The whole thermostatic adjustment is effected by the novel functioning of the hypothalamus as a heat receptor: both it and the eye (the light receptor) develop from the same tissues at the base of the brain. Finally, mammals have a behavioral response to temperature change in the skin, for they will seek out a warm place during cold weather and a cool, shady spot in the heat of the day. The effectiveness of the whole mechanism has increased during the evolution of mammals, and in Figure 2.2 we can get

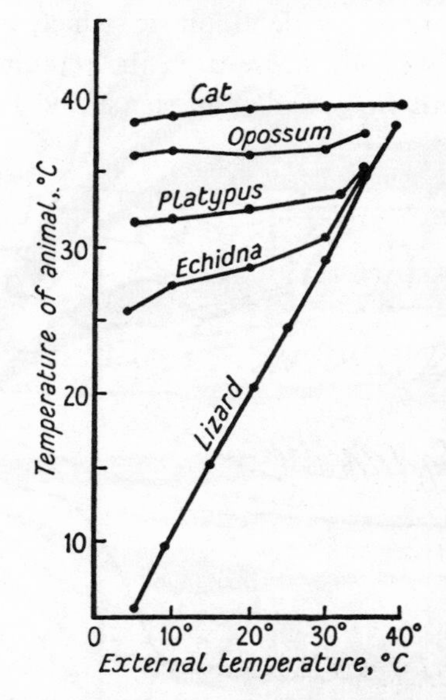

FIG. 2.2. The evolution of homoiothermy is demonstrated in the diagram. Shown are changes in body temperature of different animals during changes in the external temperature. The body temperature of the lizard varies with the environment; the three primitive mammals are some way toward a controlled body temperature, which is almost constant in the cat. (From Martin 1902.)

some idea of this progress by noting the response of different animals to change in the external temperature. Although in some mammals and a few primates the regulation of body temperature is subject to activity rhythms (see DeVore 1965, p. 293), this pattern is superimposed on the homoiothermic mechanism.

2.3 Mastication and heterodontism

THE MAINTENANCE of a constant body temperature requires a reliable source of food—as fuel—to develop not only muscular power but also heat, which is carried round the body by both the deep and the superficial blood vessels. Mammals therefore need a readily available food supply not altogether dependent on season. The need for such a food supply involved the evolution of a new kind of jaw function and tooth structure (*dentition*) to make possible a more complete exploitation of the environment.

The mammal's *heterodont* dentition, in which different teeth have different shapes and functions, arose from the reptilian *homodont* dentition, in which the teeth were similar in shape and function (Fig. 2.3).

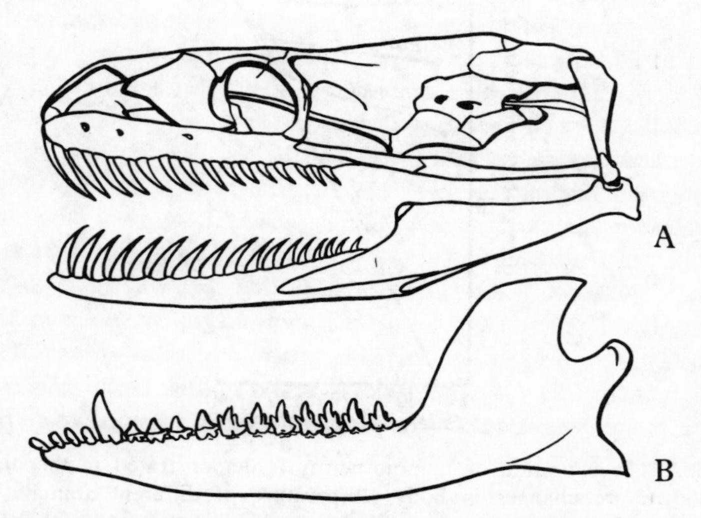

FIG. 2.3. Skull of python (a reptile) (*A*) and lower jawbone of primitive Jurassic mammal (*Amphitherium*) (*B*) drawn the same size. Note that in the python all teeth are identical in shape and function, whereas in the mammal the teeth are differentiated into incisors, canines, premolars, and molars. (Python after Romer 1956; *Amphitherium* after Simpson 1935.)

The reptile jaw and homodont dentition acted mainly as a trap, and food was swallowed whole, as is commonly seen in snakes and alligators. The teeth were often incurved and served only to grasp the prey or, in the case of herbivores, to tear the foliage. In contrast, the mammal jaw and heterodont dentition achieved much more. The jaw was strengthened and different teeth evolved specially for chewing, grinding, and cutting, with the result that the entrance to the alimentary canal served a more valuable purpose than the mere trapping of food. In particular, more kinds of food could be exploited, and the effectiveness of the digestive juices were increased by mastication (as well as by the higher body temperature). A new digestive enzyme (*ptyalin*) was introduced to the food in the mouth as saliva, so that starch digestion could begin during mastication, even before the food was swallowed. The heterodont dentition has evolved within the different orders of mammals in a remarkable variety of ways, and not least in the primates, though in that group the more extreme specializations (such as tusks) have been avoided. A reversal to a homodont condition can be seen in toothed whales, while in other members of the Cetacea, such as the whalebone whales, the teeth have been altogether lost.

The evolution of mastication and the resulting heterodontism (to be considered again in chapter 7) is of great importance to mammalian evolution. By it, mammals were enabled to exploit food resources that were not available to their reptile predecessors. The teeth of mammals were able to release nutriment from food that the digestive juices alone could not penetrate. Just as the gizzard performed this function among birds, the heterodont dentition released to mammals the full food value locked up within the hard outer skeleton of insects and in plant-food storage organs (such as nuts and other seeds). The densely packed carbohydrates and fats of seeds, the sugars and starches of underground tubers and roots, and all the plant proteins locked up in tough vegetation were released to mammals by the chewing and grinding action of their teeth, thus making possible the maintenance of constant body temperature and more or less continuous activity.

Another development, of less importance, that affected the dentition was the loss of continuous tooth replacement and its evolution into a sequence of only two successive series of teeth during the growth of the individual: the so-called "milk" teeth and the later permanent dentition. Changes in diet and tooth action probably reduced tooth loss, which in turn removed the need for perpetual tooth replacement. The milk teeth or deciduous dentition, make possible mastication when the jaws are still small and unable to carry the permanent dentition. The permanent dentition erupts during the second half of the growth period and forms an extremely strong and powerful masticatory apparatus.

2.4 Reproductive economy: prenatal

THE EARLY REPTILES had become fully adapted land animals by the evolution of internal fertilization by copulation, thus removing dependence on fresh or salt water for the external fertilization of their eggs. The reptiles had also evolved the *amniote* egg—an egg for dry land, sealed against dehydration by a shell and membranes, which allowed the growing embryo to develop in its own little aqueous environment. Reptiles, however, maintained their numbers, like more primitive animals, by the mass production of eggs rather than by the successful establishment of small numbers of offspring. Their numerous eggs were usually hidden and left to hatch alone, and the young animals had to fend for themselves from that moment. Both eggs and young were easy prey to small carnivores, especially the little mammals that were then evolving.

These evolving mammals developed a quite new kind of reproductive system, which, by cutting down losses of eggs and young, improved their overall reproductive economy. It consisted of two basic developments, each involving a whole range of new physiological and anatomical characters. The first was *viviparity* (which has occurred as a parallel adaptation in some other classes)—the retention and nourishment of the fertilized egg within the mother until a quite advanced stage of growth. The second, to be considered in the following section, was the evolution of parental care of the young after birth, for their nourishment and protection.

While such behavior as feeding affects the future of every individual separately, the reproductive activity of animals affects in particular the future generations rather than the fate of the individuals involved. In this sense, reproductive mechanisms are above all the property of the species rather than the individual. As a result, reproductive behavior tends to be innate and stereotyped and much less subject to versatility, for too much variation in reproductive behavior might endanger the survival of the species. The first requirement of biological fitness is the maintenance of the reproductive rate by regular and successful reproductive activity.

Partly for historical biological reasons and partly for the reasons discussed above, we find that the reproductive processes of mammals are primarily under control of the *endocrine* system rather than the nervous system. The endocrine system was early evolved as a method of communication between the cells and organs of an animal's body. It operates by means, not of nerves, but of *hormones,* sometimes described as "chemical messengers," which are chemical substances emitted by one group of cells and carried in the blood stream and by diffusion around the body. According to their chemical structure, they evoke a specific response in another group of cells, tuned, as it were, to receive them. In mammals the endocrine system operates in cooperation with the nervous system: the former tends to effect response in long-term life processes, such as the rate of basic metabolism and the seasonal activity of the reproductive system; the latter effects more immediate changes in the body, such as behavioral responses in general.

The part of the endocrine system that controls seasonal egg-laying and copulation in reptiles and amphibians has in mammals been expanded to control a regular cycle of sexual activity throughout the mating season and, in some species, throughout the whole year. The cycle, known as the *estrous cycle,* arises with the evolution of viviparity, and is part of the mammal's reproductive adaptations. It results from the need to follow up the growth and production of the egg or eggs (*ovulation*) within the *follicles* (egg-producing bodies) of one or both *ovaries*, by changes in the walls of the ducts through which they pass (the *oviducts*, which in higher mammals become the *fallopian tubes* and *uterus* [Fig. 2.4]). These changes are such as to allow the *implantation* of the egg or eggs after their fertilization. Implantation is the means by which the developing egg becomes attached to the specially modified wall of the uterus (*endometrium*); the egg derives nourishment for its growth from the maternal blood vessels within the endometrium. If fertilization occurs, it is followed by implantation and *gestation* (pregnancy). If not, the changes in the wall of the uterus are arrested, and new eggs develop in the ovary, later to be shed into the uterus. The complete estrous cycle can be observed only in the absence of fertilization, for implantation arrests the cycle. The complete cycle is as follows:

1 Egg develops in the ovarian follicle.
2 Egg is shed into fallopian tube and passes to uterus.
3 Endometrial lining of uterus is prepared for implantation.
4 If no implantation occurs, endometrial lining of uterus returns to normal condition.
1^2 New egg develops in ovary.

Since the egg remains in the female for only about three days after ovulation in most mammals, fertilization, if it is to occur, must take place during this period. Sexual interest of the male, culminating in copulation, is induced at this time by the particular physiological changes in the female known as *estrus,* commonly called "heat." (The word "estrus" comes from the Greek word for "sting," which sends an individual into a frenzied condition). By means of scent signals (and occasionally visual signals) the female attracts the male to copulate. It is characteristic of mammals generally that only during the period of estrus and at no other time in the estrous cycle is the female receptive or the male sexually stimulated.

In order to understand the changes that have occurred in the evolution of man's reproductive system, it is desirable to understand how the

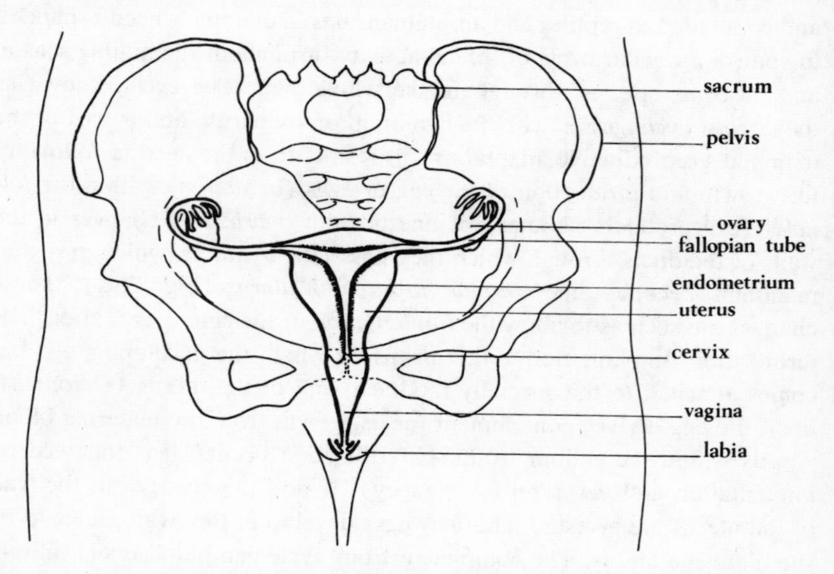

FIG. 2.4. The oviducts in the female mammal have been modified to supply nourishment to the developing embryo. The arrangement of the reproductive organs of a woman is shown in relation to the pelvic girdle; the front of the pelvis (the pubic region) is shown cut away where it would obscure the cervix. The arrangement is not strikingly different in other higher primates.

estrous cycle is controlled by a system of hormones produced by the *pituitary body,* a small endocrine gland that lies close under the base of the brain (see Fig. 8.11). In spite of intensive research, it is still not fully understood how the whole system of the estrous cycle is maintained, but at least part of the mechanism is known. Probably under the control of the hypothalamus (Fig. 8.11), the pituitary, during each estrous cycle, produces two hormones successively, which have been termed the *follicle-stimulating hormone* (*FSH*) and the *luteinizing hormone* (*LH*). Their production is shown graphically in Figure 2.5, *A*. Shed into the blood

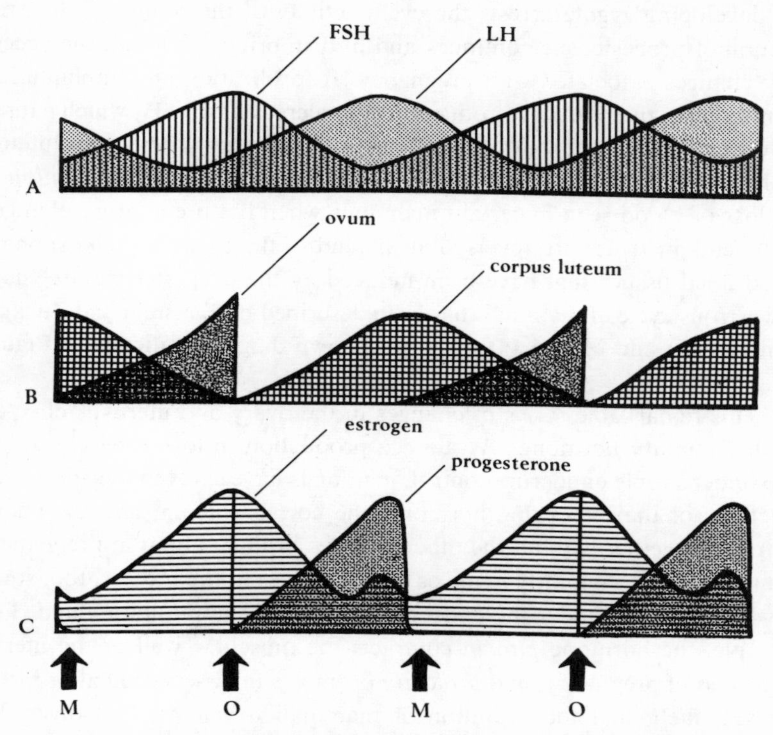

FIG. 2.5. Some of the phenomena associated with the estrous cycle: *A*, pituitary hormone levels; *B*, development of ovarian products; *C*, ovarian sex hormones; *M*, menstruation; *O*, ovulation and termination of estrus. *FSH*, follicle-stimulating hormone, *LH*, luteinizing hormone. (Adapted from Mason 1960.)

stream, the two hormones stimulate, respectively, the growth and production of one or more eggs in the follicles of one or both ovaries (FSH) and what has been called the "yellow body" or *corpus luteum* (LH), which develops in the ovary in place of the egg after it has been shed (Fig. 2.5, *B*). But the developing ovarian follicle and the corpus luteum themselves both produce hormones when they are activated by the pituitary: the well-known female sex hormones called *estrogen* and *progesterone* (Fig. 2.5, *C*). The estrogen brings on estrus or "heat" by causing activity of the scent glands and enlargement of the external female genitalia; the progesterone initiates the special changes in the endometrium that prepare it for implantation.

If fertilization and implantation take place, a hormone produced by the developing zygote arrests the cycle activity of the pituitary; the production of progesterone continues and in turn brings about all the accessory changes associated with pregnancy. If fertilization and implantation do not occur, the pituitary produces a new secretion of FSH, which causes the development of new ovarian follicles and new eggs. The corpus luteum in this case degenerates, and the progesterone level falls. *Menstruation* is the loss of blood seen in certain mammals when the preparatory changes in the endometrium are reversed at the end of the cycle, and the spongy, blood-filled tissues that have been induced by the progesterone are shed. The estrous cycle of mammals has been described by Eckstein and Zuckerman (1956) and Mason (1960). It is shown diagrammatically in Figure 2.6.

This remarkable series of changes in the ovary and uterus is effected by the pituitary hormones. While egg production in lower vertebrates is also under simple endocrine control, mammals have evolved the additional functions of the luteinizing hormone, the corpus luteum, and the endometrium, together with all the other changes brought about in pregnancy. But mammalian viviparity involves other changes in the mother, too, some of which are also under the endocrine control of the pituitary gland. For example, the hormone *pitocin* contracts the muscular wall of the uterus at the end of pregnancy, and *prolactin* promotes milk secretion after birth. We see, then, that the evolution of mammalian viviparity involves the evolution of new endocrine activity, as well as anatomical modifications, such as those we see in the oviduct.

Correlates of these evolutionary developments are the adaptations of the membranes of the *fetus* (the developing young in the womb) to form a *placenta*, the organ that penetrates the endometrium and, by diffusion, receives nourishment for the growing fetus from the maternal blood stream. Waste products also pass from the fetus by this route.

The intimate association of mother and fetus, which is typical of placental mammals, means that the egg is fully protected during its development and that the young stand a far greater chance of survival than do the young of reptiles, which are left to the mercy of predators—especially the egg-eating mammals. But the prenatal relationship between mother and fetus is not more important as a means of reproductive economy than is the postnatal relationship between mother and young.

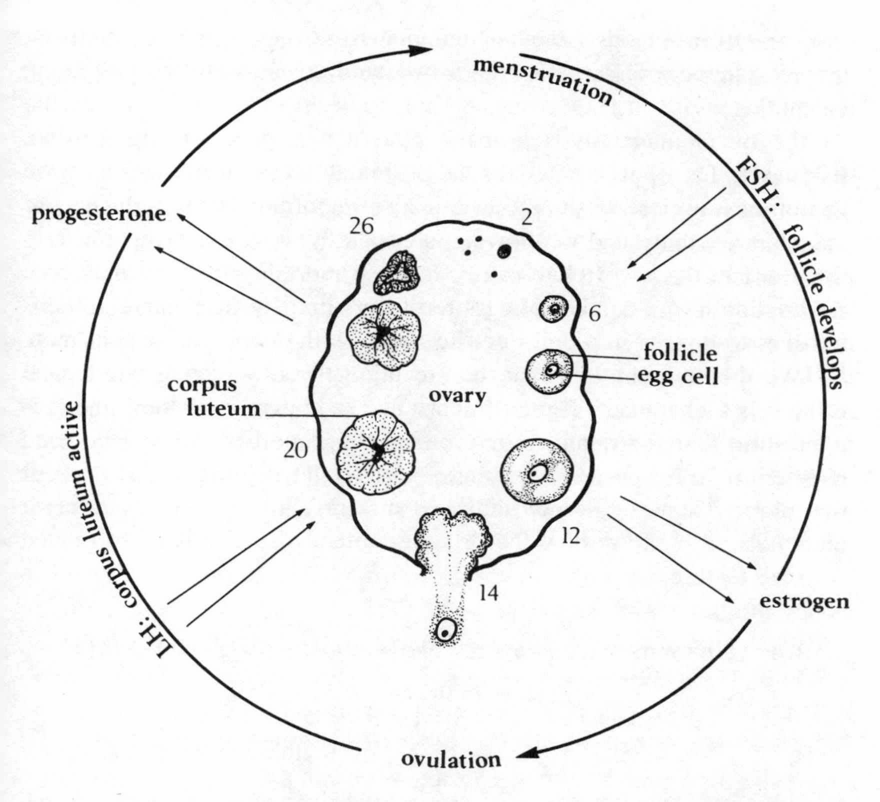

FIG. 2.6. Diagram showing the changes in the ovary that take place during the estrous cycle. The cycle in higher primates lasts approximately 28 days, and the numbers shown round the ovary give some indication of the timing of the events, counting from the first day of menstruation, when the follicle starts to grow. Menstruation has been recorded as characteristic only of the Old World monkeys, apes, and man, though it is known in a few other primates.

2.5 *Reproductive economy: postnatal*

CARE OF THE NEWBORN by one or both parents is not peculiar to mammals but has been extensively evolved among birds and is found occasionally among reptiles and in a few invertebrates. What is unique to mammals is that the "parasitic" mode of life of the offspring is continued after birth, in that they are fed by the special milk glands (*mammae*) of the mother, subcutaneous in origin and probably evolved from the sweat glands (another mammalian novelty). Milk contains all the substances necessary for the growth and development of the young (including antibodies) and may continue to be taken for long periods (the record for all mammals is probably four to five years among Alaskan Eskimos). The relative safety of the postnatal environment, in which the young remain close to the mother, is also important in the reduction of mammal mortality and in allowing, secondarily, a slower growth rate in some orders. As a correlate of this lowered mortality rate, we may note a reduction in the number of eggs fertilized annually from some millions in fishes to dozens in reptiles and to no more than one per year in man.

We shall see (in 9.7) that the exceptionally slow growth rate typical of man is a character of great importance in human evolution, and it is interesting that the quantity of protein in the mother's milk (the food most essential for growth) is quite precisely related to the growth rate of the infant. Table 2.2, which indicates the early growth rate of different mammals, also shows how the protein content of the milk is correlated with this feature.

TABLE 2.2 GROWTH RATE AND MILK PROTEIN (AFTER SIVERTSEN 1941)

Animal	*Days to Double Birth Weight*	*Protein in Milk Parts per 100*
Rabbit	6	10.4
Dog	8	8.3
Sheep	10	7.0
Pig	18	6.9
Cow	47	4.0
Horse	60	2.3
Man	180	1.9

The intimate relationship between generations resulting from suckling and caring for the young makes possible the transmission of learned behavior by observation and imitation. Behavior patterns learned by trial and error have been recognized among lower vertebrates and invertebrates, but the simplicity and origin of such behavior puts it into a different class from that of mammals. It is the transmission of complex learned behavior by imitation that makes it unnecessary to learn solely from direct experience—a dangerous process (see 1.6). Instead, the experience of generations can be assimilated in a short time with reduced danger to the young. Thus the mammals have the great advantage (shared to some extent by birds) of being able, as it were, to "inherit" (by observational learning from parents) "acquired" (i.e., learned) behavioral characters.

The "inheritance" of acquired characters makes possible a high rate of evolution of those characters, and thus the behavior patterns of mammals have evolved fast and increased in complexity in such a way as to put them in that respect at a different level from all other living organisms. Their learned behavior repertory is immense, complex, and flexible (because it is still subject to trial and error learning). It makes possible the maintenance of the population in face of even very extensive and demanding environmental changes.

2.6 *Determinants of behavior*

BEHAVIOR HAS BEEN DESCRIBED as both innate and learned (see 1.6), and we have seen that learned behavior assumes great importance among mammals. The intimate relationship between different generations brings the possibility of building up quite complex behavior patterns by imitation, without the need to learn by trial and error in each generation. Such learning is confined to animals (birds and mammals) in which parental care has evolved.

But mammals also depend on learning by trial and error, which must always be the original source of learned behavior. While in most animals activity and learning result from the need to satisfy the primary drives of hunger, sex, and self-preservation, a new determinant of behavior is found among mammals that does not satisfy these immediate needs. This new factor is a drive to investigate and explore the environment without the immediate object of satisfying one of the three prime biological needs mentioned above.

The exploratory drive is seen in the playful and exploratory activity of young mammals with the parts of the environment that provide changing and interesting feedback in connection with effort expended. Such activity leads individuals to discover how the environment can be changed and what consequences flow from the changes. Such moderate but persistent activity is possible only under the protection of parents, because its value to the individual and the species is not immediate, as are the primary drives, but is of a long-term nature. This type of behavior forms part of the process whereby the young animal learns to interact effectively with the environment and build up its perception of its surroundings (see 10.2).

The value of a computer in the solution of a very complex problem depends not only on the computer's size and complexity but also on the amount of data that is fed into it. For reliable results, a large input is a prerequisite. It is equally necessary in the evolution of the brain as a mechanism for prediction and problem-solving, for complex problem-solving depends on sufficient input data, and this need explains the evolution of an exploratory drive. The motor activity that results in the satisfaction of the primary drives of hunger, sex, and self-preservation was supplemented by the increased input that was derived from exploratory activity and is an essential correlate of the evolution of mammals into organisms with brains much larger (in relation to body size), much more complicated, and much more efficient than had been known before in the organic world.

There were two developments within the brain itself concomitant with the increased input. The first was the evolution of a greatly expanded memory store (that is, an unconscious record of experience); the second was the evolution of an expanded "computer," that is, the part of the brain that actually makes predictions and solves problems.

When we compare the mammalian brain with that of lower vertebrates, we see that it is typified by the relatively great development of what has been termed the *cerebral hemispheres,* compared with other parts (Fig. 2.7). It is in the cerebral hemispheres, and particularly in their surface layers or *cortex*, that we believe the memory store and most complex computing activities are to be found. We know the function of some of the lower parts of the brain, and they seem to be of a simpler order. For example, we have recorded how the hypothalamus controls body temperature and the estrous cycle, and it has other relatively simple functions, such as the generation of emotional response (see 8.6), the control of sugar metabolism, heart rate, blood pressure, salt and water balance, and sleep. The *brainstem* and *cerebellum* (see 8.6) control many innate reflexes relating to balance and muscular control. It is the cerebral hemispheres, however, that are concerned not so much with executing a program of bodily activity and maintaining it as with initiating such activity with a "decision" (unconscious, except occasionally in man) as to what activity should be undertaken next, on the basis of the sensory input and accumulated experience that is memorized.

Just which parts of the cerebral hemispheres are mainly concerned with memory will be discussed later (see 8.6). Investigations into which parts of the hemispheres are concerned with decision have achieved poor results, which has led to the theory that it is the activity of the cerebral

FIG. 2.7. Sections (not to scale) through the longitudinal axis (termed "medial" or "sagittal") of the brain of a fish, reptile, rabbit, and man, showing the difference in size of the cerebral hemispheres in relation to the older parts of the brain. For names of the different parts see Figs. 8.8 and 8.11. (From Magoun 1960.)

hemispheres as a whole, rather than that of any part, that determines most mammalian behavior (the so-called "law of mass action"). Quite extensive damage to the human brain may be sustained in some parts without any serious impairment of function, leading us to deduce that the problem-solving and predicting activities of the brain are not truly comparable to those of a computer, which cannot function after the mutilation of any part.

In summary, we see the evolution of the cerebral hemispheres of mammals into a mechanism somewhat similar in function to a large computer with an extensive memory store. Unlike a computer, the brain operates to some extent with nonlocalized systems of activity; like a computer, its value to its user in complex prediction is directly related to the information content of the input. The mammalian brain functions therefore not only to control the mechanisms that maintain the steady state of the body, but also to store the memory of an enormous range of experiences. Input is maintained by the highly evolved sense organs which bring to the animal an enlargement of experience. On the basis of such experience the brain is able to make predictions about the likely course of external events and on this basis to initiate novel patterns of adaptive behavior.

2.7 *Man a mammal*

WE HAVE DISCUSSED four evolutionary novelties that characterize the placental mammals, each one a whole complex of characters. Two of them, homoiothermy and heterodontism, contribute in an obvious way to make possible the constant metabolic level of the organism in the face of environmental change. Following our discussion of evolutionary progress (see 1.8), we see that insofar as they bring improvements in the internal homeostatic state of the organism, they are progressive in the direction of man and in the sense defined by Herrick (1946). We have already indicated their importance to the early mammals, in the advantages that these new character complexes gave to their bearers in the reptile-dominated environment of the Cretaceous period.

Clearly, any such advances in internal homeostatic mechanisms increase the chances of survival, not only of the individual, but of the whole species, because each individual is more likely to survive and reproduce itself. But the advantages to the species of the second two characters are perhaps more obvious. The improvements in reproductive economy that result in a much lower wastage of eggs and young mean an increase in biological efficiency as a whole for the species: fewer eggs need to be produced, and less energy and living substance is devoted to that end. The trend of a lowered rate of fertilized egg production reaches its apogee in *Homo sapiens*.

Our fourth evolutionary novelty, the exploratory drive, is a behavioral development, concomitant with the evolution of a larger brain, which by making possible a better knowledge of the environment increases the chances of survival of the individual and so of the species. The ability to assimilate and utilize this knowledge is selected in the genotype, but its transmission between generations is made possible by imitation within the close mother-infant relationship. This transmission of behavior patterns in an extra-genetic manner is the seed from which culture was eventually to grow (see chapter 10).

It is not necessary to stress the importance to man of these complexes and all their structural correlates. Man, with his remarkable ability to survive every sort of climate, to live off almost every sort of food (7.2), has a wonderfully efficient reproductive system (9.1 and 9.2) and a body of transmitted behavior (10.1) that is far beyond anything known in the rest of the animal kingdom. This vast body of transmitted behavior of different kinds he owes mostly to the extraordinary knowledge of the environment that he has gained by experience. His perception of the environment, which gives him that enriched experience, he owes to the exploratory nature which he shares with other mammals. Improved homeostatic mechanisms and a new behavioral relationship with the environment made mammals the potential ancestors of *Homo sapiens*. Man is a mammal and could belong to no other class of vertebrates.

SUGGESTIONS FOR FURTHER READING

For an outstanding text on the biology of mammals see J. Z. Young, *The Life of Mammals* (Oxford: Oxford University Press, 1957). For a simple taxonomic survey of the Mammalia see M. Burton, *Systematic Dictionary of Mammals of the World* (London: Museum Press, 1962). Useful introductions to the social behavior of animals include W. Etkin, *Social Behavior and Organization among Vertebrates* (Chicago: University of Chicago Press, 1964) and N. Tinbergen, *Social Behavior in Animals* (London: Methuen, 1953).

3

Primate Radiation

In each great region of the world the living mammals are closely related to the extinct species of the same region. It is, therefore, probable that Africa was formerly inhabited by extinct apes closely allied to the gorilla and chimpanzee; and as these two species are now man's nearest allies, it is somewhat more probable that our early progenitors lived on the African continent than elsewhere.

CHARLES DARWIN
The Descent of Man, 1871

3.1 The origin of primates

A GLANCE at Figure 2.1 will show that the mammals that gave rise to the primates (and indeed to all other placental mammals) are believed to have been of a kind that would now be classified as Insectivora. The central position of the order Insectivora in the figure is intended to indicate that they have diverged least from the ancestral mammalian stock that gave rise to the whole radiation of mammals.

Some of the fossil insectivores from the Paleocene period are very similar to some living genera of this order, and in particular to the genus *Sorex,* the little shrew (Fig. 3.1). Though many of the Cretaceous and Paleocene insectivores were rat-sized, the shrew and its relatives are very small animals, and the one illustrated is only 4.5 inches long, including tail. The insectivores are insect-eaters, as their name implies, but, like the shrew, the majority supplement their diet with other small animals, seeds, and buds; they are in fact omnivorous. The shrew is an active, nervous, nocturnal animal and eats its own weight in food daily. A large food supply is apparently necessary to maintain a creature so active and so small, since small animals have a much larger surface area in relation to their weight, through which heat may be lost, than do larger animals.

The shrews and other ground-living genera have exploited the dense environment of the forest floor, where, living almost exclusively upon the ground, they are able to find the practically continuous food supply that

FIG. 3.1. Common European shrew (*Sorex araneus*), with a body about 3 inches in length, is very similar to the North American species *S. cinereus*.

they require. As can be seen from Figure 3.1, the shrew has a long snout and sensitive tactile *rhinarium* (the sensitive area of naked skin at the end of the muzzle, familiar in dogs). They snuffle through the grass after seeds and minute animals. To their small size and dense habitat they owe their obscurity, which has certainly contributed to their survival by protecting them from the larger carnivorous reptiles and mammals. The adaptations of small size and nocturnal activity have obviously given these lively and omnivorous creatures an immense advantage over their reptile competitors. Yet their adaptations did more than this, for not all of them remained in obscurity. The moles went underground; the African otter shrews and American marsh shrews (among others) have become aquatic; other insectivores took to the trees and evolved into tree shrews, now recognized to be closely related to the earliest and most primitive primates. From some of the arboreal shrews evolved the bats (Chiroptera). Still other ground-living forms evolved at the same time into the whole range of great terrestrial mammals (Fig. 3.2).

Our special interest lies in the tree shrews, a group that invaded the trees, climbing with their sharp claws and using their long tails to balance themselves. From animals not unlike them, the order Primates is believed to have evolved, and finally man himself.

3.2 *The primates*

THE DIAGNOSTIC CHARACTERS of the order Primates, whose origins it is claimed can be traced to the Late Cretaceous (Van Valen and Sloan 1965), are not easy to define and describe in a simple list. But their characters, though numerous, are easily understood as adaptations to the arboreal environment in which the primates lived. Rather than present a list of characters at once, it is better to consider first the primate adaptations as a whole and then the characters that evolved as their basis.

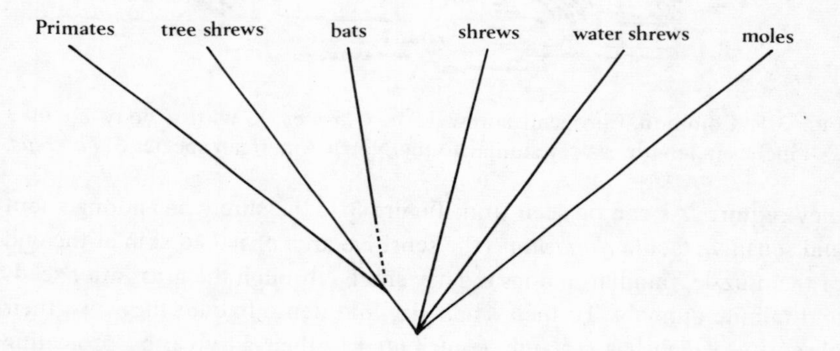

FIG. 3.2. Simplified diagram showing part of the radiation of the primitive ground-living shrew-like placental mammals.

In the face of competition from other forest-living creatures, the primates evolved the ability to climb trees. Tree-climbing was necessary to exploit the forest environment to the full, to gather fruit and insects, as well as to escape from predators on the ground. The environment was a difficult one, densely filled with objects of different kinds and demanding accurate mobility and keen sensory awareness. Since scents were unreliable in identifying the exact direction of food or enemies and were of little help in negotiating trees and undergrowth, efficient vision became the great need of these forest dwellers. Primates therefore became characterized (1) by the evolution of limbs adapted for tree-climbing and (2) by eyes adapted for much-improved vision. Almost every character we normally associate with primates is part of one or the other of these two functional complexes, and from this point of view they fall into place very neatly. Let us consider each complex in turn.

1. Tree-climbing involved the following adaptations of body and limbs:
   a) Retention of a generalized limb structure and *pentadactyly* (five fingers) with free mobility of limbs and digits, giving support to the body from any direction, and with limb movement not restricted to one plane.
   b) Development of mobile digits capable of grasping, with sensitive pads supported by nails (which replaced claws in primate evolution) and palmar surface with extensive friction skin.
   c) Retention of a tail as an organ of balance (or as a "fifth limb" in some New World monkeys).
   d) Evolution of upright body posture and extensive head rotation.
   e) Increase in body size—ultimately limited by arboreal environment.
   f) Evolution of nervous system to give precise and rapid control of musculature.
2. The following sensory adaptations were also selected:
   a) Enlargement of the eyes, which increased the amount of light and detail received.
   b) Evolution of an improved retina with increased sensitivity to low levels of illumination and to different frequencies (that is, to color).
   c) Eyes that look forward, with an overlapping visual field that gives three-dimensional perception of the environment, accompanied by the mechanisms associated with three dimensional interpretation and analysis of input to the brain.
   d) Development of a bony protection for the large eyes.
   e) Reduction (as a corollary of enhanced visual sense) of the apparatus involved in the sense of smell, the olfactory lobes, and protective bony snout or muzzle.

The foregoing two functional complexes reflect only part of the whole pattern of primate biology, but the most important part. One complex, the developed visual sense, is concerned with receiving information about the environment; the other is concerned with its active exploitation. The brain processes information from the first and instructs the second. Their interaction is an example of the intensification of total metabolic and nervous activity typical of the primates. The brain, the mechanism for analysis and prediction, is the necessary correlate of the evolution of either sensory or motor activity. Development of both these functions goes hand in hand with development of the brain.

The sight of gibbons moving through a forest canopy is a beautiful instance of precise motor action under accurate visual control. The speed and accuracy of their movement depends on a brain that can assimilate many factors of weight, distance, and wind, and process them to give immediate and precise motor instruction. It is to the exploitation of this dense, complex, and demanding environment that we owe the development of a brain that is unique in the extent of its memory bank, in the quantity of the data it processes, and in the speed of its operation. It is accompanied by nervous and muscular mechanisms that are precise and flexible in operation.

The structural characters listed here will be referred to again in the following chapters when we consider how they have contributed to man's own form. We shall in this chapter consider first the classification of the order Primates and how the different groups within the order have developed different locomotor adaptations. Then, we shall consider some aspects of the evolution of the nervous system that have accompanied these important adaptations. Finally we shall review some of the Primate fossils that evolved into modern man.

3.3 Primate classification

AMONG LIVING PRIMATES we find a wide range of fifty-seven genera of different appearance, size, and habitat. This varied order, which extends from a prosimian the size of a mouse to man himself, forms a series that suggests a trend in evolutionary development—a suggestion that is in some parts supported by the fossil remains available for study. Groups perhaps not very dissimilar to different ancestral stages in human evolution appear to have survived surprisingly unchanged from early times, since they can often be identified as close in morphology to their ancestral forms. The situation has been diagrammatically presented by Le Gros Clark (1971), and his figure is reproduced here as Figure 3.3. Five main terminal kinds of living primate, the tree shrew*, lemur, tarsier,

* Whether the tree-shrews, called Tupaioidea, should properly be classified as Primates is at present under discussion. Whatever their taxonomic status, their close relationship to primate ancestors is not in doubt.

monkey, ape, and man, are shown in Figure 3.4. These particular kinds have survived because they became successfully adapted to particular ecological niches; intermediate forms have been lost.

It is a remarkable, rare, and happy fact that we have this series of forms still living, for they can help us visualize certain stages of human evolution. Of course, we can never be certain of the interrelationships of living forms in past time, but, *on the basis of the evidence available at present*, a working hypothesis of the kind shown in Figure 3.3 forms a popular interpretation of the evidence. It seems unlikely, however, that human ancestry passed through a stage that would be identified as a tarsier, though the evidence from that remote period is still very limited. The fossil forms lack many of the characteristic features of their living representatives and the scheme merely suggests a way of visualizing past ancestral species. The actual character of the forms marked *E* and "Hominidae" in the figure and how they evolved are the subject of this book.

A formal classification of the order Primates is given in Table 3.1. Six of the superfamilies listed have been illustrated in Figure 3.4; the other two are not important to this study. We shall be concerned mainly with the so-called "higher" primates or Anthropoidea: the monkeys, apes, and man, which are most closely related to ourselves (see Tables 3.2 and 3.3). The lower primates (Prosimii) are of less immediate relevance.

We see from the classification that there are two groups of monkeys: the Ceboidea, or New World monkeys, confined to Central and South America, and the Cercopithecoidea, or Old World Monkeys, which are spread throughout the tropical regions of the Old World, in Africa and Asia. (These two groups, which have striking and important differences, are also sometimes termed the *platyrrhine* and *catarrhine* monkeys, respectively.) The third superfamily of the Anthropoidea, the Hominoidea (apes and men), was also confined to the Old World until man entered America and Australia in geologically recent time. The Hominoidea are clearly much more closely related to the Old World monkeys than to the New World monkeys, as both geographical and anatomical evidence make clear. The New World monkeys became isolated at an early stage in primate evolution, and they probably have had a separate prosimian ancestry. Therefore, we shall refer very little in this book to the New World monkeys. Though they are a successful and fascinating group, we must confine our attention here to the Old World forms, which may lie near to man's ancestral lineage.

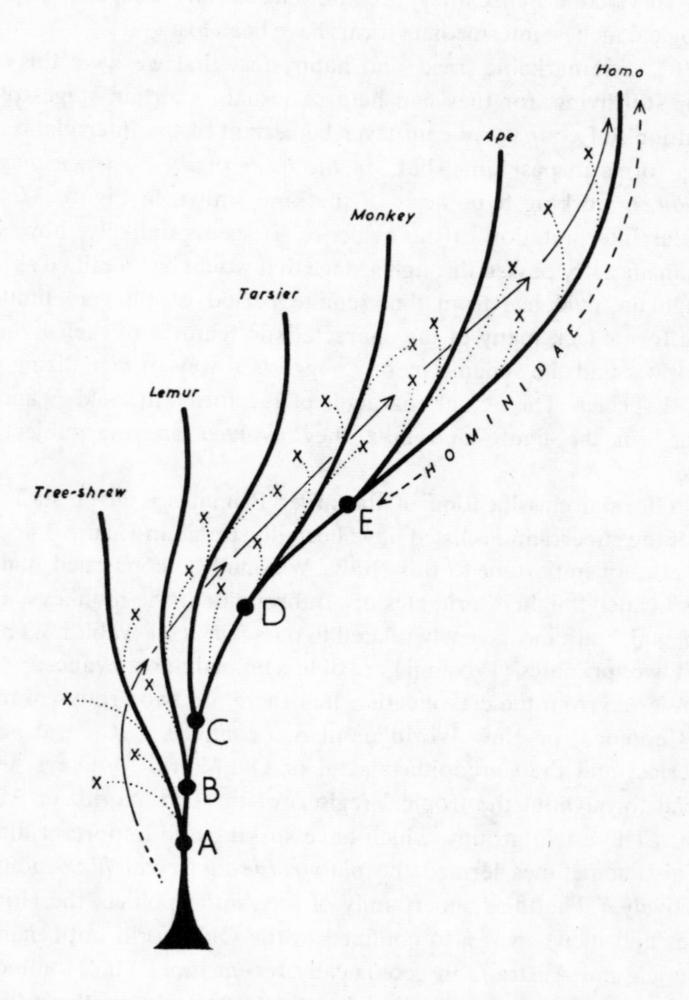

FIG. 3.3. Diagram showing that the living primates form a series from tree shrew to *Homo*, a series that suggests a general trend of evolutionary development. From comparative studies of such types it is possible to postulate the probable linear sequence of human ancestry. This sequence leads from the basal primate stock through hypothetical transitional stages, *A–E*, at which the different ramifications of the primates are presumed to have branched off. Fossil primates approximating to the postulated ancestral stages are represented by crosses. (From Le Gros Clark 1971.)

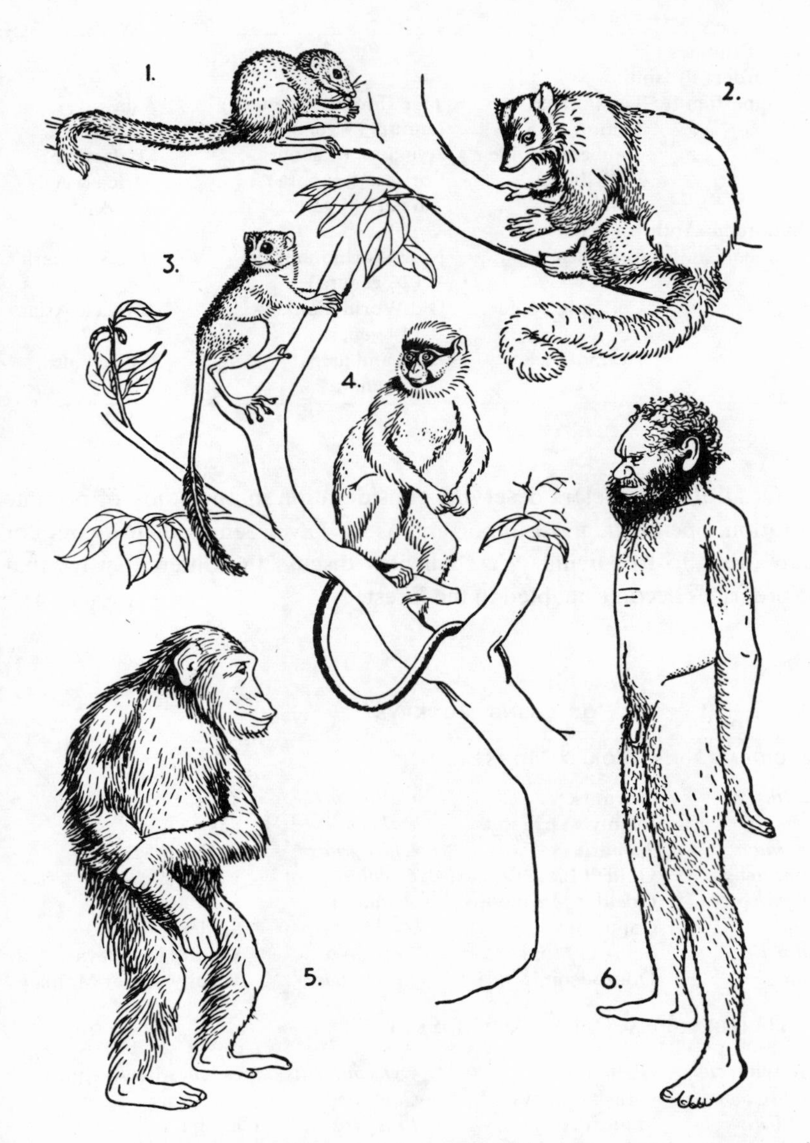

FIG. 3.4. Some living members of the order Primates, representing a series that, broadly speaking, links man anatomically with some of the most primitive of placental mammals: *1*, tree shrew; *2*, lemur; *3*, tarsier; *4*, Old World monkey; *5*, chimpanzee; *6*, Australian aboriginal. (From Le Gros Clark 1971.)

TABLE 3.1 CLASSIFICATION OF THE PRIMATES SHOWING NUMBERS OF LIVING GENERA

Order: Primates			
Suborder: Prosimii			
Superfamily:	Tupaioidea	Tree shrews (5 genera)	Asia
	Lemuroidea	Lemurs (9 genera)	Madagascar
	Daubentonioidea	Aye-ayes (1 genus)	Madagascar
	Lorisioidea	Lorises (5 genera)	Africa & Asia
	Tarsioidea	Tarsiers (1 genus)	S.E. Asia
Suborder: Anthropoidea			
Superfamily:	Ceboidea	New World monkeys (16 genera)	C. & S. America
	Cercopithecoidea	Old World monkeys (14 genera)	Africa & Asia
	Hominoidea	Apes and men (5 genera)	Worldwide

Of all the characters deserving consideration in any study of primate evolution, one of the most important, as we have seen, is their locomotor adaptations to tree-living. We shall now discuss the different ways that the order has become adapted to the forest.

TABLE 3.2 GENERA OF LIVING MONKEYS

CEBOIDEA: NEW WORLD MONKEYS			
Callithrix	Marmosets	*Callicebus*	Titis
Cebuella	Pygmy Marmosets	*Pithecia*	Sakis
Saguinus	Tamarins	*Chiropotes*	Bearded Sakis
Leontideus	Golden Lion Tamarins	*Cacajao*	Uakaris
Callimico	Goeldi's Marmosets	*Alouatta*	Howlers
Cebus	Capuchins	*Ateles*	Spider Monkeys
Saimiri	Squirrel Monkeys	*Lagothrix*	Woolly Monkeys
Aotus	Douroucoulis	*Brachyteles*	Woolly Spider Monkeys
CERCOPITHECOIDEA: OLD WORLD MONKEYS			
Cercopithecus	Guenons	*Cynopithecus*	Celebes Black Ape
Erythrocebus	Patas Monkeys	*Colobus*	Guerezas
Mandrillus	Mandrills	*Presbytis*	Langurs
Cercocebus	Mangabeys	*Nasalis*	Proboscis Monkeys
Papio	Baboons	*Simias*	Pagai Island Langurs
Theropithecus	Geladas	*Rhinopithecus*	Snub-nosed Langurs
Macaca	Macaques	*Pygathrix*	Douc Langurs

(From Napier & Napier 1967.)

FIG. 3.5. The baboon (*above*) is a quadrupedal monkey that has become terrestrial. The gorilla (*below*) is an ape with an arm-swinging ancestry, and when he walks quadrupedally he takes his weight on his knuckles. Note that the gorilla's arms are much longer than his legs. (Not drawn to scale.) (By permission of the trustees of the British Museum [Natural History].)

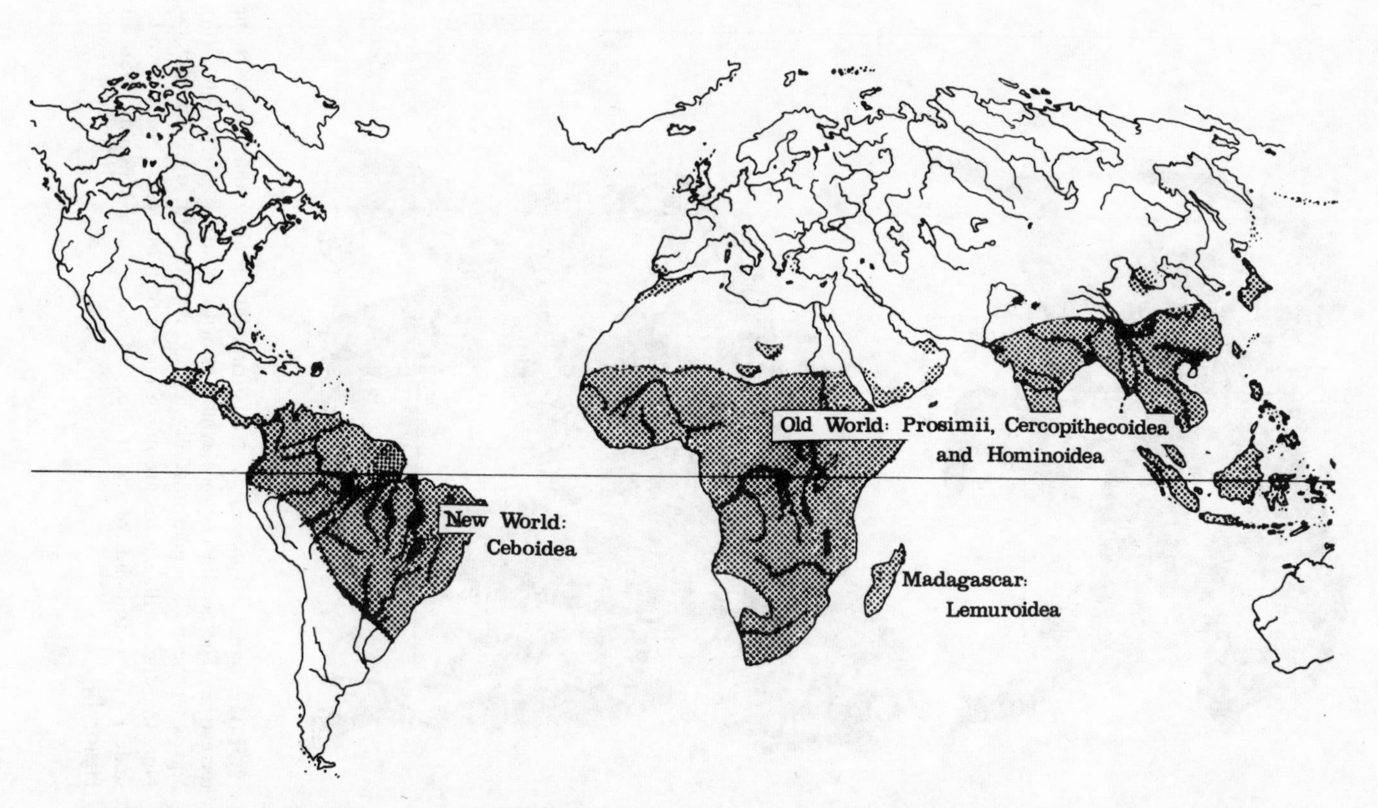

FIG. 3.6. World map showing the approximate distribution of the three geographical radiations of non-human primates.

TABLE 3.3 CLASSIFICATION OF THE HOMINOIDEA AND HOMINIDAE

Family	Oreopithecidae	
Genus	*Oreopithecus*	Swamp apes
Family	Pongidae	
Subfamily	Hylobatinae	
Genus	*Hylobates*	Gibbons
	Symphalangus	Siamangs
Subfamily	Dryopithecinae	
Genus	*Dryopithecus*	Oak apes
	Gigantopithecus	Giant apes
Subfamily	Ponginae*	
Genus	*Pongo*	Orang–utan
	Pan	Gorilla & Chimpanzee
Family	Hominidae	
Genus	*Ramapithecus*	Early Ape-men
	Australopithecus	Ape-men
	Homo	Men

* In the subfamily Ponginae, I have chosen the classification in which both chimpanzees and gorillas are included in the one genus *Pan*, rather than separated into the two distinct genera, *Pan*, and *Gorilla*, respectively. Their relationship is close, and they are together quite distinct from *Pongo* (orang-utans).

3.4 *Primate locomotion: prosimians and monkeys*

THE MODE OF LOCOMOTION of a group of animals is necessarily closely adapted to the ecological niche they occupy. Different species of primates occupy different niches and this factor, coupled with the differences in body size of the different species, results in considerable differences in locomotor behavior and in anatomical adaptations. Differences in these adaptations are broadly reflected by the taxonomic groupings of the primates, and they will be considered under taxonomic headings.

1. *Prosimians.* Since the primates were evolved from ground-living forms, we can be sure that the earliest mode of locomotion in the group was running up the trunks of trees and along the branches on "all fours." This mode of locomotion would have involved minimum arboreal specialization in the first place. From our knowledge of the living prosimians,

however, we can identify two general kinds of adaptation to arboreal life that have replaced this primitive state.

Vertical clinging and leaping adaptations were probably the first to appear in prosimian evolution. Relatively long and powerful hind legs are characteristic of many small prosimians (such as *Tarsius* or *Galago*) as well as some larger forms (such as *Indri*). This development has given these creatures the ability to leap, like a small kangaroo, from tree to tree, over enormous distances in relation to their body size. Thus they can avoid predators, both by avoiding the ground and by leaping if they are attacked in the trees. This evolution of the hind legs is accompanied by a movement of the center of gravity to a position near the hind legs, and by the adoption of a resting posture with the trunk held vertically. They usually cling to a vertical rather than a horizontal branch. This locomotor adaptation has been called vertical clinging and leaping (VCL). It is an arboreal specialization, but it has the result that if such an animal crosses the ground at speed it proceeds by hopping, like a kangaroo, rather than quadrupedally.

There is evidence that the ancestors of the Old World monkeys and apes evolved from an Eocene prosimian with VCL locomotor adaptations. Although today the monkeys do not carry this extreme locomotor adaptation, they and all other Old World higher primates are characterized by the erect trunk that we see in the VCL prosimians, at least when at rest. The specializations of the VCL complex may be summarized as follows (Napier and Walker 1967).

a) A considerable lengthening of the hindlimbs in relation to the trunk. This lengthening includes the femur, tibia and fibula, and in *Tarsius* and *Galago* the tarsal bones are also lengthened.

b) Powerful muscles are associated with the hind limbs. These developments bring the center of gravity back over the hind quarters.

c) A large foot carries a divergent big toe and reduction of digits II and III. This allows the animal to grasp bigger branches.

d) The vertebral column and its relation to the head reflect the vertical posture.

e) The tail is usually long and carries thick fur (but it is absent in *Indri*). Where present, its function is probably to give aerodynamic control during flight, and to control the position of the center of gravity.

A second adaptation seen among prosimians contrasts strikingly with VCL. A group of four genera move by very slow quadrupedal climbing (SQC), with neither leaping nor running. Typical of this adaptation is the slow loris (*Nycticebus*) which moves without shaking a branch, and relies on its stillness both for stalking its prey of insects, lizards, and birds, and for protection from predators. SQC specializations are summarized as follows (Walker 1969).

a) The lorises and other slow climbers have very robust limb bones and musculature adapted for slow, powerful and prolonged contraction. The legs are only slightly longer than the arms.

b) The digits are long, but the tarsal footbones are short.

c) Hip and ankle joints show great mobility, allowing suspension in strange postures by any or all the limbs.

d) Tail absent or very short.

In the case of both VCL and SQC the size and mode of locomotion reflects the animal's adaptation for feeding and for predator avoidance in a particular type of arboreal habitat.

2. *Old World Monkeys.* Old World monkeys in general are larger than the living prosimians. They often live in forest areas and their greater size allows them to run along the larger branches and leap from tree to tree without the extreme specializations we find among prosimians. Their large size gives them not only increased mobility, but also protection from the smaller animals that prey upon the prosimians.

This larger size does, however, present certain problems. The vegetable food in the forest (fruit, leaves, and buds) is found mainly at the ends of branches and these parts of the trees are not very safe for exploitation by a heavy animal. The tendency, therefore, among monkeys and apes, is to spread the weight among a number of different branches by raising the arms above the head and hanging from above, and at the same time getting support from below. Another problem relating to large size is the difficulty of balancing when walking along horizontal branches; swinging underneath the branch gives greater stability during locomotion. Arm-swinging, suspension from above, is therefore an adaptation that enables the larger monkeys to travel through the smaller branch zones of the forest canopy and to exploit it for food (Avis 1962). Arm-swinging seems to be an evolutionary concomitant of size increase coupled with an herbivorous arboreal diet.

Monkeys are versatile creatures and they all use their arms above their heads at least occasionally, but certain groups seem to use this means of locomotion more regularly than others. On the basis of observed behavior, some authors (such as Napier and Napier 1967) have published classifications of primate locomotor patterns. The Old World monkeys can in this way be divided into three groups: 1) branch running and walking; 2) branch running, arm-swinging, and leaping; and 3) terrestrial running and walking. In the first group we find monkeys that use their arms above their heads with much less frequency than those in the second group. The best example of the arboreal quadrupedal (group 1) monkeys is the genus *Cercopithecus* (the guenon, vervet, and other common African monkeys). The best known of the arm-swingers (group 2) are the genera *Colobus, Presbytis,* and *Nasalis* (the common colobus of Africa, the langur, and the curious proboscis monkey of Asia.) It is, however, important to remember that monkey locomotor behavior is very flexible, and that it is adapted to the particular environment in which a population lives and to the particular locomotor demands made on each individual. There is great variation in behavior within genera, within species, and

FIG. 3.7. Baboon sleeps in a tree, sitting upon his ischial callosities. (Drawn from a photograph in Washburn 1957.)

within populations and troops. Locomotion may also vary with the age and sex of monkeys as well as with environmental variables. Any generalizations about primate locomotion must be considered in this light.

Quadrupedal arboreal locomotion (group 1) requires certain specializations. Some of these are found in the prosimians, but they are more clearly seen in the monkeys:

a) The hand and foot have long flexible digits, with independently mobile thumb and big toe which are opposable for gripping branches.

b) The hindlimbs are longer and more powerful than the forelimbs.

c) The resulting backward shift in the center of gravity is associated with a sitting position when resting, with the trunk more or less vertical (Fig. 3.7).

d) The tail is retained as an aid in balance.

The arm-swinging forms (group 2), whose locomotion is to a considerable extent suspensory, are characterized as follows:

a) The arms are lengthened in relation to the legs, but they are still just shorter than the legs; this gives greater leverage to the arm and shoulder muscles.

b) The fingers are lengthened, and the thumb is usually used not in opposition to the fingers but alongside them. It is somewhat reduced in length, and in *Colobus* it is lacking or reduced to a stump.

c) The shoulder blade and associated musculature show modifications associated with increasing the power of the arm to lift the body from above, as well as support it from below. The freedom of movement of the arm above the head is also improved. (See chapter 6.)

All living primates can cross open ground, and some Old World monkeys have become adapted to spending a lot of time on the ground (group 3). Again, there is a whole spectrum of behavior, from forms that sometimes come to the ground (such as the vervets) to forms that are always terrestrial (such as the hamadryas baboon). Monkeys that are quadrupedal in trees are preadapted to terrestrial locomotion, but a few species have developed locomotor specializations to life on the ground, and these require some consideration.

The Baboon (*Papio*) is classified into two species. The common savanna baboon which extends from W. Africa to Ethiopia and to the Cape of South Africa (and is called the chacma, yellow, olive and Guinea baboon in different regions) is best considered a single species (*P. cynocephalus*) with a variety of geographical subspecies. This baboon is the most extensively studied of all monkeys, and being in most of its range terrestrial, like ourselves, is of particular interest to us (Fig. 3.5). It is a large and powerful monkey and spends most of the day on the ground in the sparsely wooded savanna areas of its range. It feeds on grass shoots, seeds, and fruit, often from a sitting position. At night it climbs into trees (or occasionally, cliffs) for protection from predators. It walks quadrupedally and takes its weight on the fingers of its hands (*digitigrade*) and on the toes and palms of its feet (*plantigrade*). This mode of support increases the effective length of the forelimbs and gives the monkey greater height and improved vision, as well as greater speed when running. The thumb and fingers, however, are relatively shorter than in arboreal monkeys; they are no longer required so much to grasp branches, but rather to act as props.

The second species is the hamadryas baboon (*P. hamadryas*). This monkey is closely related to the savannah baboon and differs from it in relatively superficial respects. It lives in the even drier environment of eastern Ethiopia and is adapted to sub-desert conditions where trees may be absent. It sleeps at night on cliffs for protection from predators.

Besides the baboons, two other ground-living monkeys should be mentioned. The patas monkey (*Erythrocebus patas*) is probably the terrestrial monkey with the most striking locomotor adaptations. It is distributed along the sub-desert steppe country between the southern edge of the Sahara and the tree savanna which lies to the south. The patas sleeps in trees but finds its food on the ground during daylight, like the baboons. Its main difference is in its lighter build and its adaptation for speed. The monkey has long legs in relation to trunk length, and this gives it great speed which it uses to escape from predators. It runs digitigrade, like the baboon, but its fingers are even further shortened and strengthened. Like the baboon, it often stands bipedally to improve vision, but it uses a powerful tail to form a tripod with the hind legs, like a kangaroo.

A fourth monkey, the Gelada (*Theropithecus gelada*), probably carries more adaptations to terrestrial life than the other forms, and they do not lie only in its locomotor organs. These monkeys live on mountain slopes at high altitudes in Ethiopia, between 6,000 and 16,000 feet. Like the baboons, they are powerful, digitigrade animals, and they normally remain on the ground at all times.

All these four species of monkey have particular anatomical and social adaptations to terrestrial life to which we shall return (chapter 9). Here we are concerned only with their locomotor adaptations. We may summarize this section by noting the following characters of group 3 monkeys:

a) The front limbs are digitigrade, and the fingers shortened. An opposable thumb is retained, however, as the hand is used for feeding. The finger-thumb grip is quite precise.

b) The limbs tend to be nearly equal in length, and in the patas monkey they are long in relation to body length.

c) A bipedal stance is common to improve vision, and the patas uses the tail for support when bipedal.

d) The sitting posture is common and usually used in feeding. The Gelada feeds continuously in the sitting posture.

3. *New World Monkeys.* A range of variation between quadrupedal and arm-swinging monkeys is also present among the New World species. In fact, arm-swinging, as a small-branch adaptation, is taken further among some New World forms than in the Old World species. Most New World monkeys have powerfully developed tails and in four species these are fully prehensile, so that they act like a fifth limb in spreading the animal's weight. The best known is probably the spider monkey (*Ateles*). Their tails are not only long and very strong, but have a well developed sense of touch (with special friction skin on the ventral surface) as well as a good sense of position. The monkeys can hang by the tail alone while feeding, and almost all arm-swinging is accompanied by use of the tail. As a result, these monkeys depend less than the Old World species on leaping between trees.

The whole arboreal locomotor spectrum among the New World species is broader than that found in the Old World monkeys, though there are no terrestrial species. Our knowledge of these monkeys is not, however, as good as that of the Old World species, and a detailed discussion of their locomotion would be irrelevant. It is of interest to note, however, that the New World arm-swingers have taken the structural adaptations we have discussed even further than we find in the Old World: two species have arms longer than legs, and no thumb at all.

3.5 Primate locomotion: apes and men

1. ASIAN APES. The trend toward arm-swinging reaches its most advanced form among the Asian apes. In Southeast Asia there are two groups: the smaller gibbons (the gibbon *Hylobates* and the very closely related Siamang *Symphalangus*) and the large orang-utan

(*Pongo pygmaeus*). The gibbons are the most evolved arm-swingers among the primates and move through the trees suspended by their forelimbs alone, with no support from below. They hold their feet tucked up beneath their body. Because of the degree of dependence on the arms, this type of locomotion has been called *brachiation*. The tendency that we have traced through the order of primates from almost total dependence on the hind limbs among VCL prosimians, to the arm-swinging monkeys, finally reaches its end point with almost total dependence on the long and powerful forelimbs of the gibbons (Fig. 3.8).

a) The forelimbs are greatly lengthened so that they are one quarter again as long as the hind limbs.

b) The hand is long (especially the fingers), but the thumb has not shared this increase in length.

c) Structural modifications of the shoulder and arm give even greater power and flexibility than we find among the arm-swinging monkeys.

d) The tail is lost.

e) The trunk has reduced flexibility, as it is no longer used to add to the power of a leap. The arms have taken over almost the entire locomotor role in the trees.

FIG. 3.8. The gibbon (*Hylobates*) is a skilled brachiator. When he is traveling fast from tree to tree, he carries his legs fully retracted; under these circumstances they have no locomotor function. Note the very long arms and long hands.

It is noteworthy that the gibbon is probably the only nonhuman primate that occasionally adopts a bipedal run along a branch. This is more common than a quadrupedal posture on large branches. He will also run bipedally when crossing open spaces. We shall return to this observation when we discuss the origin of man's bipedalism.

The heavily built orang-utan has many of the anatomical characters we associate with brachiation, but he is perhaps better described as a climber. While he regularly uses his arms above his head, he usually takes weight on his feet too, which grip branches below and to his side. Although the forest habitat of the orang (now found only in Borneo and Sumatra) is somewhat different from that of the gibbon, it appears that the differences in their locomotor behavior and adaptations are due more to their difference in size than any other single factor. The gibbon is a small and fast moving animal, while the orang is heavy and usually slow moving; he has no predators but man. When resting in the trees, he usually holds onto a branch above his head; although he rarely brachiates, his long arms probably supply more locomotive power than his legs. He sometimes walks bipedally on the ground, but more frequently he walks quadrupedally, with clenched fists and feet, though occasionally his hands are used *plantigrade* (Tuttle 1967). The orang is primarily an arboreal animal and has no stereotyped locomotor pattern for terrestrial use. His arboreal adaptations include:

a) Great relative elongation of the arms, as in the Siamangs; both have relatively longer arms than even the gibbons.

b) Great elongation of the fingers, but reduction of the thumb to a short, almost vestigial, digit (see Fig. 6.13E).

c) Great mobility of the shoulder and wrist, with independent finger control.

d) Powerful development of the shoulder, arm and finger musculature (Tuttle 1969).

2. *African Apes.* The Asian apes are less closely related to us than the African apes, so we must discuss their locomotor adaptations in some detail. The chimpanzee and gorilla (*Pan troglodytes* and *P. gorilla*) are closely related and have many features in common. The most striking difference is one of size: gorilla males may weigh up to 400 lbs. with a mean of about 340 lbs., while chimpanzees average only 110 lbs. (the orang male averages 160 lbs.). They live in neighboring territories of the rain forests of central Africa, though there may be some overlap northwest of the Congo river. There are at least two subspecies of gorilla, the best known of which are the Mountain gorilla and Lowland gorilla; the former is adapted to high altitude woodland with small trees, the latter to dense tropical rain forest. The chimpanzee is usually divided into two species, *Pan troglodytes,* the common chimpanzee, and *Pan paniscus,* the pygmy chimpanzee or bonobo. *Pan troglodytes* is sometimes divided

into three subspecies, the western, central and eastern types, though their morphological differences are very slight. (Fig. 3.9.)

Variations in anatomy, locomotor behavior, diet, and social behavior throughout these species of the genus *Pan* are not yet well known, as only limited field work has been done in this part of Africa. It is, however, possible to make some generalizations on the basis of our limited existing knowledge.

The genus *Pan,* as we have said, occurs in species of three different sizes, and locomotor behavior, feeding, and diet vary accordingly. The larger the animal, the more time it appears to spend on the ground. When arboreal, these species show a rather varied repertoire of locomotor patterns. The smaller species, and juvenile chimpanzees and gorillas, are often seen to brachiate like the gibbon, though less rapidly. The larger animals climb more slowly using all four limbs. Unlike the Asian apes, these species also have very well-defined terrestrial adaptations. They walk quadrupedally on the soles of their feet and the phalangeal knuckles of the hands (Fig. 3.5). Appropriately, the skin on the *outside* of the middle finger bones is tough, hairless friction skin, such as we find on the palms of our hands, and the wrists are strengthened by alterations in the structure of the small wrist bones (the carpal bones) to give better support. This curious adaptation gives extra height to the animal's head when moving quadrupedally, an advantage where the undergrowth or grasses are high, as it allows more effective vision. All species will stand bipedally when displaying or to get a better view, and the chimpanzee in particular (the best known species both in captivity and the wild) will often walk bipedally over short distances. But bipedalism has never become a major locomotor adaptation among the apes.

It seems clear that the terrestrial adaptations of the African apes are secondary, in the sense that they evolved following a period of arm-swinging suspensory locomotion. In this character, the genus may once have shown some similarity to monkeys like *Presbytis* or *Colobus.*

The brachiating gibbon and slow-climbing orang have both lost their tails during the course of their evolution, as have the African apes. This development seems to be associated with a suspensory mode of locomotion where the function of the tail as an aerodynamic organ of balance is no longer as appropriate as it is in the running and leaping quadrupedal monkeys. In a mainly terrestrial form like the gorilla, the tail is even less necessary.

The differences in locomotion and general anatomy between the

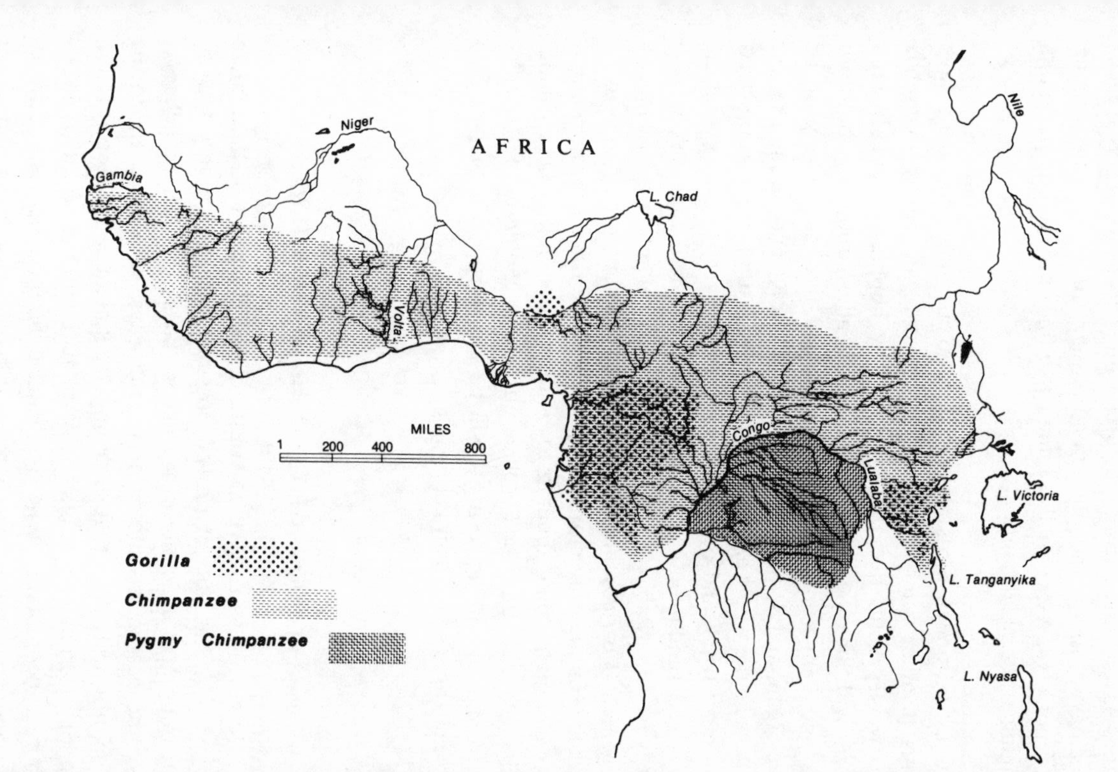

FIG. 3.9. Map of Central Africa showing the approximate distribution of the three species of *Pan*. The data for much of the area are very limited and unreliable, and these rare species may today occupy a much smaller range than that indicated. The continuity of distribution is illfounded, thanks to increased deforestation, cultivation, and hunting.

terrestrial monkeys on the one hand, and the fairly terrestrial African apes on the other, can be accounted for by considering their arboreal past. Terrestrial monkeys have evolved from quadrupedal (group 1) monkeys; African apes have evolved from creatures with suspensory arboreal locomotion such as we find in the gibbon and orang, and in the arm-swinging (group 2) monkeys. As we shall see, man also finds his early origin in this arm-swinging arboreal adaptation.

3. *Men.* Although birds are bipedal, no mammals but man have evolved his peculiar kind of alternating bipedal gait. Many mammals move on two legs but, like the kangaroo, by hopping. Probably man alone shares with the dinosaurs the symmetrical alternating walk, but both dinosaurs and kangaroos have brought the tail into use as a prop or propulsive organ, and neither group shows any equivalence to man's erect posture. Among the whole animal kingdom, man is unique in his combination of an erect trunk and a bipedal walk. In this respect, his evolution, which involved fundamental changes in structure and function, is perhaps nearly as remarkable as the evolution of marine mammals. The two hominid genera *Australopithecus* and *Homo* we know to be plains-living and bipedal. (Having only cranial fragments of the third genus, *Ramapithecus,* we can say nothing about its locomotor adaptations.) Quite fundamental changes must have taken place during the evolution of *Australopithecus* from a tree-living form. It will be our purpose in chapter 5 to trace these changes so far as it is possible.

3.6 Neural correlates of arboreal locomotion

In the functional analysis of the evolution of the human body, it has not yet proved possible, except in a very few instances, to determine in detail the neural correlates of a functional change. Studies of the descending motor nerve pathways from the brain to the muscles do, however, show that some important changes have occurred in the evolution of primates that may be correlated with the evolution of primate arboreal locomotion. Some significant differences between primates and other mammals, on the one hand, and between the lower and higher primates, on the other, have been demonstrated in these descending nerve tracts.

Among mammals generally we find a relatively direct system of nerve fibers running from the enlarged cerebral cortex of the brain to the motor *neurons* (nerve cells) of the spinal cord, which supply nerves to the muscles and effect their contraction. These *corticospinal fibers* (also called the *pyramidal system*) are exclusive to mammals and reach their greatest development in the higher primates. This phylogenetically new system (that is, one relatively recently evolved) is complementary to the phylogenetically ancient and indirect *extrapyramidal* tracts, the fibers of which arise in the brainstem rather than the cortex and reach the motor neurons via a sequence of relay neurons. The evolution of the corticospinal tracts is correlated with the evolution of the cerebral cortex (see 2.6, 8.6); recent experiments have shown that they are responsible for precise motor control and that they add to muscular activity both speed and a capacity for very precise movement. This capacity of refinement is superimposed upon that of the indirect brainstem systems of the extrapyramidal tracts that direct movement in a more general way. Figure 3.10 shows in diagrammatic form the difference between the arrangement of these two kinds of nerve tracts.

In all mammals the contraction of the muscles is brought about by nerves that find their origin in the motor neurons within the spinal cord. As has been said, these motor neurons are activated by connections from the descending nerve fibers from either the brainstem or the cortex. The connection is made by a relay neuron, the so-called *internuncial neuron*. However, it has been shown through studies on monkeys, chimpanzees, and man, that in primate evolution an increasing number of corticospinal fibers make direct connection with the motor neurons and bypass the internuncial cells (Kuypers 1964). This is a most important primate evolutionary development, and it is especially interesting that the motor neurons of *distal* muscles (farthest from the trunk) receive more direct cortical fibers than do those of *proximal* muscles (nearest to the trunk). This finding correlates nicely with the evolution of the primate hand and foot, and especially with the evolution of the human hand. Kuypers has shown that these direct connections from the cortex to the motor neurons are not established at birth but develop during early life with the coming of precise motor activity.

In the evolution of man's motor-nerve pathways, we see therefore (1) a large increase in the number of direct connections between the cortex and the spinal cord that are shown to effect muscular speed and precision and (2) the appearance during primate evolution of a direct connection between these fibers and the motor neurons, especially in connections to the distal musculature of the hand and foot. Here we

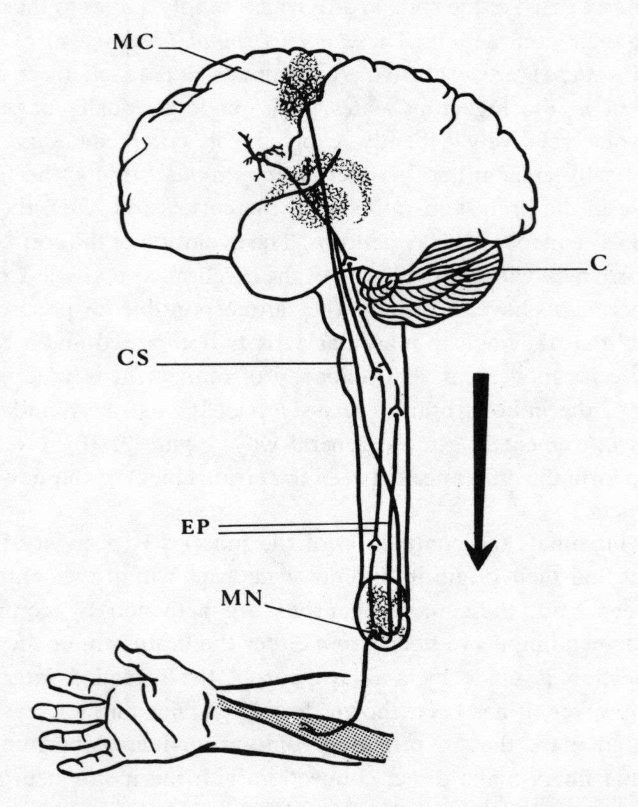

FIG. 3.10. Diagram of motor (descending) nerve fibers. *CS* labels a corticospinal fiber that in higher primates forms a direct link between the motor cortex (*MC*) and the motor neuron (*MN*) of the spinal cord. *EP* labels two of the extrapyramidal fibres that connect the cerebral cortex through a complex sequence of relay neurons in the brainstem to the motor neurons of the spinal cord. (Internuncial neurons omitted.) (Redrawn after Noback and Moskowitz 1963.)

have a clear-cut neurological correlate of the evolution of the human hand, a striking and significant organizational improvement in motor control.

Precise muscular control is not, however, achieved by fast muscular contraction. The extent of such contraction is controlled by information returned to the brain by special sensory nerve endings within every

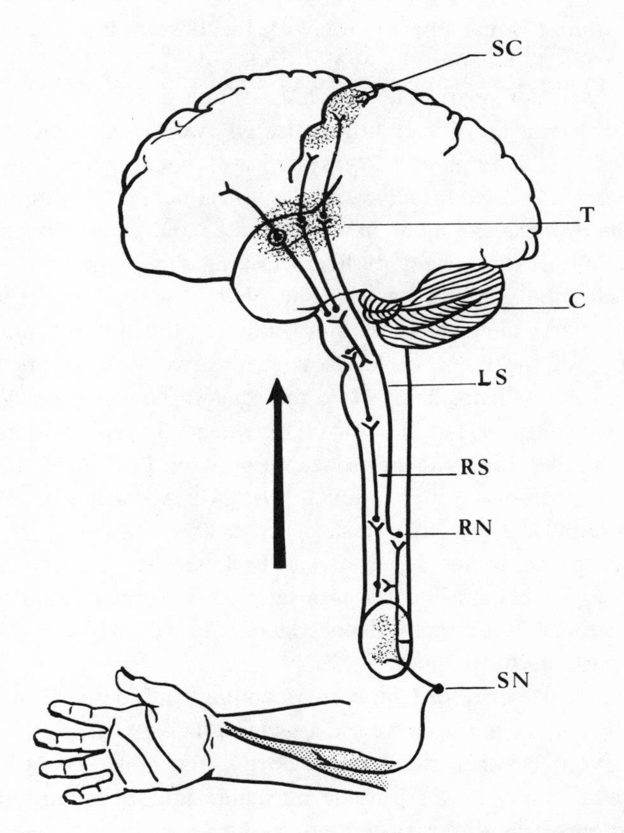

FIG. 3.11. Diagram of sensory (ascending) nerve fibers. *LS* labels a fiber of the lemniscal system, which forms a relatively direct link between the sensory neuron (*SN*) (from proprioceptors in the muscle shown here, but also from the skin) and the somatic sensory cortex (*SC*) via relay neurons (*RN*) in the brainstem near the cerebellum (*C*) and in the thalamus(*T*). *RS* labels a reticular fiber that constitutes a relatively indirect link between the sensory neuron and the cortex (Redrawn after Noback and Moskowitz 1963.)

muscle, the *proprioceptors.* The proprioceptors feed back to the brain information as to the extent of the effected contraction and cause adjustment in the frequency of the motor-nerve impulses. The precise control of muscular contraction by proprioceptors is believed to be carried out in the cerebellum (Fig. 3.11, *C*) as well as in the cerebral cortex.

Ascending fibers from the proprioceptors as well as from the sensory nerve endings in the skin (the somatic sensory nerve tracts) are therefore of almost equal importance with the descending fibers in the evolution of primate locomotion. While studies of these ascending tracts in mammals have not yet revealed such a striking novelty as that which we have recorded in the descending pathways, we do have evidence of two equivalent and complementary systems of nerve paths like those described above. The *lemniscal system* is prominent in mammals and reaches its highest development in the primates and especially in man. It is a relatively direct system of fibers linking the peripheral sensory nerve cells with the cerebral cortex. The phylogenetically older *reticular system* is, as its name implies, more indirect and diffuse and is the more primitive of the two. Nerve tracts of the two systems are shown diagrammatically in Figure 3.11. Here the sensory neuron, which lies in a small knot (*ganglion*) of nervous tissue near the spinal cord, can transmit an impulse from the proprioceptors in the muscle of the forearm to both the lemniscal and reticular systems. According to Herrick (1956), the lemniscal system is fast, specific, and analytical, the reticular system nonspecific and integrative. The latter has been shown experimentally to be responsible for arousal and wakefulness and effects a general condition of alertness. Both systems are well developed in primates, and particularly in man.

When we look, therefore, at the ascending and descending nerve pathways in primate evolution, we see a general increase in relatively fast direct connections between the cerebral cortex, the sense receptors, and the muscles. The precision of primate muscular activity is due not only to the evolution of the direct systems of nerve fibers, however, but also to a very complex feedback control mechanism, which depends on both descending and ascending nerve tracts as well as upon an increasingly complex organization in the brain.

3.7 Neural correlates of binocular vision

IN ORDER TO UNDERSTAND the evolution of primate vision, it is necessary to understand in some detail the structure and operation of the visual system as a whole.

The brain is a spatially organized and oriented organ in which the two sides are connected by nerve tracts to the opposite sides of the body.

Through these motor and sensory connections, the two sides of the brain are, as it were, concerned with the opposite two halves of the environment, the left hemisphere with the right half and the right hemisphere with the left half. The optic nerves, which carry the input to the brain from the eyes, therefore cross at a structure called the *optic chiasma*. The evolution of stereoscopic vision has, however, involved an overlap in the field of view of each eye, so the light from a single object can enter both eyes at the same time (Table 3.4). There is some overlap—some binocular vision—in many mammals, especially cats (whose eyes face forward), but in the primates this feature reaches its full development.

TABLE 3.4 APPROXIMATE EXTENT OF VISUAL FIELDS OF CERTAIN MAMMALS (WITHOUT HEAD MOVEMENT)

Mammal	*Angle between Optical Axis of Eye and Midline of Head*	*Total Visual Field**	*Greatest Width of Binocular Field*
Horse	40°	360°	57°
Dog	20°	250°	116°
Cat	8°	287°	130°
Monkey, ape, and man	0°	180°	120°

* Note that, since in the primates the eyes come to point straight ahead, the total visual field is reduced; this reduction is compensated for by free movement of the head. Data from Walls (1963).

In primate evolution the eyes have moved toward each other in the face, so in many lower and all higher primates the two fields of view overlap almost entirely. As sense organs, the eyes are no longer concerned individually and exclusively with the two halves of the environment; instead, they are each receiving information from both halves of the environment, which involves a new distribution of the optic nerve tracts. We find that in animals with stereoscopic sight the nerve fibers are so arranged at the chiasma that those from the left side of *both* eyes pass to the left side of the brain and similarly for the right side (Fig. 3.12). That is to say, a portion of the nerve fibers do not cross at the chiasma, and, as the extent of the binocular field increases over that of the monocular field, the noncrossed fibers increase over the crossed fibers until in man nearly 50 per cent of the fibers from each eye pass to the same side of the brain. Information from each eye is thus equally distributed to the two sides of the brain, which receives a double image of the environment.

It is on the basis of the slight differences in this double image that the brain is enabled to assess the relative and, with experience, the absolute distance of different objects. Visual knowledge of the relative distance from the animal of different parts of the environment makes possible three-dimensional perception. Depth is added to breadth and height.

The estimation of distance on the basis of a difference in the analysis of form (that is, a slightly different view of an object) involves a comparison of the images of the two eyes. The difference in the two images of a near object will be much greater than that in the images of a distant object. It follows that to interpret depth at a distance involves greater visual discrimination than such interpretation at a close range. This fact no doubt goes some way to explain the degree of visual discrimination available to primates, for a high level of discrimination is essential to an accurate analysis of the difference between the two images received by the brain. The eyes, it appears, have evolved as precision range-finders; the ability of man to read fine print looks like a byproduct of this adaptation.

We are only beginning to get some understanding of the process of visual analysis in the brain. Experiments with the cat (Hubel 1963) have shown that analysis begins in the *retina* and continues in the *lateral geniculate body* and then in the *visual cortex* (Fig. 3.12). Hubel has shown that the pattern of impulses from the retina is analyzed by banks of neurons for individual components of pattern, color, and movement. We must assume that the result of the analysis is then compared in some way with visual patterns stored within the brain and that such comparison may result in "recognition" of certain environmental phenomena.

In animals with stereoscopic vision the analysis must also include a simultaneous comparison of images, and the differences between them are interpreted as distance and depth in accordance with memory traces of previous visual experience. Though the complexity of this process is immense, it seems quite clear that to turn an image into a view of the environment (that is, perception), it is essential to compare the image with recorded memory of other images with their context of sensory and motor experience. Perception involves memory, and perception of the detailed kind available to higher primates (and eventually man)

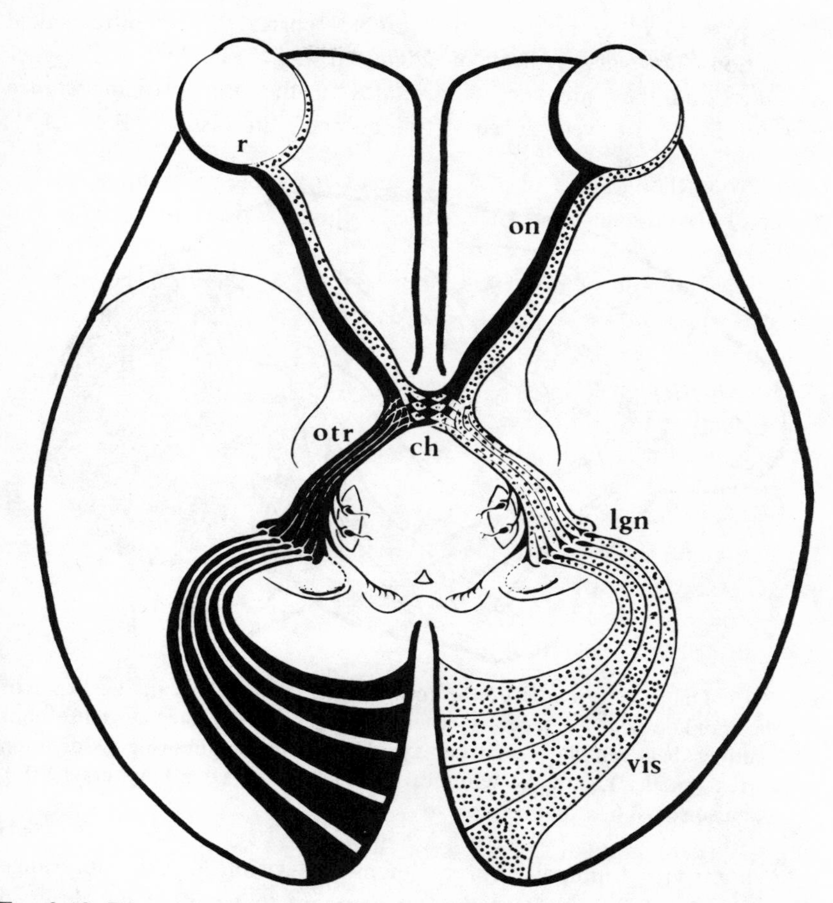

FIG. 3.12. Diagram of the human visual system: *r*, retina; *on*, optic nerve; *ch*, optic chiasma; *otr*, optic tracts; *lgn*, lateral geniculate body; and *vis*, tract to visual cortex. Fibers originating in the inner or nasal halves of the retinae cross at the chiasma so the visual cortex of each side receives signals from the same half of the field of view—that is, the opposite half. The signals, however, represent two slightly different images, and on the basis of these differences depth and distance are perceived. (After Polyak 1957.)

involves increasingly extensive analysis and increasingly extensive memory banks. As we shall see (8.6) the parts of the brain believed to function as memory banks (in particular, the *temporal lobes*) have expanded rapidly in primate evolution. Recent investigations by Kuypers *et al.* (1965) show the tremendous extent to which nerve tracts from the visual cortex pass to other areas of the cerebral cortex. These intracortical connections are very typical of advanced mammals and the higher primates, and they are surely correlates of the immense importance of vision in almost every aspect of the animal's life history (Fig. 3.13).

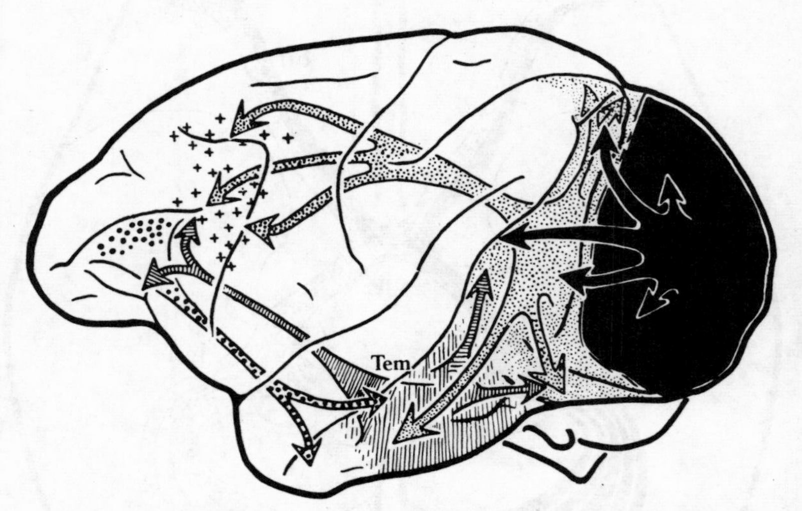

FIG. 3.13. Diagram of some of the intracortical connections in the brain of the monkey *Macaca* recently revealed by Kuypers. Note the dense connections between the visual cortex (*black*), its neighboring association cortex (*shaded*), and the temporal lobe (*Tem*). (After Kuypers 1965.) See also sec. 8.6.

We are visual animals. Three-dimensional vision gives a wide range of precise information about the environment that cannot be obtained by any other means. Because light travels in straight lines and because the nature of reflected light is determined by the chemical and physical structure of objects, very precise data about the nature of objects can be obtained at a distance. Stereoscopic vision and competent motor investigation (in particular, manipulation) make the world view of the higher primates unique both in quality and in kind. Primates alone have come to know the structure of the environment in terms of both pattern and composition. They see the environment as a collection of objects rather than merely as a pattern, and the recognition of objects is, as we shall see (9.2), the beginning of conceptual thought. With the coming of stereoscopic vision, the primate can begin to perceive non-spatial abstractions, and this advanced perception finds its foundation

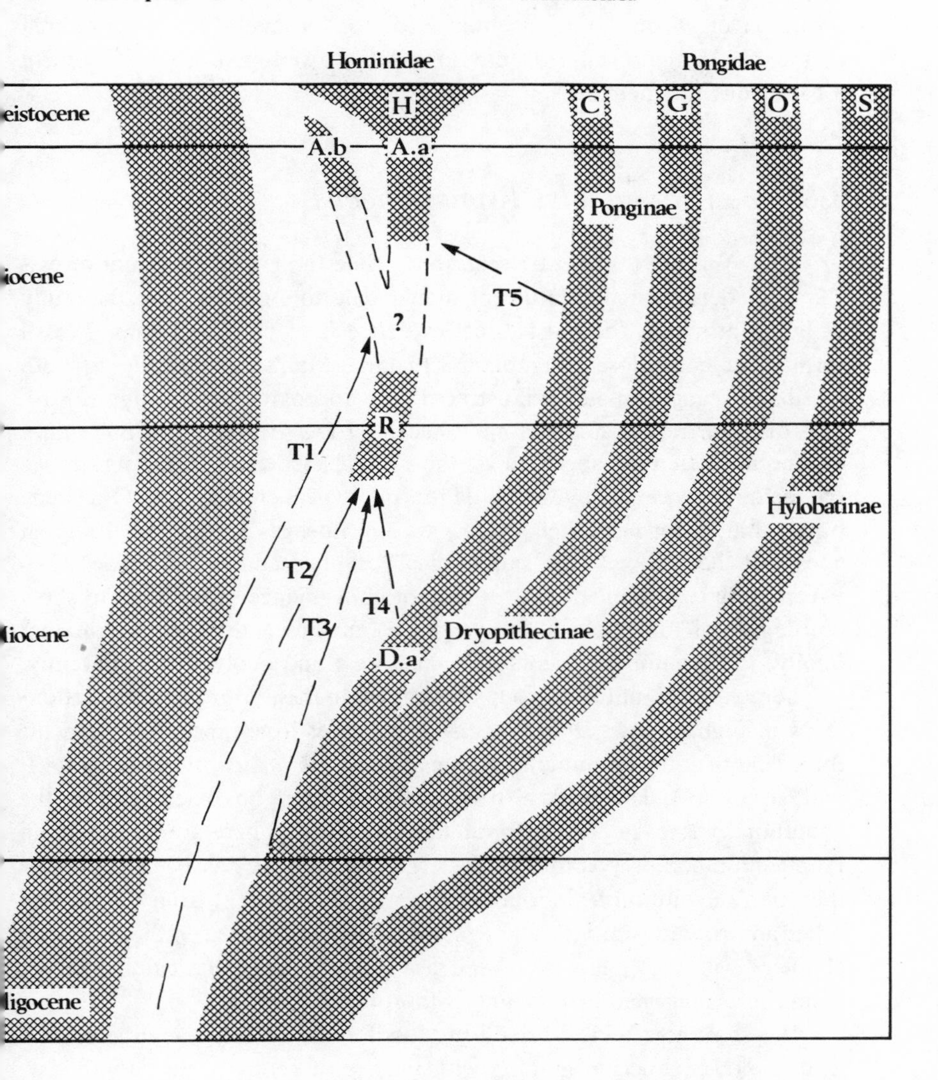

FIG. 3.14. Diagrammatic representation of lineages suggesting the course of the evolution of the Old World Anthropoidea. The origin of the different groups is not known with certainty and may date from the Oligocene or earlier. Important genera are labeled as follows: *Homo* (*H*), *Australopithecus* (*A*), *Ramapithecus* (*R*), *Dryopithecus* (*D*), Gorilla (*G*), chimpanzee (*C*), *Pongo* (*O*), and the gibbons *Hylobates* and *Symphalangus* (*S*). *T1–T5* are some possible hypothetical origins of the Hominidae.

in combining the analytic functions of the two sides of the brain. The evolution of vision is one essential basis for the evolution of an animal that came to understand its environment so well that it could control it for its direct benefit.

3.8 Fossil hominids: Ramapithecus

WE HAVE ALREADY seen that, while the greatest amount of evidence for the nature of man's ancestors comes from the study of living primates, the most direct evidence is to be found among fossil forms that bear close resemblance to him. Man's relationship to both fossil and living primates is illustrated very approximately in Figure 3.14. The diagram represents no finally accepted fact of evolution but rather a hypothesis that is supported by the present evidence. In the figure, we see the monkeys evolving with little morphological change from Oligocene times, and, from near their origin, we see lineages leading to the great apes and the gibbons (the subfamilies Ponginae and Hylobatinae). Between these two groups (drawn here for convenience rather than to show the degree of morphological divergence) is shown the aberrant human family, the Hominidae, which left the forest and evolved so differently.

Some important fossil groups are shown in their approximate relationships in Figure 3.14: *Dryopithecus,* a genus of fossil apes (*D*); and the three genera of the Hominidae, *Ramapithecus* (*R*), *Australopithecus* (*A*), and *Homo* (*H*). The figure also shows five different possible routes for the evolutionary lineage of the Hominidae (*T1–T5*). There are of course an infinite number of possible routes; for example, it has been suggested that man's evolution from a prosimian was independent from that of any other anthropoid primate. The evidence to be considered in the following chapters will show, however, that any other lineage basically different from those suggested here is highly improbable.

We shall now review briefly the fossil evidence for human evolution in its time sequence, beginning with the oldest genus of the Hominidae, *Ramapithecus.* The inclusion of this genus in the Hominidae is in fact still controversial, but since a majority of workers recognize this form as a hominid, we shall include it here. The fossils of *Ramapithecus* fall into two groups, each of which has the status of a species: *R. wickeri* from East Africa, and *R. punjabicus* from the India-Pakistan border. *R.*

wickeri is the oldest known hominid and is represented by the fragments of a single individual: an incomplete upper dentition in two fragments of the maxillary bone, and part of the left side of the lower jaw or mandible (Fig. 3.15). They were found by the paleontologist L. S. B. Leakey in 1961 at Fort Ternan in Kenya* and the upper dentition was described by him as hominid in character.† The lower jaw fragment from the same site was later recognized to belong to the same individual. The volcanic deposits at Fort Ternan which overlie the bones are estimated to be 14 million years of age (Evernden and Curtis 1965) and so fall into the Late Miocene period.

This is a fossil of great importance, bearing characters of both man and ape. On superficial examination it seems very ape-like, far more so than any other hominid, but the total morphological pattern is not that of an ape and varies from it quite specifically in the direction of man. The features which distinguish it from a pongid are its broad, shortened, flat, compressed molar teeth which are more or less of equal size and generally man-like (see 7.5 and Fig. 7.7). The canine teeth, which are big in apes but small in men, are here ape-like in shape but more the size of human canines, and the first premolar is just half-way between that of an ape and a man (for details of the teeth, see 7.6–7.9). The front teeth are relatively small as in men and unlike those in apes, and the face is clearly shorter and deeper than the face of an ape. In summary it is fair to say that these jaws are close to what might be predicted for an intermediate form between ape and man.

Long before the discovery of *R. wickeri*, fragments of another group of fossils had been found in India and Pakistan, in the Siwalik Hills. The first specimens were found about 1910, but only later were their manlike features recognized by Lewis (1934) and Simons (1964). These fossils are distinct from the Fort Ternan jaws and are more manlike. This fits in well with the more recent age assigned to them which is somewhat less than fourteen and possibly nearer nine million years BP (before the present), and puts them on the Miocene/Pliocene boundary (see Fig. 3.14.) The age estimate is at present based on faunal correlation, that is, the similarities of the fauna accompanying the fossils to those of other

* For the location of this and other sites mentioned in this chapter, see Fig. 3.23.

† Originally named *Kenyapithecus wickeri* by Leakey, the specimen was later found to be similar to the North Indian species described below, so it was placed in the same genus as the latter, *Ramapithecus*.

well-dated sites. It does not have the authority of an absolute age determination by the potassium/argon method.*

This material, now called *R. punjabicus,* has been listed and described by Simons (1968), and consists of five mandibular and two maxillary fragments with teeth from different individuals. It differs from the older African form by carrying the incisors in such a way that they form an almost vertically oriented cutting edge, as in man. (In the African form they are rather procumbent, that is, set in the jaw at a projecting angle.) There is no canine known, but the size of the socket suggests it was small, and the fragments as a whole have much in common with later hominids (Fig. 3.16). A later find consists of a single molar tooth from the Pliocene site Ngorora, in Kenya, which is believed to fall into this genus.

The hominid characters of *Ramapithecus* are perhaps most clearly

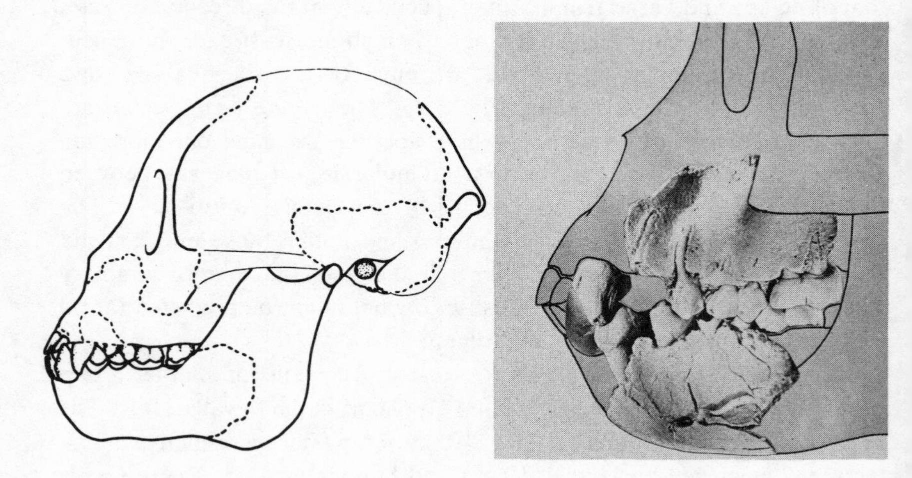

FIG. 3.15. On the left, a reconstructed drawing of *Dryopithecus africanus.* Dotted lines indicate the edges of the bone fragments; fine lines indicate the reconstructed shape of the missing parts. On the right a photograph of the left side of the fragmentary jawbone of *Ramapithecus wickeri.* Note the relative size of the canines. (Photo courtesy Peter Andrews.)

* Techniques for measuring the age of volcanic rocks by analyzing the ratio of radioactive substances in them have enabled us to get reasonably reliable dates for a number of fossil-bearing sites. The most useful method, known as K/A (potassium/argon) dating, covers the period with which we are most concerned, as it can be used on volanic ash and lava as old as 100 million and as young as one million years of age. There is no radiometric dating technique available at present that has proved of much value for more recent periods, until we reach about 40,000 years BP. The Carbon 14 method can be used from that time to the present.

seen when the genus is compared with the contemporary genus of fossil apes, *Dryopithecus* (Figs. 3.15 and 7.7). Probably the best known of these fossil pongids are the East African specimens, previously called *Proconsul* (Clark and Leakey 1951, Napier and Davis 1959). *Dryopithecus africanus* is known from its skull, teeth, and forelimb, and because of its value as a related comparative fossil, is referred to frequently in this book. In those parts which can be compared with *Ramapithecus* it contrasts very clearly, and in its teeth it carries most of the features of the living great apes (*Pan*), to the ancestor of which it probably lies quite close. The remains of *D. africanus* come from Miocene deposits of about twenty million years BP (Bishop 1964), but more recent specimens of other species of *Dryopithecus* have been found in Africa and Eurasia up to about eight million years BP.

In summary, most paleontologists agree that on the basis of the known evidence (which is weak) the fossils of *Ramapithecus* should be classified in the family Hominidae. It is possible that *Ramapithecus* will yet prove to be more ape-like or more monkey-like than it now appears on the basis of the known fragments. The hypothesis we select for man's ascent

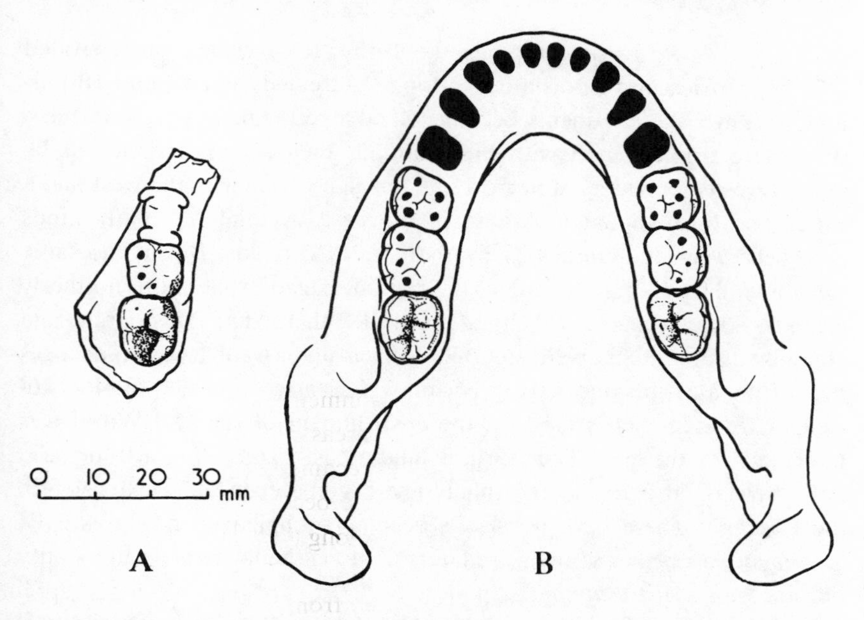

FIG. 3.16. Mandibular fragment of *Ramapithecus* (*A*) and mandible of a later hominid from Swartkrans (reconstructed) (*B*). Note the similarity in size and shape of teeth. (After Simons 1964, and Robinson 1953.)

will depend on the interpretation we place upon these fragments (Fig. 3.14). Their structure and significance will be further discussed in chapters 7 and 11. It should be added that biochemists have claimed that similarity in blood chemistry between man, gorilla, and chimpanzee exclude a long independent lineage such as this early fossil hominid implies (Wilson and Sarich 1969). Others, such as L. S. B. Leakey, have gone to the opposite extreme and recognized hominids from even more remote times, such as *Kenyapithecus africanus* of the early Miocene, but the hominid nature of these very early finds is not generally accepted. Because any fossil specimen less human than *Ramapithecus wickeri* would be indistinguishable as a hominid, we can safely put this specimen near the origin of man's distinct lineage. On this basis we might suppose that the Hominidae became separated as a lineage at least fifteen million years ago.

3.9 *Australopithecus*

DURING MIOCENE times, the Tethys sea—which had separated Africa and Eurasia for so long—retreated, and around 18 million years ago, the continents became joined together in a single land mass. Both were tropical, and with high rainfall, they became connected by richly forested corridors, a narrow one through Gibaltar in the west and a wider one through Saudi Arabia in the east. Animals of many kinds passed between the continents by the eastern corridor. But at the same time, a cooling and a lowering in the rainfall began in the more northerly latitudes so that a more arid climate prevailed there. The Alps, Urals, and Himalayan mountains were uplifted. Africa underwent tectonic changes too; rifting and volcanic activity began, which lasted well into the Pleistocene. A fundamental change in the environment of the Old World was underway; by the late Miocene, grassland areas began to replace forest in both Africa and Eurasia, and this biome became even more extensive in the Pliocene. These new grasslands became occupied by newly evolved grazing quadrupeds and their predators. Among the latter, we can identify our ancestor, *Australopithecus*.

This genus, descendant in all probability from *Ramapithecus,* is represented by fossils which occur mainly in Africa, with later members found in Southeast Asia. Unlike the fossils of *Ramapithecus*, which have been found only in forest environments, *Australopithecus* is invariably associated with grassland and savanna animals, often with some evidence of woodland as well.

The first African finds of *Australopithecus* were made in 1924 and were recognized as a missing link in man's ancestry by the anatomist Raymond Dart. They came from the site named Taung, in the Republic of South Africa. Further finds in the region of Johannesburg followed in the 1930's and since 1945 many more fossils have been added from this very fossiliferous region. Fossils were also found in Java in the 1930s and later, which can be placed in this genus, and quite recently a very large number have been discovered in East Africa (Tanzania, Kenya, Ethiopia). More finds are announced every year.

The fossils that fall into this genus represent two species that are not successive as those of *Ramapithecus* but represent two contemporary lineages. One lineage, *Australopithecus africanus*, leads on toward man, while the second, *Australopithecus boisei*, appears to become extinct. Fossils from the first species range from about 5.5 million to 1.5 million years BP; those from the second from around 3 to 1.2 million years BP. The genus *Ramapithecus* was, we believe, ancestral to the genus *Australopithecus*, but the latter split into these two lineages, only one of which led to man.*

Australopithecus africanus† is thus very well represented in the fossil record, being better known than any hominid species except *Homo*

* When we use the term species here, we have in mind something different from the biospecies described in 1.4. Fossil species are properly called chronospecies, which implies that they have a dimension of time as well as spatial and morphological variability. They represent a segment of an evolving lineage, a biological species extended through a long period, with the variation due to evolution added to what we would expect of any population at a given point in time (Fig. 3.17). The main difficulty that arises from the concept of the chronospecies is that successive chronospecies in a single lineage are continuous and not discrete; there is no real natural boundary between them. Those that represent two contemporary lineages can easily be distinguished by the methods usually employed to recognize biospecies, but where they succeed each other in a single lineage, the boundary must be artificially drawn by the taxonomist. The result is that many different opinions exist as to the best place to draw these boundaries; the same fossils are given different names by different authors and their nomenclature becomes confusing. The classification and nomenclature used here (Table 3.5) is the most modern and the simplest available; its main features are discussed and justified elsewhere (Campbell 1972).

† Under the terms *Australopithecus africanus* I am including fossils previously designated as *Australopithecus robustus, Plesianthropus transvaalensis, Paranthropus robustus* and *P. crassidens, Homo habilis, Telanthropus capensis, Meganthropus africanus, Meganthropus palaeojavanicus* and *Homo modjokertensis.*

sapiens. The most important collections of fossils come from sites east of Lake Rudolf in Kenya and near Johannesburg in South Africa, but by taking all the evidence into account we can get a good idea of the variation present in the species in both space and time. Spatial variation is due to environmental differences in different ecological zones (such as we find among modern human races), while temporal variation represents the evolution of these races from an earlier *Ramapithecus*-like form to a creature which eventually becomes *Homo*.

The best preserved skeleton belonging to this species comes from the Sterkfontein site near Johannesburg. There is a nearly complete skull (without mandible), parts of the scapula and humerus, some rib fragments, nine vertebrae from the thoracic and lumbar regions, the sacrum, pelvis, and a poorly preserved femur. Other parts of the skeleton of this species are known from fragments of other individuals from this and other sites. Some idea of the quantity of material is given in Table 3.6, and the reconstructed skeleton is shown in Fig. 3.18 alongside that of modern man.

From the remains we can get some idea of the size of these creatures, though, as we might expect, it varied a great deal. In the first place, the jaws show it to be larger than *Ramapithecus*, which was a small creature, as well as more manlike than the latter in the form and arrangement of the teeth. We do not, however, have a complete arm or leg from any one individual, so we do not know the proportions of the parts of the limbs (see 6.3) or the proportions of the body as a whole, arm to trunk or leg (5.1). Estimates of stature based on single leg bones vary from 4′3″ to 5′6″, which suggests that as a whole they were shorter than their successors, *Homo*. Evidence of the overall size and proportions of *A. africanus* are of great interest, but since they require the discovery of complete skeletons, they may elude us for some time. Meanwhile we can attempt to summarize the individual anatomical features of this species.

The skull of *A. africanus* is distinctly ape-like in appearance, though much more lightly built than that of the large African apes. It somewhat resembles that of the pygmy chimpanzee. While it is smaller than a chimpanzee skull, it has a brain as big as that of a gorilla; cranial capacity ranges from 435 to 815 cc. As a result the braincase is relatively large and well-rounded. Though smaller than that of a large chimpanzee, the face is relatively massive; the jaws are large and the brow-ridges (supra-orbital torus) are well developed. The big jaws have no chin, but a parabolic dental arcade and dentition like that of modern man is present. The

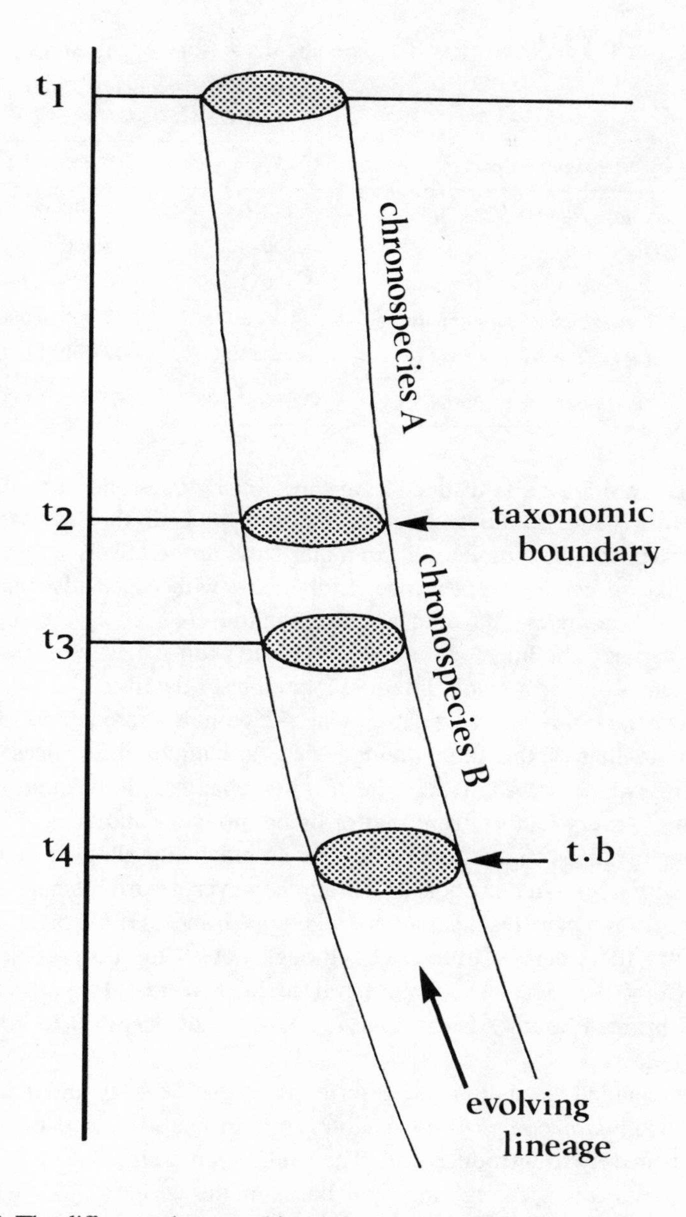

FIG. 3.17. The difference between biospecies and chronospecies are here shown diagramatically. An evolving lineage is divided by time lines at time t2 and t4 into two chronospecies, A and B. Biospecies can be recognized at any point in time (e.g., t3) and the term is used to describe the present day status of the lineage (t1).

TABLE 3.5 APPROXIMATE CHRONOLOGICAL RANGE OF HOMINID SPECIES

Hominid Species	CHRONOLOGICAL RANGE IN THOUSANDS OF YEARS	
	Known	*Assumed*
Homo sapiens	250–0	300–0
Homo erectus	1,300–400	1,300–300
Australopithecus africanus	5,500–1,300	6,000–1,300
Ramapithecus punjabicus	c. 9,000	12,000–6,000
Ramapithecus wickeri	c. 14,000	18,000–12,000?
Australopithecus boisei	3,000–1,200	6,000?–1,000?

canine teeth are quite distinct from those of the apes and very similar to our own, being spatulate like the incisors, and all the front teeth are much reduced. The molar and premolar teeth are relatively large, but still man-like in general appearance. Individuals with especially heavy jaws (usually the males) have an ape-like sagittal crest along the top of the skull to carry the huge jaw muscles, but the skull is better balanced than in the apes and as a result differs in a number of details.

The anatomy of the post-cranial skeleton is more obviously man-like than that of the skull, though there are minor differences in most features which we shall describe in later chapters. It is quite different in almost every feature from that of living monkeys and apes. The implications of this is that *A. africanus* was an efficiently adapted bipedal animal with a human dentition. His brain, however, was still small, and with his large jaws gave his head an ape-like appearance. Of his place in man's ancestry there can be little doubt, though not all the geographical races known of this species are ancestral to later forms. In particular, the small-brained South African race is possibly not ancestral to later races of *Homo*.

A detailed discussion and description of the anatomy and adaptations of *Australopithecus africanus* follows in later chapters, and our purpose here is merely to introduce him. The genus spans some 4½ million years and shows much change in form, between the earliest races which approximate *Ramapithecus* and the later and better known races that blend into his descendant *Homo erectus*.

The successful adaptation of *Australopithecus africanus* to woodland and grassland savanna evidently allowed this species to expand rapidly

TABLE 3.6 AN APPROXIMATE SUMMARY OF THE FOSSIL EVIDENCE FOR MAN'S EVOLUTION, EXCLUDING THE VERY NUMEROUS SPECIMENS, BOTH FOSSIL AND SUB-FOSSIL, OF *Homo sapiens*

Skeletal remains	*Ramapithecus*		*Australopithecus*		*Homo*
	wickeri	*punjabicus*	*africanus*	*boisei*	*erectus*
Crania			11	2	12
Fragmentary Crania			41	8	17
Mandibles			26	3	8
Fragments of mandibles and maxillae	2	10	110	16	21
Fragments of axial skeleton			14		2
Pelves			9		1
Humeri			10	3	3
Radii and Ulnae			10		1
Femora			17	8	14
Tibiae and Fibulae			7	3	2
Hand bones			19		1
Foot bones			23	2	
TOTAL	2	10	297	45	82
Isolated teeth	1	2	186	29	50

All fossilized skeletal remains are imperfect, and listing in this table implies only that some part of the named bone has been recovered. The totals do not represent numbers of individual fossil hominids, but numbers of fossil bone fragments. Besides the isolated teeth listed, some 850 further teeth are present in the fossil jawbones listed above, which brings the total to well over 1000 permanent and deciduous teeth.

over a wide geographical area. The evidence suggests that he spread from East Africa through southern Asia to Java, which he had reached by two million years BP. Ancestral populations must have left Africa by the northeast corridor long before this. The fact that we have no fossils

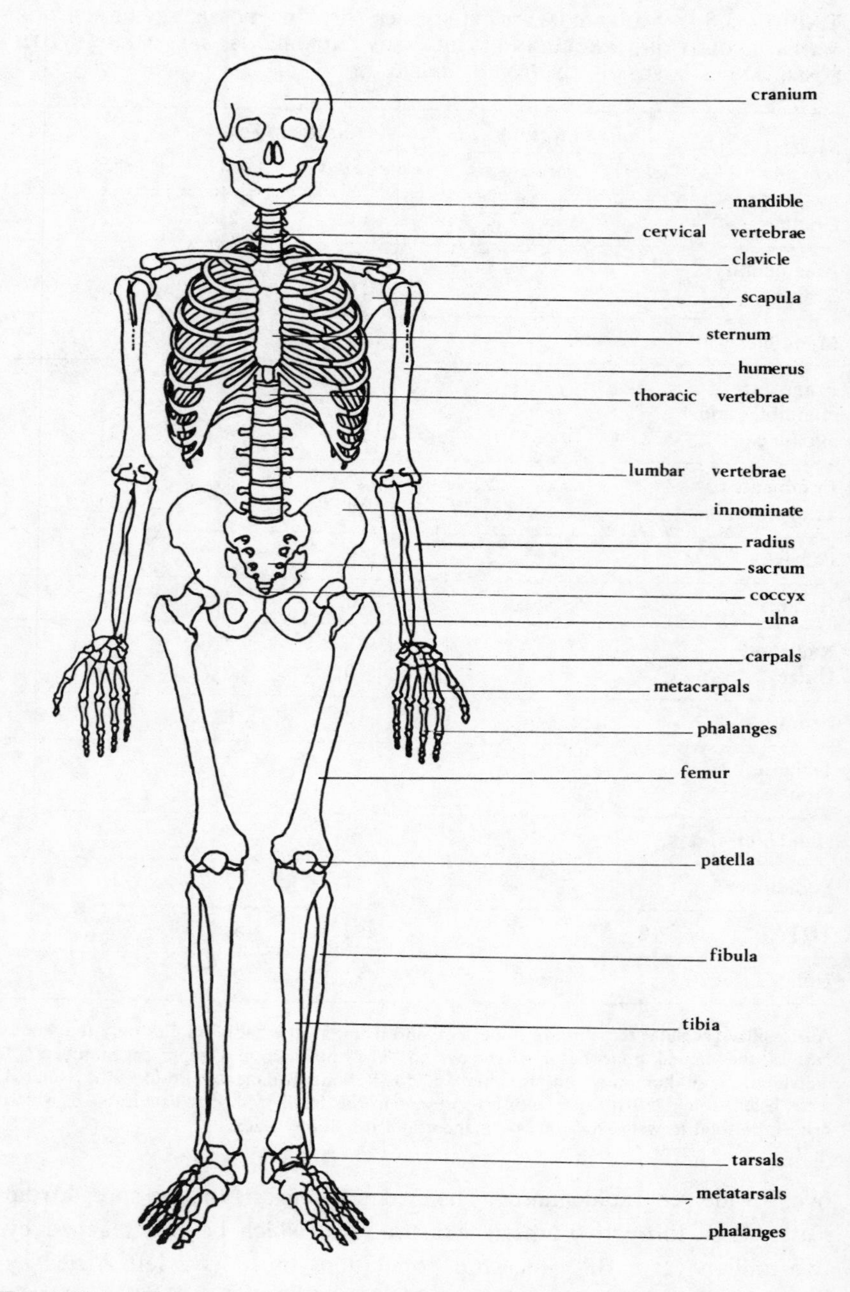

FIG. 3.18A. The human skeleton with the most important bone labeled.

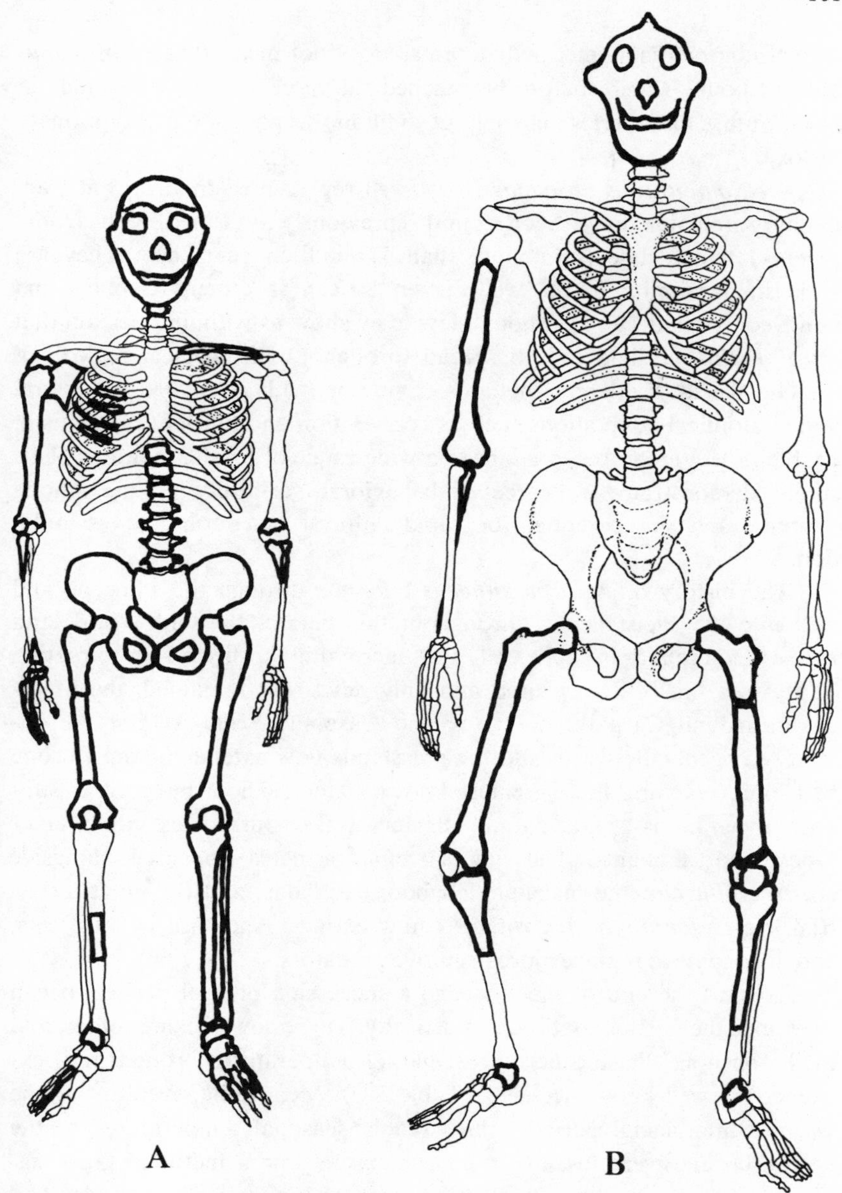

FIG. 3.18B. Reconstructed skeletons of *Australopithecus africanus* (*A*) and *Australopithecus boisei* (*B*) with those fossil bones which have been recovered drawn with heavy line. The skeletons are reconstructed from fragments of many different individuals, and in the case of B, the reconstruction is highly imaginative, in the absence of much fossil material.

from India or other intermediate areas does not mean that *A. africanus* did not flourish there before he reached the more remote woodlands of Java. Future discoveries may help us to fill in this part of our evolutionary history.

Australopithecus africanus is not well represented in Java, but fragmentary remains have been found (previously classified with *Homo erectus*) which date from more than 1.9 million years ago. They are surprisingly similar to the well-known african specimens (Tobias and von Keonigswald 1964). Though few, they show us without question that *Australopithecus* had indeed spread throughout the tropical regions of the Old World. Evidently, methods of hunting, with associated behavioral and anatomical adaptations such as cooperation and efficient bipedalism, enabled these creatures to adapt to a wide range of local conditions. This surely demonstrates a degree of behavioral versatility unique among animals, and the potential for rapid cultural invention and assimilation.

The history of *Australopithecus* takes us through the Pliocene and well into the Pleistocene. Throughout this period, the Old World land masses continued in upheaval and increasing aridity. The retreating Tethys sea finally disappeared and only relict seas remained, the Mediterranean and Carpathian (Black and Caspian) Seas. At first the climate was generally warm and the grasslands now extended from Europe to China, reaching their greatest known extent. These huge areas supported vast herds of grazing animals ancestral to our horses, cattle, antelopes, and elephants. The first savanna primates appeared alongside *Australopithecus*, the macaques, baboons, geladas, and the great terrestrial ape *Gigantopithecus*, which is now extinct. Ancestral hyenas, cats, and dogs came to replace more primitive predators.

Toward the end of the Pliocene a succession of cool periods began to change the northern latitudes drastically. These now became colder, and in the ensuing Pleistocene the oscillating temperatures brought into existence the well-known ice ages (Table 3.7). Meanwhile, even during the warmer interglacial periods, the equable seasonal temperatures of the early Pliocene were lost and cold winters became a factor of great significance for the evolving flora and fauna. As we shall see, they also had a profound effect on the evolution of *Homo*.

Before we leave this short introduction to *Australopithecus* it is appropriate to refer to that close relative of our ancestral species, *Australo-*

*pithecus boisei.** This group of early hominids, of which fossils have been found only in East Africa, responded to the grassland environment not as an adventurous tool-using predator, but more like the other higher primates mentioned above, as a thoroughly orthodox herbivore. His teeth underwent exactly the kind of evolutionary changes that we associate with herbivores; that is, a vast increase in the grinding surface of the molars and a reduction of the front teeth. That he is an *Australopithecus* is obvious at a glance (Fig. 3.18), but it is also clear that his adaptation was different and involved a commitment to the mastication and digestion of tough vegetable food. His molars are vast, bigger than those of any other

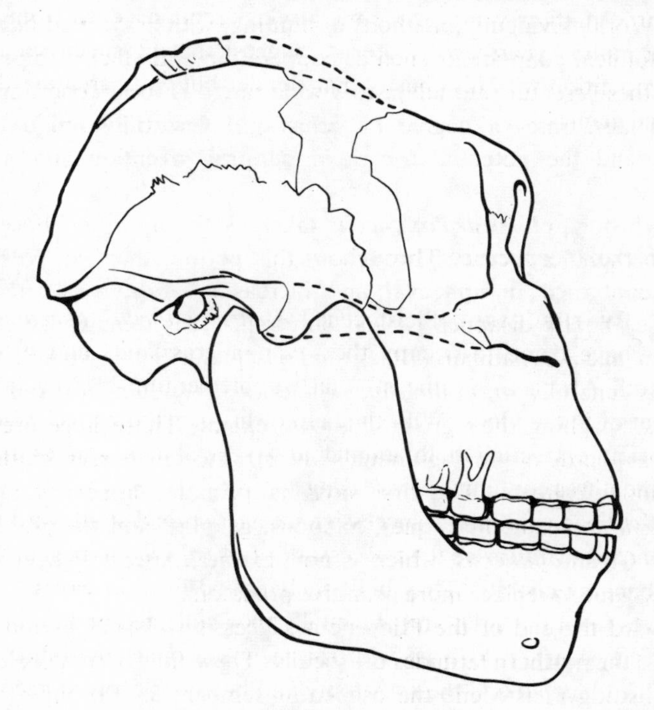

FIG. 3.19. The skull of *Australopithecus boisei* carries exceptionally large jaws and teeth, and the temporal muscles meet at the top of the cranium to form a crest. This drawing is of the first specimen found at Olduvai Gorge in 1959 and believed to date from about 1.75 million years BP. (Dotted lines are reconstructed and the drawing of the mandible is based on another specimen.)

* Here I am including fossils previously designated *Zinjanthropus boisei.*

primate, and his jaws are the most efficient in terms of the power brought to bear on these grinding teeth. Like many of the other herbivores he is a big animal, far more heavily built than his relative, *A. africanus* (Fig. 3.19).

Beyond this we know very little about him. The few bones of his skeleton we possess do not tell us for sure if he was bipedal or quadrupedal, though it is possible that he was a poorly evolved biped who dropped onto his front legs when stationary. But he had a brain of about 530 cc and brought all the advantages of advanced primate nature to his new environment. His distinct appearance in the fossil record begins about three million years ago, and his last remains date from a little over one million years ago. Theorists suggest that he succumbed in too close competition with his cousin, that slender but witty creature that was to become man.

3.10 Homo erectus

BY THE EARLY Pleistocene, *Australopithecus africanus* must have come to occupy the savanna grasslands and woodlands from South Africa to Southeast Asia, supplementing his vegetable diet with meat of many kinds. With the help of his developing technology, his efficiency at extracting both animal and vegetable resources from the environment was continually increasing.

As time passed, and the successive ice ages spread southward through north temperate Eurasia, the successors of *A. africanus* began adapting to the changing climate. They built shelters and lived in caves to protect themselves from the cold, and eventually they collected and domesticated fire. As they learned to cope with this new seasonal environment, their style of life changed and their anatomy and behavior evolved accordingly. Their success as hunters implies a well developed social structure and extensive cooperation. During the warm interglacial periods they expanded northward and eventually learned to survive the cold winters of the north temperate zone.

Fossils from this period are placed in our own genus, *Homo*, but designated *Homo erectus* to distinguish them from modern man. The earliest are found in Africa (Olduvai Bed II) and Java (Sangiran and Trinil) and are about one million years of age. Soon after this we have skulls and jaws of slightly more modern appearance from many sites in the Old World. These date from between 600,000 and 300,000 BP; the most valuable come from Algeria (Ternifine), Germany (Heidelberg), Hungary (Vértesszöllös), and China (Choukoutien (Table 3.7). They reach latitude 41° in China and 49° in Germany. By 750,000 years BP, hearths are regularly found in caves and other occupation sites, together with stone tools and bones.

As we might expect, the fossil remains of *Homo erectus* show much geographical variation in their general appearance and size. Unfortunately we do not have as big a collection of fossils of this species as of *A. africanus* (Table 3.6); nevertheless we can assess their anatomy fairly precisely. Although the skull still looks primitive, even somewhat apelike, the post-cranial skeleton was becoming very close to that of modern man—so close in fact that individual bones are often indistinguishable from those of *Homo sapiens*. Compared with his ancestor *A. africanus, Homo erectus* seems to have been taller and much more heavily built. In comparison to modern man, he was a powerful, muscular, stocky, thickset creature with perhaps slightly longer arms and shorter legs. He was, however, almost certainly as efficient at walking and as well balanced as we are.

While that thickset body might have been taken for an unusually robust modern, his head was quite different. He still carried clear traces of the large face and powerful jaws of *A. africanus,* and although his brain was now much larger (close in size to our own), his skull was heavily built, thick, and carried a powerful musculature to support itself and the jaws (Fig. 3.20). The total effect was to retain a head still clearly distinct from that of modern man (Fig. 3.22).

The earliest finds of this species date from 1891, when the Dutchman Eugene Dubois unearthed the first really ancient fossil skull ever found, at Trinil in Java, amid much controversy. He called it *Pithecanthropus erectus* ("erect ape-man"), and that is the origin of the name of this species. Dubois did not find any cultural remains with his discovery and seriously underrated the level of cultural achievement of these people. Today we have extensive evidence from sites further north, especially at Choukoutien, near Peking in China, where there is a vast cave that was occupied for perhaps 70,000 years by a band of hunters who left us their hearths, their tools, their food remains and their own bones in profusion.

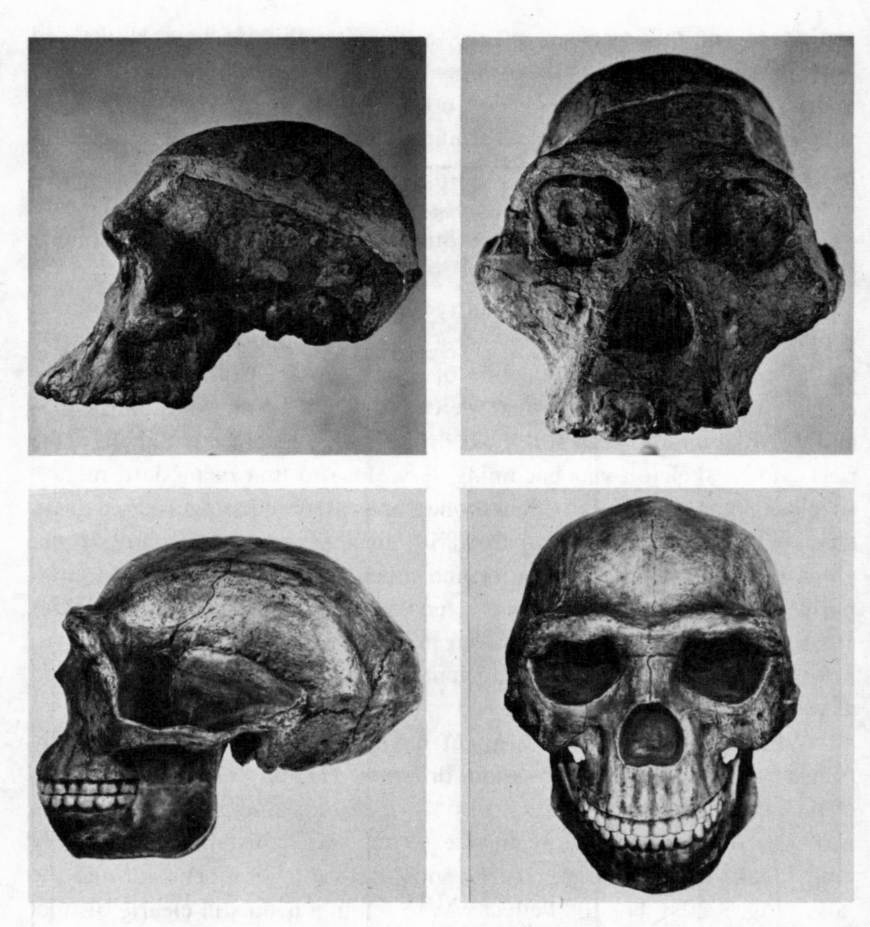

FIG. 3.20. A well preserved cranium of *Australopithecus africanus* from Sterkfontein (above); and a reconstruction of the skull of *Homo erectus* from Peking (below). (Above, by courtesy of the Transvaal Museum; below by courtesy Trustees of the British Museum [Natural History].) Not reproduced to scale.

Somewhat similar sites are known from Europe (Torralba, Ambrona, Terra Amata and Vértesszöllös) and from Africa (Ternifine and Olduvai Bed II). We now have a fairly clear picture of the adaptations and style of life of *Homo erectus*. His achievements were considerable: he mastered fire, and survived the northern winters by his superior technology and remarkable skill as a hunter.

For convenience we assign the species name *Homo erectus* to our

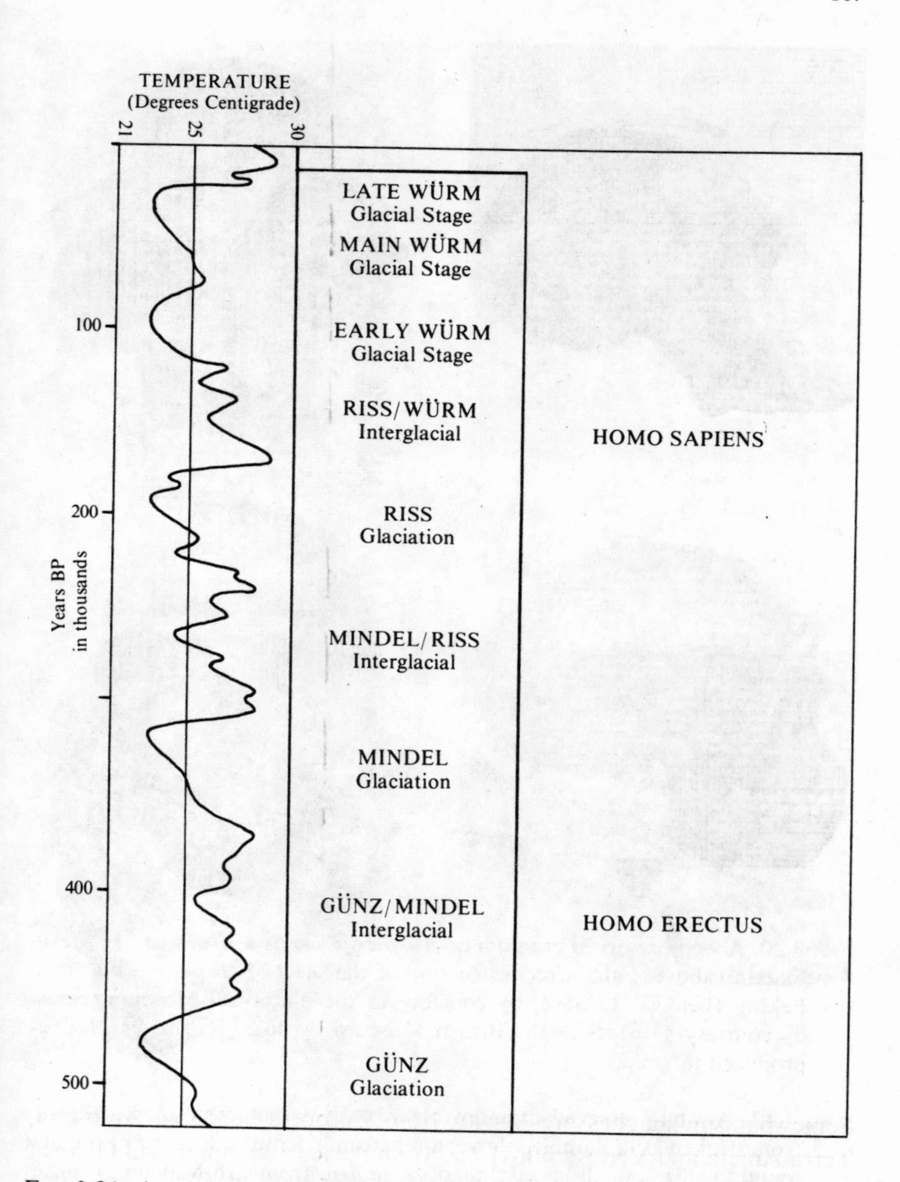

FIG. 3.21. Apparently random fluctuations recorded in the mean annual temperatures in the North Atlantic can be interpreted as a series of ice ages, as indicated here. The absolute dates associated with the ice ages are not clearly established, but the dates shown here are quite widely accepted. The species of *Homo* are indicated on the right side. For further details and discussion, see Oakley (1969).

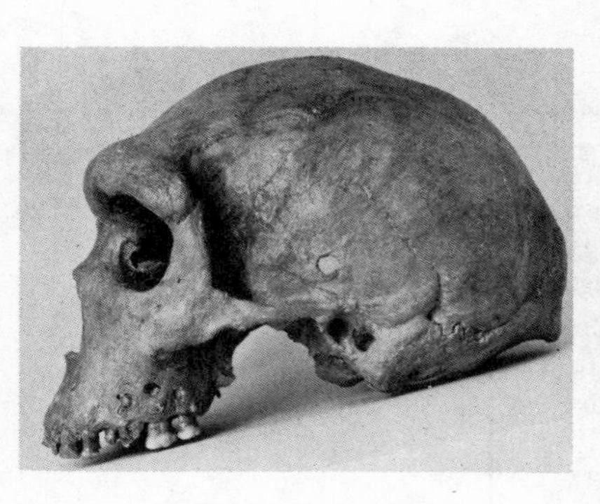

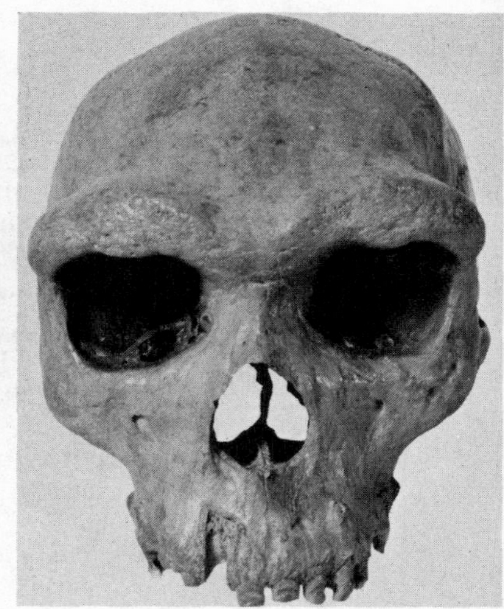

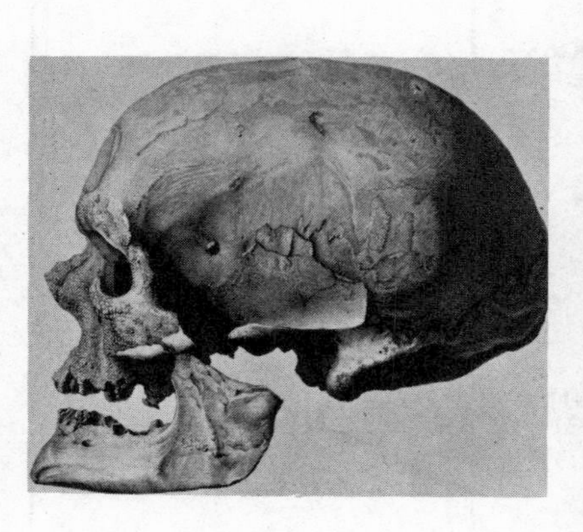

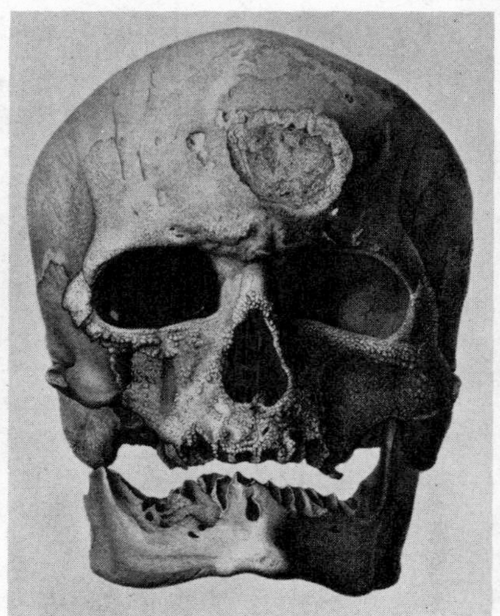

FIG. 3.22. A well-preserved cranium of an early large-jawed *Homo sapiens* from Broken Hill, Zambia, above; a nineteenth century drawing of the first found fossilized small-jawed skull of *H. sapiens* from Cro-Magnon, France, below. (Courtesy Trustees of the British Museum [Natural History].) Not reproduced to scale.

ancestral hominid lineage during the period 1.3 to 0.3 million years BP. After that time our ancestors become almost indistinguishable from ourselves.

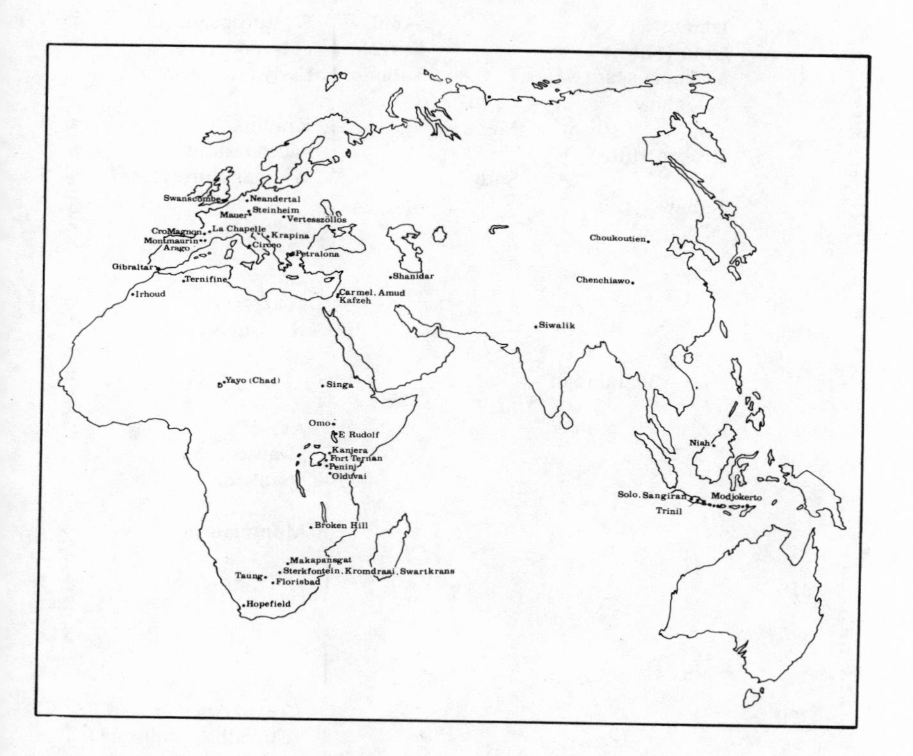

FIG. 3.23. Map of the Old World showing the locality of some of the important sites mentioned in the text and listed in the Appendix to this chapter.

TABLE 3.7 RELATIVE CHRONOLOGY OF SOME MIDDLE AND UPPER PLEISTOCENE FOSSIL HOMINIDS AND IMPORTANT PREHISTORIC SITES (IN PARENTHESIS); C = COLD, W = WARM

		Africa	*Asia*	*Europe*	*Climatic Stage*
	0				
		Anatomically modern man			
		Irhoud	Skhul	W. European	C Würm
		Omo (Kibish)	Kafzeh	Neandertals	
Upper Pleistocene	50	Makapansgat (CH)	Tabun	Early Neandertals	
		Saldanha			
			Mapa	Krapina	
		Broken Hill		Saccopastore	
	100		Solo	Weimar-Ehringsdorf	W Riss/Würm
		Rabat			
				Quinzano	
				(Lazaret)	
	150			La Chaise	
		Sidi Abderahman			C Riss
				Arago	
	200			Swanscombe	
				Steinheim	
				Montmaurin	
Middle Pleistocene	250				W Mindel/Riss
	300			(Terra Amata)	
				(Torralba/Ambrona)	
					C Mindel
	350	Ternifine			
	400		Choukoutien	Vértesszöllös	W Inter-Mindel

3.11 Homo sapiens

BY ABOUT 300,000 years ago, the people living in Eurasia and Africa were practically modern in appearance if not in culture. Their brain size was well within the range of modern man and their bodies were indistinguishable from ours, though some of the early populations may have been rather more robust. Only their heads still looked strange, with long skulls and heavily built faces and jaws. Through the evolution of *Homo sapiens* during the last 300,000 years, we see the final reduction of the jaws and the appearance of man's chin. As the jaws became smaller, the whole face shrank and receded under the brain case, so that it became surmounted by a vertical forehead. This changed the balance of the head, and the long narrow skull of *Homo erectus* became the rounded skull of many modern populations.

During this final period, which brings us to the present day, the succession of ice ages continued with two more major advances separated by warm interglacial periods (Table 3.7 and Fig. 3.21). Man's response to the ice was now to become his greatest cultural achievement. By about 70,000 years ago, man had made his last climatic conquest and developed the ability to survive the year-long cold of the arctic biome. The first evidence of this extraordinary feat comes from France, where the Neandertal people survived for some 10,000 years in a climate as rigorous as many Eskimos enjoy today. We do not yet know the whole story of their adaptation, but survive they did, to demonstrate that man was indeed the most adaptable creature on earth.

For the earliest period of *H. sapiens'* time on earth, the fossil record is extremely sparse. We have four substantial finds dated between 300,000 and 200,000 BP. They all come from Europe: Montmaurin and Arago (France), Swanscombe (England) and Steinheim (Germany), reaching north to a latitude of 52° (Fig. 3.23). The last three sites have yielded us skulls of great value (Howell 1960); here we find considerable variation in the extent of reduction of the jaws, as well as the usual variability in size, robusticity and cranial capacity. The impression given is that jaw reduction was proceeding faster in some areas (northern Europe) than in others (southern France).

Four later sites which belong to the next 100,000 year period have yielded fragmentary remains (Table 3.7). It is not until the final 100,000 years of our history that the fossil record supplies us with good evidence of the most recent stages of man's evolution, and even in this last period, we find the story is unclear. For its first half (100,000–50,000 BP), the known sites carry fossil men who still have relatively large jaws though they are modern in most other respects. These are the Neandertal, Rhodesian, Shanidar and Solo people (see Appendix to this chapter). Then between 50,000 and 30,000 years BP we see the big-jawed people disappearing and being replaced by small-jawed people like ourselves. The earliest men of this type are not generally intermediate but are like ourselves in *all* respects; they come from South Africa (Border Cave, Natal, c. 60,000 BP) and Southeast Asia (Niah, Sarawak, c. 40,000 BP) and somewhat later from Europe (Dolni Vestonice, Czechoslovakia, c. 30,000 BP) (Fig. 3.22). The only fossils that are clearly intermediate are those from Es-Skhul and Djebel Kafzeh in Israel, which date from about 50,000 years BP.

This transformation from a large-jawed "Neandertaloid" to a small-jawed modern type has always been something of a mystery. In many areas the replacement apears to be so sudden that an invasion of modern forms from elsewhere has been postulated (e.g., in France), while in other places the two forms seem to have overlapped in time (e.g., Southern Africa). Only in Western Asia are truly intermediate specimens to be found, but it may be pure chance that they have not been uncovered elsewhere. The simplest hypothesis on the basis of the present evidence is that modern forms first appeared in western Asia or northern Africa and soon spread south, moving later to eastern Asia and finally to Europe. We might expect the big-jawed forms to have survived longer in outlying areas. We do not at present have enough evidence to determine whether the big-jawed populations were genetically swamped by the invaders, were exterminated by them, or evolved into their successors. Just one thing is certain: the genes of the big-jawed people survived longer in Southeast Asia than elsewhere, and have made a substantial contribution to the aboriginal people of Australia, who have the largest jaws and teeth among living peoples.

The search for the origin of modern man is today being pursued in many continents. In a few years we may have better data with which to test our hypotheses.

3.12 Summary

WE OURSELVES are primates, and the origin of our nature may be traced quite precisely to the adaptations of the primates that evolved in response to the forest environment. The special problems associated with that environment demand sensory awareness, precise mobility, and a brain able to make accurate and immediate prediction. We owe our nature primarily to the challenge of that environment. In the following chapters of this book, an attempt will be made to show how each aspect of man's physiology and behavior evolved in response to the challenge either of the forest or of the totally different open environment of the plains, in which he later achieved his manhood. The majority of mammals have evolved for tens of millions of years in one environment alone, antelopes on the grasslands, whales in the sea, primates in the forests; man's move from the forest to more open country made us finally into what we are and is the story of our own evolution.

The human family, the Hominidae, includes those living and fossil forms that are more like modern man than like other living primates. This assessment of similarity is the general basis upon which the classification of fossil remains must be made. However, we know enough of the processes of organic evolution to realize that, by convention, families should be differentiated not by slight anatomical differences but by a distinct complex of adaptations. The adaptations that distinguish the hominids, apes, and monkeys are primarily locomotor and have been discussed above; the Hominidae are quite strikingly distinct from all other primate families as a result of their terrestrial bipedalism. They have accordingly been defined by Le Gros Clark on the basis of their morphological pattern, and the following has been modified from his summary definition (1964, p. 119). The meaning of these statements and the terminology will become clear in the following chapters.

Family Hominidae: a subsidiary radiation of the Hominoidea distinguished from the Pongidae by the following evolutionary trends.

> Progressive skeletal modifications in adaptation to erect bipedalism, shown particularly in a proportionate lengthening of the lower extremity, and changes in the proportions and morphological details of the pelvis, femur, and foot skeleton related to the mechanical requirements of erect posture and gait and to the muscular development associated therewith.

Preservation of well-developed pollex, ultimate loss of opposability of hallux.
Relative displacement forward of the occipital condyles and restriction of nuchal area.
Consistent and early ontogenetic development of a pyramidal mastoid process.
Diminution of canines in phylogeny to a spatulate form interlocking slightly or not at all and showing no pronounced sexual dimorphism, with disappearance of diastemata.
Replacement of sectorial first lower premolars by bicuspid teeth (with later secondary reduction of lingual cusp).
Alteration in occlusal relationships, with the result that in later forms, at any rate, all the teeth tend to become worn down to a relatively flat, even surface at an early stage of attrition.
Development of an evenly rounded dental arcade.
Marked tendency in later stages of evolution to a reduction in size of the molar teeth and in subnasal prognathism.
Progressive acceleration in the replacement of deciduous teeth in relation to the eruption of permanent molars.
Marked expansion (in the terminal products of the hominid sequence of evolution) of the cranial capacity, associated with reduction in size of jaws and area of attachment of masticatory muscles and the development of a chin.

Within the Hominidae, we have described and recognized three genera, and the characters of these will now be summarized in more detail:

Ramapithecus: A genus of the Hominidae distinguished by the following characters:

Masticatory apparatus absolutely smaller and mandible shallower than that of *Australopithecus*.
Face shorter than in *Australopithecus* but not so short as in *Homo*.
Incisors and canines less reduced compared with cheek teeth than in *Australopithecus*.
Incisor procumbency considerable; maxillary canine morphologically resembling the smallest of the Indian species of *Dryopithecus*, but in size no bigger than the canine of *Australopithecus*.
No diastema; first premolar non-sectorial with small second cusp.
Molar teeth simple as in other Hominidae, having thick enamel and flat broad grinding surfaces.
Parabolic dental arcade; mechanics of jaw function as in other hominids, adapted for powerful crushing and side-to-side chewing.

Australopithecus: A genus of the Hominidae distinguished by the following characters (modified from Le Gros Clark 1964, p. 168).

Cranial capacity ranging from about 435 to 815 cc.
Relatively thin-walled cranium with strongly built ectocranial superstructures and robust supraorbital ridges.

A low sagittal crest in species and individuals with heavy masticatory apparatus.
Occipital condyles well behind the midpoint of the cranial length but on a transverse level with the auditory apertures.
Nuchal area of occipital bone restricted and downward facing, as in *Homo.*
Consistent development of pyramidal mastoid process but less striking than in *Homo.*
Jaws variable in size but often massive.
Chin absent but inner symphyseal surface strengthened by heavy mandibular torus.
Dental arcade parabolic in form with no diastema.
Moderate sized spatulate canines wearing down from the tip only.
Relatively large premolars and molars.
Anterior lower premolar bicuspid with subequal cusps.
Pronounced molarization of first deciduous molar.
Progressive increase in size of permanent lower molars from first to third.
Limb skeleton (so far as is known) conforming in its main features to that of modern man but differing in a number of details, such as relatively smaller sacroiliac articulation and acetabulum.
Blades of the ilia splayed widely and birth canal relatively narrower than in *Homo.*

Homo: a genus of the Hominidae distinguished by the following characters (based on Le Gros Clark 1964 and Leakey *et al.* 1964, p. 6):

A large cranial capacity with a mean value of more than 1,100 cc. but with a considerable range of variation, from about 775 cc. to almost 2,000 cc.
Masticatory apparatus variable in size but relatively smaller than the neurocranium when compared with *Australopithecus* and much reduced in modern *H. sapiens;* accompanying changes involve flattening of face.
Forward movement of occipital condyles in relation to skull as a whole.
Reduced masticatory musculature with temporal ridges never reaching the midline of the cranial vault, their height being variable.
Nuchal area relatively smaller than in *Australopithecus*, variable, and in *Homo sapiens* much reduced.
Skull more rounded, and muscular attachment areas in general variable in size but less well marked than in *Australopithecus,* and in modern man almost imperceptible.
Supraorbital torus variably developed, being enlarged in *Homo erectus,* though with reduced postorbital constriction compared with *Australopithecus;* torus much reduced in some subspecies of *Homo sapiens.*
Dental arcade evenly rounded, usually with no diastema.
Mandible less deep than in *Australopithecus* but variable in its horizontal stressing. from an internal torus in *Homo erectus* to an external chin in many subspecies of *Homo sapiens.*

Canines relatively small, with no overlapping after the initial stages of wear.
Molar teeth variable in size, with a relative reduction of M3.
Forelimb shorter than hindlimb.
Pollex well developed and fully opposable, so the hand is capable of a precision grip.
Hindlimb skeleton seems not to be very variable and is fully adapted to bipedal locomotion, as in modern man.

As we have proposed, the story of our evolution can best be told by studying in turn the evolution of each functional complex of characters according to its changing function. In treating each functional complex, we shall compare man with some of the living monkeys and apes in order to establish his relationship to them. Where fossils are available, we shall then consider the evidence from that source in order to establish in more detail the morphology and chronology of the Hominidae.

In that framework we shall study the special adaptations that distinguish man from other primates—adaptations to life in open country. The movement from the forest was a remarkable and rather mysterious event. Some authors see it as the direct result of a lowered rainfall, which restricted the extent of the tropical forests and caused increased population pressures. Others see it simply as resulting from the normal competitive pressures of organic life, which favor the exploitation of all exploitable environments. The point to note, however, is that if a population is to exploit an environment different from that of its ancestors, it must to some extent be adapted to the new environment already, or, to use a common term, it must be *preadapted*. Of course, many animals are preadapted in varying degrees to environments not their own, and this is what is meant by a generalized species. All quadrupedal monkeys are in their locomotor character preadapted to terrestrial life in a way that jumping prosimians and full brachiators are not. It is certain that man himself must have been preadapted to terrestrial life, otherwise he could not have evolved into a fully terrestrial animal. The extent of preadaptation and the kinds of selection pressure involved will be discussed in chapter 11. Our purpose in the next and following chapters is to examine man's adaptations in some detail and in a historical setting so that it may be possible at a later stage to make an extrapolation to his origin.

SUGGESTIONS FOR FURTHER READING

The most convenient reference book on the living primates is J. R. and P. H. Napier, *Handbook of Living Primates* (New York and London:

Academic Press, 1967). For a simple account of primate anatomy, see A. H. Schultz, *The Life of Primates* (New York: Universe Books, 1969). On the general features of the nervous systems of mammals consult J. Z. Young, *The Life of Mammals* (Oxford: Oxford University Press, 1957). For the evolution of the vertebrate visual system consult S. Polyak, *The Vertebrate Visual System* (Chicago: University of Chicago Press, 1957). The most compact text on fossil primates is E. L. Simons, *Primate Evolution* (New York: Macmillan, 1972), and the best on fossil hominids is D. R. Pilbeam, *The Ascent of Man* (New York: Macmillan, 1972). Some of the most important fossil Hominidae are briefly described in M. Day, *Guide to Fossil Man* (New York: World, 1965). For a complete catalog of Hominid fossils refer to K. P. Oakley, B. G. Campbell, and T. I. Mollison, *Catalogue of Fossil Hominids* (3 vols.; London: British Museum [Nat. Hist.], 1967–1974). For a detailed review of recent work in human paleontology, see F. Clark Howell, "Recent Advances in Human Evolutionary Studies," *Quart. Rev. Biol.* 42 (1967): 471–513. This article is updated and reprinted in S. L. Washburn and P. Dolhinow (eds.), *Perspectives on Human Evolution,* Vol. 2 (New York: Holt, Rinehart and Winston, 1972). For a recent discussion of hominid taxonomy see B. G. Campbell, "A New Taxonomy of Fossil Man," *Yearbook of Physical Anthropology,* Vol. 17 (1973).

Appendix to Chapter 3

FOSSIL EVIDENCE FOR HUMAN EVOLUTION: SOME IMPORTANT FOSSIL REMAINS OF EARLY MAN

Country	*Locality*	*Remains*	*Approximate Age in Years B.P.**
HOMO SAPIENS			
Europe:			
Belgium	Spy, Province de Namur	† Parts of 2 male skeletons and tibial fragment	35,000–70,000
Britain	Swanscombe, Kent	Occipital and parietal bones	150,000–250,000
Czechoslovakia	Predmosti, Moravia	29 skeletons	c. 26,000
France	Arago, Pyrenées	† Skull, 2 mandibles and fragments	c. 200,000
	Cro-Magnon, Dordogne	Five adult skeletons	20,000–30,000
	Fontéchevade, Charente	Frontal bone of 1 individual; calotte of second	70,000–150,000
	La Chaise, Charente	Fragments of 16 individuals	c. 150,000
	La Chapelle-aux-Saints, Corrèze	† Male skull and nearly complete skeleton	35,000–45,000
	La Ferrassie, Dordogne	† Parts of 6 individuals, 1 nearly complete (male)	35,000–55,000
	La Quina, Charente	† Parts of female skull and skeletal fragments	35,000–55,000
	Montmaurin, Haute-Garonne	† Mandible and teeth	c. 250,000
	Regourdou, Dordogne	† Mandible and skeletal fragments	c. 45,000

Germany	Ehringsdorf, nr. Weimar	† Parts of female cranium and 2 mandibles plus remains of child's skeleton	60,000–120,000
	Neandertal nr. Dusseldorf	† Parts of male skull and skeleton	35,000–70,000
	Steinheim nr. Stuttgart	Female cranium	150,000–250,000
Gibraltar	Forbes Quarry	† Female cranium	35,000–70,000
Greece	Petralona	† Cranium	35,000–70,000
Italy	Circeo, Latina	† Cranium of male individual; part of mandible of another	35,000–70,000
	Saccopastore nr. Rome	† Female cranium and parts of male skull	60,000–100,000
Yugoslavia	Krapina, nr. Zagreb	† Fragmentary remains of c. 13 individuals	30,000–45,000
Asia:			
Iraq	Shanidar, N. Iraq	† 5 crania with skeletal fragments and remains of two others	45,000–70,000
Israel	Amud	† Almost complete skull and skeleton	35,000–50,000
	Djebel Kafzeh	7 crania, 4 with skeletal remains and fragments of 2 other individuals	c. 40,000
	Galilee	† Parts of cranium and fragments of other individuals	c. 70,000
	Mount Carmel:		
	Cave of Mugharet es-Skhul	† 7 crania with skeletal remains and fragments of 3 others	35,000–40,000
	Cave of Mugharet et-Tabun	† Skull and skeleton of female, mandible; fragments	40,000–45,000
Java	Ngandong (Solo)	† Eleven calvariae and two tibiae	50,000–150,000
Sarawak	Niah Cave	Cranium	c. 40,000

FOSSIL EVIDENCE FOR HUMAN EVOLUTION:
SOME IMPORTANT FOSSIL REMAINS OF EARLY MAN—Continued

Country	*Locality*	*Remains*	*Approximate Age in Years B.P.**
Africa:			
Ethiopia	Omo (Kibish)	2 calvariae and skeletal fragments	50,000–100,000
Kenya	Kanjera	Fragments of 4 crania	c. 60,000
Morocco	Jebel Irhoud	† Two crania	c. 40,000
South Africa	Border Cave	2 calvariae, mandible and skeletal fragments	50,000–90,000
	Florisbad	Calotte with facial bones	c. 39,000
	Hopefield (Saldanha)	† Calotte	c. 45,000
Sudan	Singa	Cranium	c. 20,000
Zambia	Broken Hill	† Cranium and skeletal fragments	60,000–120,000
HOMO ERECTUS			
Algeria	Ternifine	Parietal and 3 mandibles	c. 400,000
China	Choukoutien	Skulls, teeth and skeletal fragments of 38 individuals	350,000–400,000
Germany	Mauer (Heidelberg)	Mandible	c. 500,000
Hungary	Vértesszöllös	Cranial and mandibular fragments	c. 400,000
Java	Sangiran	4 calvariae, calotte and isolated teeth	c. 1.0 m
	Trinil	Calotte and femur	c. 1.0 m
Tanzania	Olduvai (upper Bed II)	Cranium, calotte	c. 1.0 m
	(Bed IV)	Pelvis, femur	c. 400,000

AUSTRALOPITHECUS AFRICANUS

China	Kungwangling	Calotte	c. 1.9 m.
Ethiopia	Omo (Shungura)	Calotte, ulna and many isolated teeth	1.9–3.0 m.
Java	Sangiran	2 calvariae, 6 mandibular fragments	c. 1.9 m.
Kenya	East Rudolf	Calvaria, many mandibular and skeletal fragments	1.3–2.9 m.
South Africa	Kromdraai	Part cranium, mandibular and skeletal fragments	1.5–2.0 m.
	Makapansgat	2 calvariae, mandibular and skeletal fragments	2.0–3.0 m.
	Sterkfontein	2 crania, many cranial, mandibular, and skeletal fragments	2.0–3.0 m.
	Swartkrans	Cranium, many cranial, mandibular, and skeletal fragments	1.5–2.0 m.
	Taung	Child's cranium and mandible	0.7–0.9 m.
Tanzania	Olduvai (Beds I–II)	3 calvariae, 3 fragmentary calottes and other fragments	1.5–1.8 m.

AUSTRALOPITHECUS BOISEI

Ethiopia	Omo (Shungura)	Many isolated teeth and mandibular fragments	1.9–3.0 m.
Kenya	East Rudolf	3 calvariae, many mandibular and skeletal fragments	1.3–2.6 m.
Tanzania	Olduvai (Bed I)	Cranium	c. 1.7 m.
	Peninj	Mandible	c. 1.2 m.

FOSSIL EVIDENCE FOR HUMAN EVOLUTION: SOME IMPORTANT FOSSIL REMAINS OF EARLY MAN—Continued

Country	*Locality*	*Remains*	*Approximate Age in Years B.P.**
RAMAPITHECUS PUNJABICUS			
India	Siwalik hills	Mandibular and maxillary fragments	9.0–12.0 m.
Pakistan	Siwalik hills	Mandibular and maxillary fragments	9.0–12.0 m.
RAMAPITHECUS WICKERI			
Kenya	Fort Ternan	Mandibular and maxillary fragments	c. 14 m.

* For further data see Oakley (1969) upon which many of these dates were based.

† Large-jawed forms of *Homo sapiens* (by implication when jaws not present).

The most important sites are shown on the map of the Old World, Figure 3.20.

For a full catalog of fossil Hominidae see Oakley, Campbell, and Mollison (1967–1974).

Note: The science of chronology is in a state of rapid development. Many of the dates given may require revision in the near future. Those about which there is most doubt belong to the species *Homo erectus*.

4

Body Structure and Posture

4.1 Body structure and locomotion

A VAST MAJORITY of the visible and gross features of the vertebrate body are related to the function of locomotion. Primitive locomotion in fish was achieved by a wave-like movement of the vertebral column in a lateral plane. The trout moves its tail from side to side, and the waves of muscular contraction pass forward up the animal's body. In primitive fish the fins are used mainly as stabilizers and for steering, but in later forms they may be used to propel the fish, especially the paired fins—the *pectoral* fins behind the head, the *pelvic* fins nearer the tail.

The first fish to climb from the water used their pectoral and pelvic fins as props for their bodies, and this came to initiate the process whereby the locomotor function was eventually transferred from the vertebral column to what became the limbs. The paired pectoral and pelvic fins were attached to the backbone or vertebral column by three paired bones of the same name, later to become the shoulder and pelvic girdles. The lateral muscles at the sides of the vertebral column (the *myotomes,* which effected the wave motion of the body) were then greatly reduced, and the limb musculature developed in their place. An intermediate stage can be seen in the newt (Figs. 4.1 and 4.2), which uses both its myotome muscles and its limbs for propulsion.

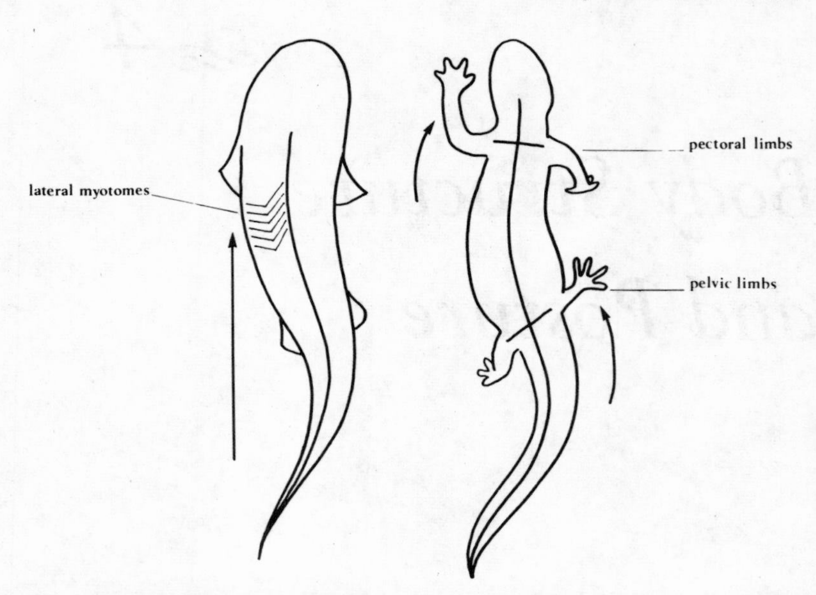

FIG. 4.1. Locomotion in the fish and newt. Note that the same wave motion is used by both animals, but in the newt it is reinforced by the action of the pectoral and pelvic limbs, which evolved from the paired fins in the fish.

As the limbs took over the locomotor function in terrestrial vertebrates, the vertebral column assumed a quite new function—that of supporting the body above the limbs. The flexibility so necessary for swimming was replaced by a stiffness necessary to carry the head and body. The soft parts already *ventral* to (on the underside of) the vertebral column in fish became suspended from it in vertebrates, and the head became supported by a cantilever. The resulting structure was compared by D'Arcy Thompson (1942) to that of a cantilevered bridge (Fig. 4.3) and can be understood as a weight-bearing compound girder. The important muscles are now ventral and *dorsal* to (that is, below and above) the vertebral column, no longer lateral. This change in the evolution of vertebrates is most important and is clearly demonstrated in the secondarily aquatic whales, which evolved from terrestrial quadrupeds and which propel themselves in the sea by moving their tails up and down rather than laterally as do the fish.

In the terrestrial quadruped, therefore, the vertebral column is under compression—the *body* of each vertebra (and the *intervertebral discs*) carry the load, and the *supraspinous* ligaments are under tension, while the

FIG. 4.2. The locomotion of the newt on land still combines features of fish and quadruped locomotion. Note that the limbs are splayed sideways; they cannot support the body, which is dragged along. The musculature is still lateral, not dorsal and ventral as in evolved quadrapeds.

nuchal muscles of the neck support the head (see Fig. 4.4). The bending moment (that is, the force tending to cause bending) in the vertebral column owing to the weight of the body and head is reflected in the length of the *spinous processes,* which, together with the oblique *interspinous ligaments* and muscles, constitute the web of the compound girder. The spinous processes are rarely so greatly developed over the pelvic girdle as over the pectoral because in most vertebrates the weight of the tail is less than that of the head, though exceptions to this are known (dinosaur, kangaroo) (Fig. 4.5). Further stressing over the center section of the girder between the limbs is achieved by an inverted parabolic girder of which the *sternum* and abdominal muscles form the ventral tension member (see Fig. 4.4).

Bridges, however, are designed for rigidity, and the anatomy of bridges is only comparable to that of animals, insofar as animals require rigidity in their skeletons. Flexible structures are of a different order of complexity and do not so clearly follow the simple mechanical principles of the bridge-builder. That is why the cantilever structure is seen more clearly in the large slow-moving animals than in the small, active forms. Flexibility means that the bony members must be jointed to allow movement, and the spinous processes of the vertebrae must be short enough not to touch each other when the body is flexed. In this case, the girder is supported by ligaments under very high stress; it has gained flexibility at the cost of strength.

The second factor to be recalled when studying skeletal architecture is that when the loads due to gravity are very slight, as in small animals,

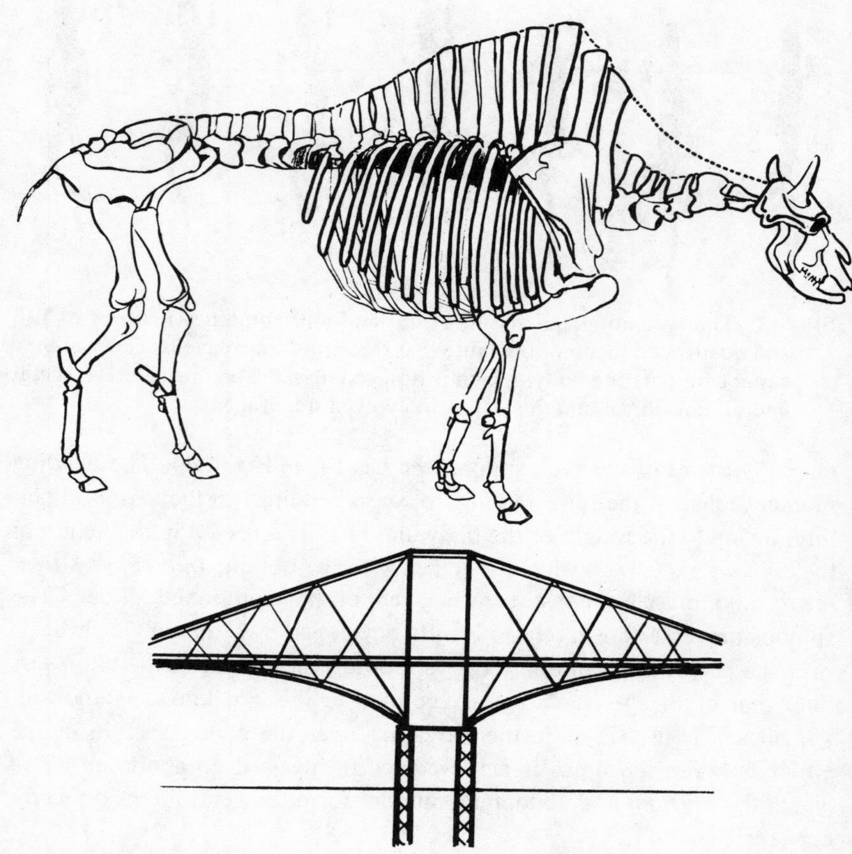

FIG. 4.3. The bison and the cantilever bridge are constructed according to the same mechanical principles. The nuchal muscles (*dotted line*) in the bison, and the upper horizontal member of the steel bridge, take the tension due to gravity; the vertebral bodies and the lower horizontal member of the bridge withstand the compression. The vertical and diagonal struts separate the horizontal members and form the webbing. The nerve cord and road are suspended within this webbed compound girder. (After D'Arcy Thompson 1942.)

the structures associated with locomotion will mask those designed to withstand the force of gravity.

In this chapter we shall trace the evolution of posture from the quadrupedal to the erect, which, as we have mentioned, is found among prosimians as well as in man. The vertebral column, which began in fish as a flexible rod under intermittent compression (when the myotomes contract) is, in quadrupeds, the compression member in a weight-bearing compound girder. In man it becomes a vertical weight-bearing flexible rod, held erect by "rigging," like the mast of a yacht. We shall see that this change of function is accompanied by change in form.

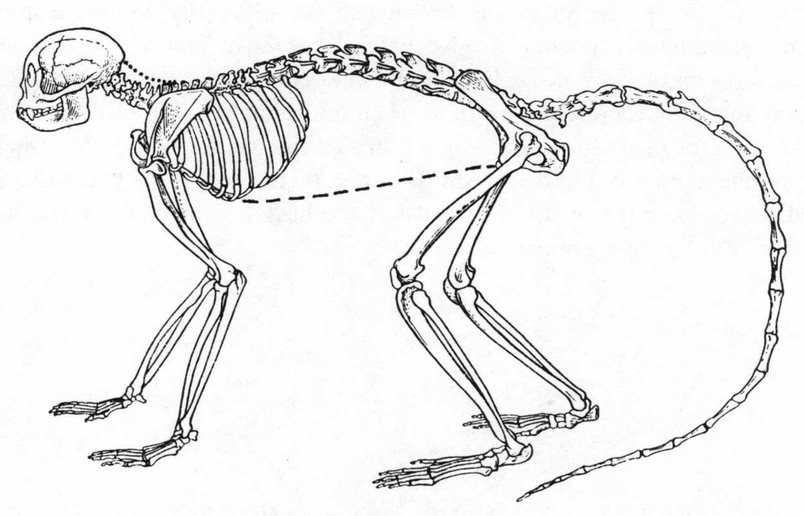

FIG. 4.4. Architecture of a quadrupedal monkey (*Cercopithecus*). Note that the nuchal muscles (*dotted*) are still important but that the head is much lighter than in the bison so the cervical spines are much shorter. The trunk is supported by an inverted parabolic structure under tension—the sternum and abdominal muscles (*interrupted line*). (After Le Gros Clark 1971.)

4.2 Center of gravity

DIFFERENCES in the form of the vertebral column among quadrupedal animals are due primarily to differences in the center of weight of the animal, better known as the center of gravity. This center will move forward or backward along the backbone depending on whether there is more body weight in the head and shoulder region or in the hind quarters and tail region. In the bison, judging by the skeleton shown in Figure 4.3, the majority of weight lies over the forelimbs, and that is where the center of gravity lies. In contrast, the dinosaur and kangaroo, for example, both have far more weight over the hindlimbs, and that is the feature which allows them semierect posture with tail support, so that they can collect vegetation with their forelimbs (Fig. 4.5). A parallel trend can be seen in the evolution of the primates. Here the center of gravity has moved back for two reasons. The first is that the musculature and length of the hindlimbs has increased, especially in the leaping and grasping prosimians, to give extra locomotive power. The second reason is that the tail has enlarged into an organ to maintain balance when jumping. It is also necessary that, when an animal is propelled forward, the force of propulsion should pass through the center of gravity; otherwise the animal would revolve in the air. It therefore follows that a center of gravity near the organ of propulsion, the back legs, will add to the stability of the jumping prosimian.

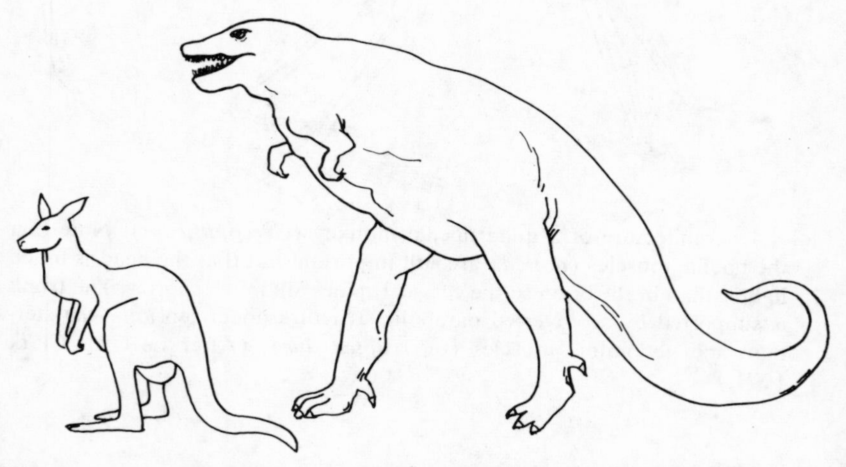

FIG. 4.5. The marsupial kangaroo and the 20-foot reptile *Tyrannosaurus* from the Upper Cretaceous are both bipedal but gain support from a heavy and powerful tail. In each, the spinous processes of the vertebrae reach their greatest development over the hindlimbs. Note the greatly reduced forelimbs. (Not drawn to scale.)

It is for these reasons that in the early stages of primate evolution we see the center of gravity well back over the hindlimbs. This development makes practical the sitting position (Fig. 4.6) and frees the forelimbs for manipulation, just as we see in the squirrel and kangaroo. It also makes possible the later stages of human evolution, when man became able to stand and run on his hindlimbs. The forest adaptation of a low center of gravity was a necessary prerequisite for this final evolutionary development. In man, while the tail was lost, the low position of the center of gravity was enhanced by still further development of the hindlimbs, somewhat reduced musculature in the forelimbs, and some changes in the form and position of the trunk and abdomen.

FIG. 4.6. The ring-tailed lemur (*Lemur catta*) achieves a comfortable sitting position, but the forearms help the balance. Note the immense tail with heavy fur. (From Le Gros Clark 1971.)

4.3 Evolution of the vertebral column and thorax

THE ANTHROPOIDEA generally, adapted as they are for speedy arboreal locomotion, have evolved very flexible vertebral columns compared with most quadrupedal plains mammals. Flexion of the backbone has been added to that of the legs to give extra propulsive power in jumping, as well as to absorb the shock of impact. Flexibility was achieved by some small modifications of the vertebral column, which are most clearly seen in the quadrupedal forms:

1. There are well-developed intervertebral discs that act as elastic buffers and allow tipping movement of the vertebrae.

2. The spinous processes are relatively short with the result that extensibility is not restricted by their proximity.

3. The *thorax* (the rib cage and its musculature), which gives the kind of support to the trunk that is achieved by a parabolic girder, stiffens the vertebral column along the thoracic region. In quadrupedal primates (such as macaca) the *cervical* (neck) and *lumbar* (waist) regions of the vertebral column are relatively long, and these regions in particular allow bending movements of the body (see Fig. 4.7, *A*).

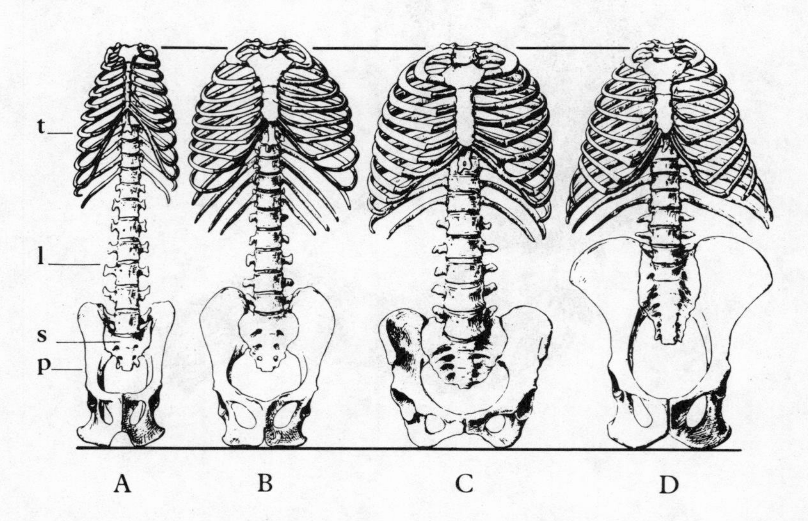

FIG. 4.7. Trunk skeletons of the macaque (*Macaca*) (*A*), the gibbon (*Hylobates*) (*B*), modern man (*C*), and the chimpanzee (*Pan*) (*D*) drawn the same size. Note the different forms of the thorax (*t*), pelvis (*p*), and sacrum (*s*) in these species and the different relative lengths of the lumbar region (*l*). (From Schultz 1950.)

In arm-swinging and particularly in brachiating forms, however, this flexibility is somewhat lost, since the vertebral column plays a far smaller part in locomotion. We find, for example, that three of the nine lumbar vertebrae found in quadrupedal monkeys become thoracic in gibbons (Washburn 1963b) (Fig. 4.7, *B*). The numbers of such vertebrae are shown in Table 4.1, where it can be seen that in this respect man is stiff-backed and morphologically near to the brachiators and semi-brachiators. It is, however, interesting that the stiffening has proceeded further among the great apes than it has in man. As a whole, these data suggest quite strongly that man evolved from a brachiating primate. But other fundamental changes have occurred in the spine and thorax during human evolution that were adaptations to man's fully erect posture.

TABLE 4.1 AVERAGE NUMBERS OF THORACIC AND LUMBAR VERTEBRAE

PRIMATE	*THORACIC VERTEBRAE*		*LUMBAR VERTEBRAE*	
	Mean	*Range*	*Mean*	*Range*
Papio	10	9–11	9	9
Hylobates	13	12–14	6	5–7
Great Apes	13		4	
Man	12	10–12	6	5–7

1. The size of the vertebral body increases at the pelvic end of the vertebral column (Fig. 4.8) because the forces of compression are no longer constant along the trunk as they are in a quadruped with a horizontal backbone. As the force of compression increases from head to pelvis because of the weight of the animal, so the cross-section of the vertebral body, which bears it, must also increase (Schultz 1953). The proportions of the vertebral body also change (Schultz 1960), and the weight distribution on the vertebrae themselves changes in order to carry the vertical forces of compression.

2. The spinous processes are more or less equally developed along the vertebral column because the bending moment—the tendency to bend—is no longer restricted to the points of support, the limb girdles, but is now more evenly distributed. For example, the gorilla has well-developed cervical spines which anchor the nuchal muscles, but they are reduced in man, since the head is more or less balanced on top of the vertebral column (Fig. 4.9).

3. The inverted parabolic girder effect of the sternum and abdominal muscles to stiffen the backbone becomes less important. The sternum tends to lie nearer to the vertebral column, its segments become fused (Schultz 1930) and broaden, and the form of the thorax is wider and less deep (Fig. 4.10), bringing the center of gravity nearer the vertebral column. Schultz (1960) has studied the angle between the "neck" and "body" of the ribs and finds that among the monkeys the angle varied from 89 to

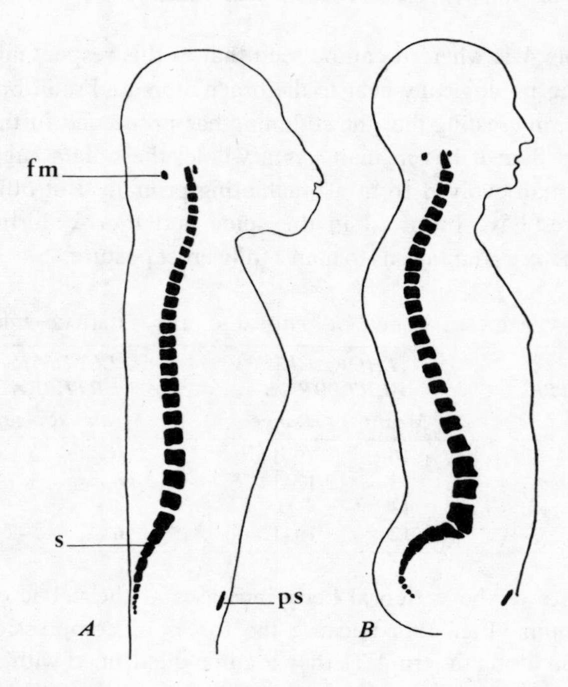

FIG. 4.8. Medial sections through a chimpanzee (*A*) and a man (*B*) showing the curvature of the backbone and form of the vertebrae. *fm*, base of skull; *s*, sacrum; *ps*, pubic symphysis. (After Weidenreich 1939–1941.)

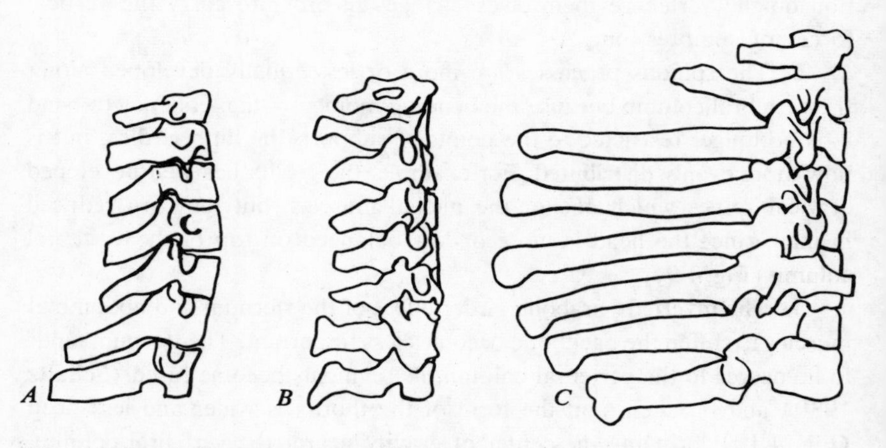

FIG. 4.9. Cervical vertebrae of modern man (*A*), *Hylobates* (*B*), and *P. gorilla* (*C*). The length of the spinous processes of the vertebrae is proportional, not to the amount of weight transmitted down the vertebral column, but to that carried by the nuchal muscles. (Not drawn to scale.)

100 degrees, while among the more erect apes and men it varied from 50 to 65 degrees (Fig. 4.12).

4. Increased weight transmission through the pelvis and legs has resulted in an enlargement of the *sacrum*, that is, the "sacralization" of more vertebrae at the base of the spine. Among the lower primates 2–4 vertebrae are sacralized, while among the great apes and man they average 5.5 and 5.2, respectively (Schultz 1930). The sacrum is also relatively wider and on the internal surface more convex (Figs. 4.7 and 4.8).

We see from these data that man has a relatively stiff backbone, which suggests his origin among arm-swingers or brachiators, and he no doubt benefited by this preadaptation in the evolution of his fully erect bipedal posture. In achieving that posture, the vertebral column evolved to form a vertical weight-bearing rod in place of a compound girder, which means that the vertebral bodies, which transmit the weight to the pelvic girdle, have increased in size downward, and the spinous processes, which supported the webbing of the girder, have become much reduced. The thorax

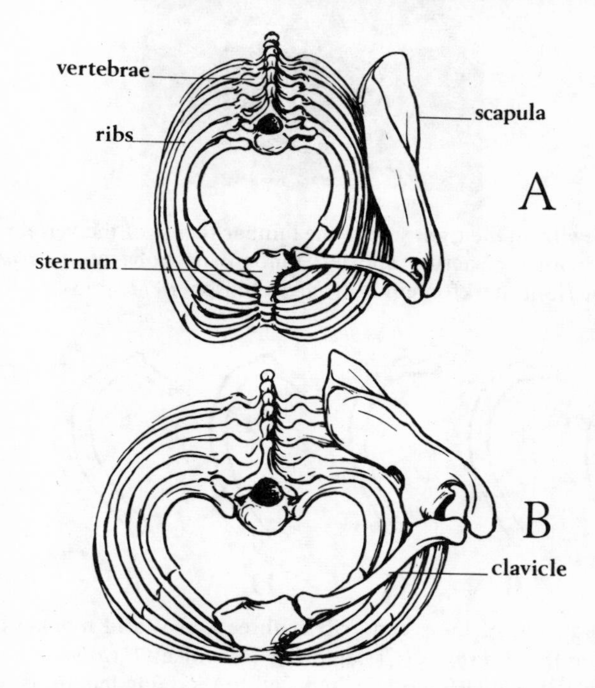

FIG. 4.10. View from above of the thorax and shoulder girdle of *Macaca* (*A*) and modern man (*B*) drawn to the same anterior-posterior depth. Note the very different shapes of the thorax and the different positions of the shoulder girdle. (From Schultz 1950.)

FIG. 4.11. View from the right side of the lumbar region of the vertebral column together with the sacrum and left innominate bone of *Australopithecus africanus* from Sterkfontein. (From Robinson 1972, p. 318.)

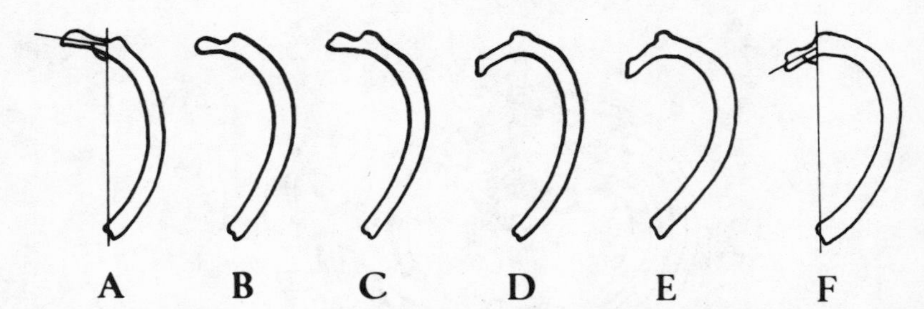

FIG. 4.12. Drawings of the second rib of three Old World monkeys (*Macaca* [*A*], *Papio* [*B*], *Presbytis* [*C*]), and two Pongidae (*Hylobates* [*D*] and *Pan* [*E*]), together with *Homo* (*F*), reduced to the same length. Note that the different form of the ribs reflects the different form of the thorax in these groups. The angle between the "neck" and "body" of the rib is shown for *Macaca* and *Homo* and is quoted in the text. (Redrawn after Schultz 1960.)

has flattened to allow the center of gravity to move back toward the vertebral column which effectively lies within the thoracic area (Fig. 4.10). The weight of the body is transmitted to the legs via the pelvic girdle, and the changes here will be discussed in chapter 5.

The evidence of man's ancestry derived from the vertebral column is important and is supported by the evidence of the fossilized vertebrae of *Australopithecus*, which show most of the man-like features associated with an upright posture, though (as might be predicted) the bodies of the lumbar vertebrae are still relatively small (Robinson 1972). Known fossils of early *Homo sapiens* populations have vertebrae similar to modern man (Straus and Cave 1957), and, as we shall see, there is much evidence that erect posture was the first necessary preadaptation to be evolved before man moved from the forest to the plains.

4.4 The head and neck

WE SHALL CONSIDER the functions of the head itself in chapters 7 and 8. Our purpose in this section is to consider the head not for itself but rather in its relation to the rest of the body. Just as semierect and erect postures have affected the form of the cervical vertebrae, so they have also affected the position of the head in relation to the spine and the connecting musculature.

We have seen how in such quadrupedal forms as the bison or monkey the head is supported by a webbed girder consisting of the cervical vertebrae under compression and the nuchal muscles and ligaments under tension (Figs. 4.3 and 4.4). It is clear that as a more erect posture was evolved, the strength required in the nuchal muscles was reduced, since more of the weight was carried directly by the vertebrae.

At the same time, we have already mentioned that in primate evolution generally the muzzle was reduced and the brain enlarged. This development resulted in the center of gravity of the head itself moving more nearly over the point of pivot of the head upon the vertebral column. This evolutionary trend is shown in Figure 4.13, where different skulls are aligned by their *occipital condyles,* the convex bone surfaces with which the skull pivots upon the first cervical vertebra. Here we clearly see the center of gravity moving back until it is almost exactly over the occipital condyles. It has not yet achieved a perfect balance and probably will not do so, because there are no powerful ventral muscles present to support the

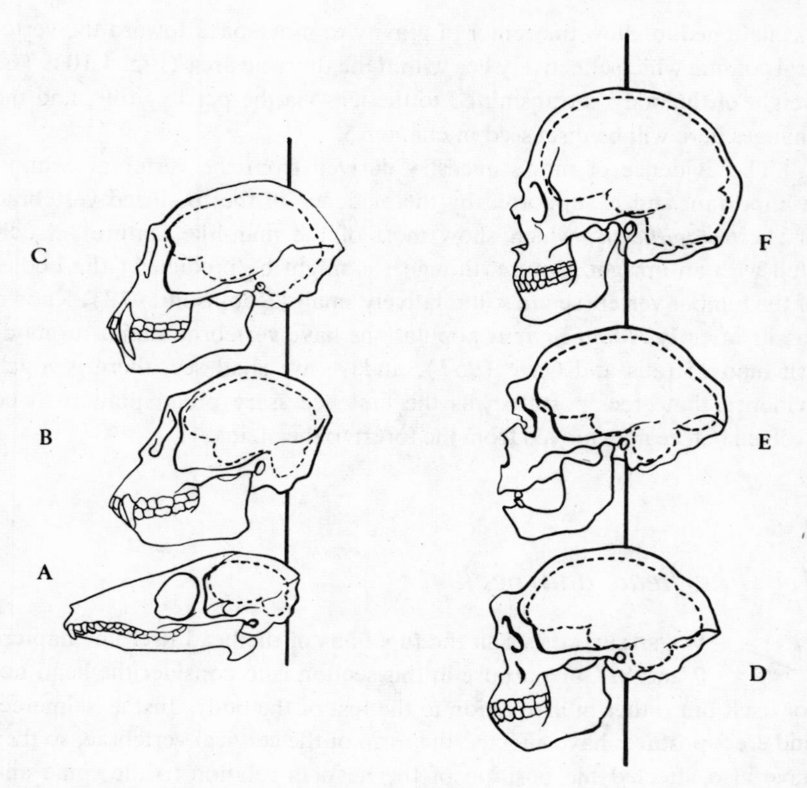

FIG. 4.13. Skulls of various primates aligned upon the point on which they pivot on the spine, the occipital condyles. Note that in passing from the lower to the higher primates, the center of gravity moves back; the brain expands, the jaws recede. *Tupaia* (*A*), *Cercopithecus* (*B*), *Hylobates* (*C*), *Homo erectus* (*D*), *Homo sapiens neanderthalensis* (*E*), modern man (*F*). (Not drawn to scale.) (After DuBrul and Sicher 1954.)

skull from the front and stop it from tipping backward. Those that are ventral to the occipital condyles (the *longus capitis* and *rectus capitis anterior*) arise very close to the occipital condyles and so exert only slight leverage on the head. The condition of modern man is seen in Figure 4.14, where the weight of the head, ventral to the occipital condyles, is balanced by the nuchal muscles, dorsal to this pivot.

Le Gros Clark (1950) has proposed a most useful means of assessing the extent to which this movement of the center of gravity has occurred in different primates: by measuring the position of the occipital condyles in relation to the rest of the skull. The *condylar position index* is calculated

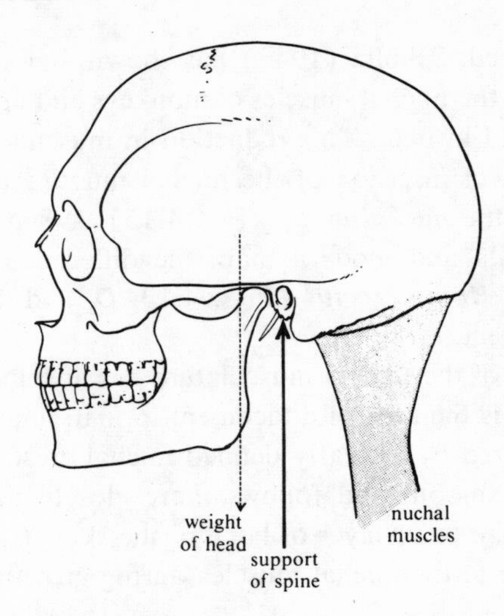

FIG. 4.14. In the course of human evolution the center of gravity of the head has moved nearer to the point of pivot upon the spine—the occipital condyles. In modern man it approaches this point but is still slightly anterior to it. The weight is still balanced by the now rather reduced nuchal neck muscles.

TABLE 4.2 THE POSITION OF THE OCCIPITAL CONDYLES

Primate	*Mean Condylar Position Index*
Dryopithecus africanus	30
Pan gorilla	24
Pan troglodytes	23
Pan paniscus	35
Australopithecus africanus	39
Homo sapiens (modern)	77–81

Condylar position index measured in Frankfurt plane; $\frac{CD}{CE}$ (see Fig. 8.13). Data from Le Gros Clark (1950), Ashton and Zuckerman (1951), and Davis and Napier (1963).

according to the simple geometrical basis described in Figure 8.14, and some figures for this index are recorded in Table 4.2.

Note that the increase of the index has begun in *Australopithecus,* but has not at this stage progressed very far in the human direction. The index numbers may be compared with the drawings in Figure 4.13. We shall return to this trend in chapter 8.

This double change, in posture and in the head's center of gravity, has had the effect of removing work from the nuchal muscles, which have in turn been reduced. Schultz (1942) has shown that six times as much power is needed in the nuchal muscles of monkeys and apes than is needed to support the head in man. This reduction in musculature has been reflected in the area of insertion of the nuchal muscles upon the back of the skull (termed the *nuchal area*) (Fig. 4.15). When we compare, for example, the gorilla and modern man, the difference is very striking. The fossil skull of *Homo erectus* (Figs. 4.13, *D,* and 4.15, *B*) is intermediate in this respect.

The reduction of the nuchal musculature has affected the form of the skull, for not only is the area of attachment in man much reduced, but it is no longer bordered by a clearly defined nuchal crest. The back of the skull is relatively smooth and follows more closely the outline of the brain, for in man the outer layer of bone of the skull (the outer "table") responds very little to the nuchal muscles during growth (Figs. 4.13 and 4.15).

It has been mentioned that there are no powerful ventral muscles in the neck to keep the head from tipping back—a problem probably unique to man. He can tip his head back beyond the point of balance, and he must be able to pull it forward again. The forward movement of the head is made possible not only by the *longus capitis* and *rectus capitis anterior* mentioned above, but also by the *sternomastoid muscles,* which, lying on each side of the neck and inserted upon the *mastoid process* of the skull, and the sternum and clavicle at the top of the thorax, are individually able to rotate the head upon the vertebral column (Fig. 4.16). Their importance to the primates is very great, since this group is dependent on rotary head movement to maintain effective all-round vision. When in man the center of gravity of the head moves back so near to the occipital condyles (the pivot of the head), the sternomastoid muscles come to counteract any backward forces upon the head by contracting together. In Figure 4.16 we see that, viewed from the side, their insertion is more or less in line with the occipital condyles.

Krantz (1963) has shown that the sternomastoid muscles obtain effective leverage to pull the head forward only because in the evolution of man an enlargement and lengthening of the mastoid processes has occurred. This lengthening alone makes it possible for the sternomastoids to raise the head after it has tipped back, because only by this lengthening is the insertion point of the muscles brought forward beyond the point of pivot when the head is thrown back (Fig. 4.17). As it is, the leverage is very poor indeed, a fact that the reader may check by throwing his head

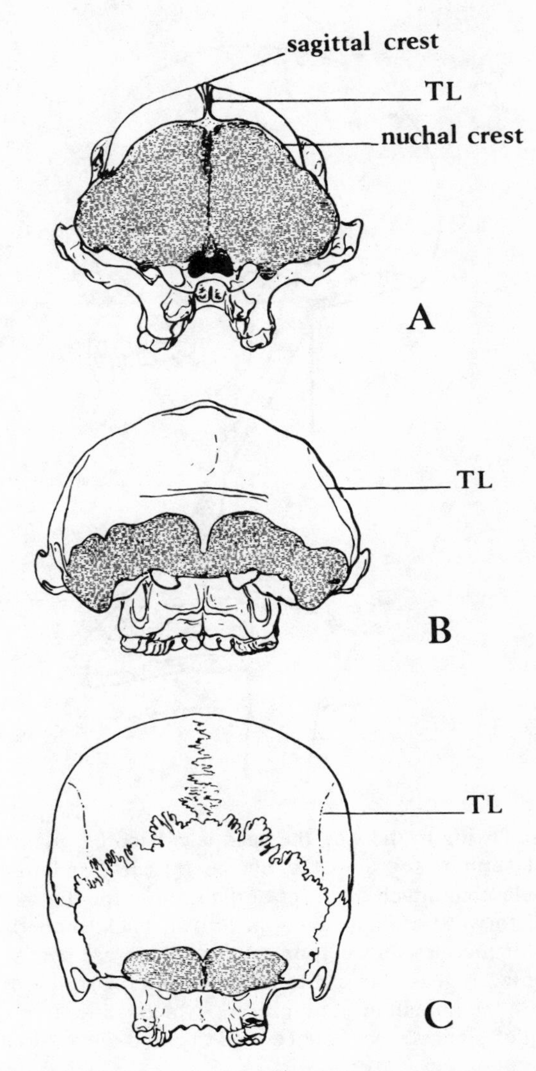

FIG. 4.15. Posterior view of skulls of *P. gorilla* (*A*), *Homo erectus* (*B*), and modern man (*C*). Note the decrease in size of the nuchal area (*shaded*). *TL*, temporal line. (After Weidenreich 1941.)

well back over the back of his chair and slowly raising it again, with his hands on the sternomastoids. The enlargement of the mastoids in human evolution is thus a direct result of man's erect posture and of the forward movement of the occipital condyles in relation to the skull as a whole.

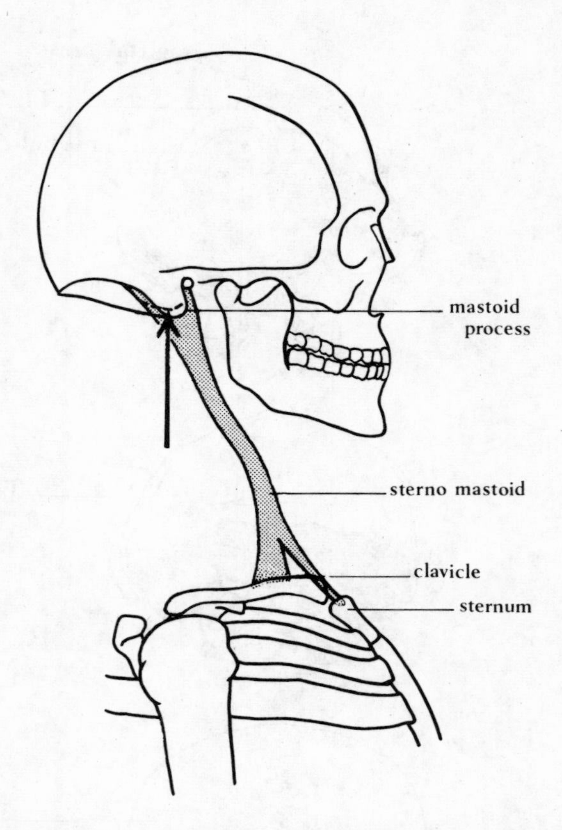

FIG. 4.16. Owing to the way the head is carried by quadrupeds, these is sufficient support for it in the nuchal muscles, and there are no powerful muscles that attach the front of the skull to the thorax or neck. To pull the head forward after it has been thrown back, modern man relies on two rotator muscles, the sternomastoids, which are contracted together. Sufficient leverage is obtained to pull the head forward only by a slight elongation in the human mastoid process, which comes to project forward of the point of pivot, shown here (dotted) behind the mastoid process.

In this section we have seen how man comes to have a reduced nuchal musculature, a reduced nuchal area, and elongated mastoid processes. All these developments are facets of a double adaptive complex involving changes in the olfactory region and brain, which caused a movement in the head's center of gravity and so in the relative position of the occipital condyles. The fully erect vertebral column of evolving man tended to tip the balance still farther, with the result that the center of gravity came to be very little forward of the occipital condyles themselves.

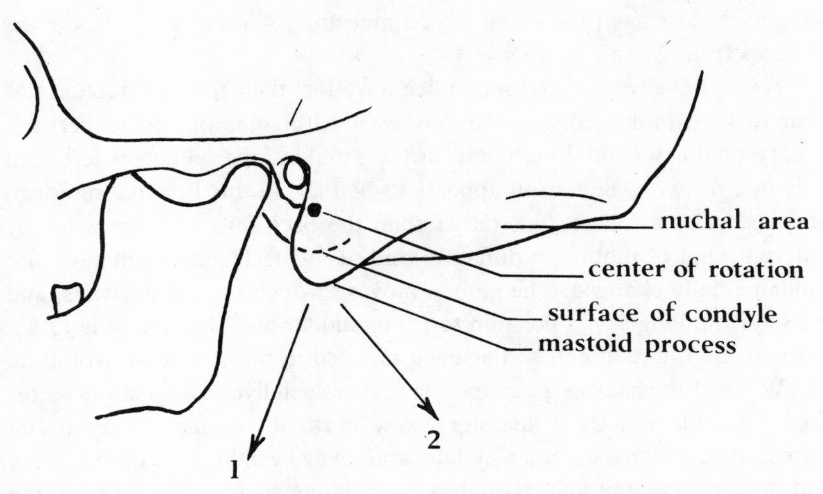

FIG. 4.17. The mechanical advantage bestowed by the mastoid process is apparent only when the head is moved from its normal position (when the sternomastoid muscle lies in direction *1*) and is tilted back, when the sternomastoid pulls in the direction indicated as *2*. Under these circumstances the mastoid can be seen to give some leverage about the center of rotation of the skull upon the spine. A smaller mastoid would reduce to nil the leverage available. (After Krantz 1963.)

4.5 The tail

AS WE HAVE SEEN, the tail is used by many primates for balance in jumping, and it has evolved in such a way as to bring the center of gravity back over the hindlimbs. It is also used in many groups as a rudder and elevator (in the same way as the tail of a bird or airplane), to adjust the position of the body as it passes through the air so that the animal will arrive with its limbs in a position to grasp on impact. This function accounts for the thick hair that the tail sometimes carries, which increases its aerodynamic effect (Fig. 4.6). We have also noted that in the New World monkeys the tail has evolved as a grasping organ—a fifth limb—so it adds to their mobility as well as to their stability.

The tail is also used in other ways by terrestrial monkeys, among which it tends to be reduced. It acts as a support for the baby riding on its mother's back like a jockey, as a device for signalling (like a dog's), as a fly-whisk for the ano-genital region (as in cattle), or it may be almost completely lost as in the stump-tailed macaque. Among macaques alone it varies from 27 to 3 vertebrae.

As we have seen, the Hominoidea have lost their tails. Reduction has gone furthest in the gibbons, the most evolved brachiators. From perhaps twenty or thirty caudal vertebrae in the monkey, the gibbon is left with only one or two. The reason appears to be that, in this brachiating form, propulsion by the front legs rather than the back legs involves a totally different kind of motion, a different kind of flight. The aerodynamics are fundamentally changed. The gibbon moves through the air sideways, and the tail is no longer in a position to act as rudder and elevator (Fig. 3.8). It serves no function and so has been reduced in the course of evolution.

Whether the heavier great apes were ever such lively brachiators seems doubtful, although their ancestors were certainly smaller than present forms. But, in any case, as sedate and heavyweight brachiators, they had no use for a tail and, though it is not quite so far reduced as in the gibbon, all visible signs have been lost.

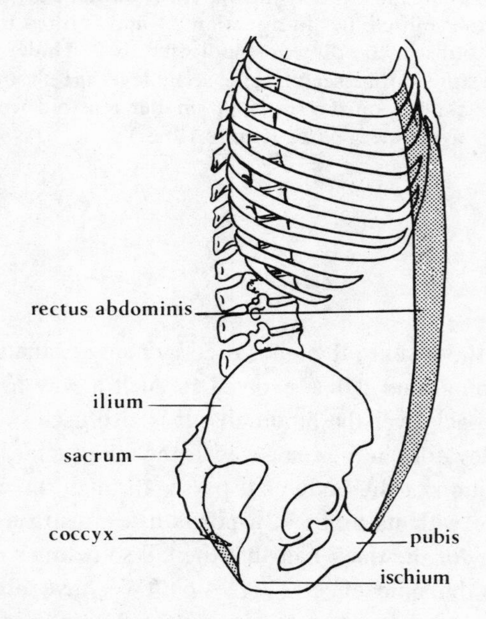

FIG. 4.18. Diagrammatic side view of the pelvis and associated musculature, which form a basin and support the viscera in man. Note that the coccyx, the remains of the tail, has an important function as a supporting structure —a totally new use for a tail.

In man the tail is reduced to from three to five small vertebrae known as the *coccyx*. This reduction may be due to fast brachiation, sedate brachiation, or simply to man's bipedal locomotion. We do not know when the tail was reduced, and at present no coccygeal vertebrae are known from *Australopithecus*. Reduction of the tail in modern man has not gone so far as in the gibbon, which may well be because it has assumed a totally new function arising from man's erect posture. The tail vertebrae, the coccyx, are now curved ventrally and help form the basin-shaped structure that carries the viscera. Ligaments run from the forward-pointing coccyx to the *ischium*—the lower part of the pelvis.

Figure 4.18 shows how the viscera in man are carried in a sack-shaped cavity: the vertebral column at the back and the *rectus abdominis* muscles in front. At the base, the sack is rounded off by the curved bones of the pelvis, the sacrum, and the coccyx and the *sacrotiderous* ligaments. This viscera-carrying function of the pelvis is again something new and is associated with bipedal locomotion. We shall consider the complex evolutionary changes in these bones in the next chapter.

4.6 Summary

IN THIS CHAPTER we have seen how the arboreal life of the lower primates has influenced the evolution of the vertebral column in a number of ways:

1. The center of gravity moves to the back legs.
2. The tail assumes special functions, as a balancing rudder and elevator in Old World monkeys and as a fifth limb in New World forms.
3. Erect sitting and climbing posture develops.
4. The vertebral column maintains considerable flexibility.

This process leads to the form of many modern arboreal monkeys without important modification, but a number of new trends have appeared in the descendent groups of higher primates:

1. A reversion to terrestrial quadrupedalism: all trends are reversed (e.g., *Papio*).
2. The evolution of brachiating locomotion: the tail is lost and the flexibility of the spine is reduced, the thorax is flattened and the posture is erect (*Hylobates*).

3. The evolution of terrestrial bipedalism: the posture is fully erect, the function of the vertebral column is greatly modified as a vertical weight-bearer, the thorax is flattened to bring the center of gravity over the pelvis, the tail is modified to form a floor to the pelvis, and the relation of the head to the neck approaches balance (*Australopithecus, Homo*).

We have seen in this chapter how the different locomotor adaptations discussed in sections 3.4 and 3.5 have fundamentally affected the form of the vertebral column, the ribs, and the neck. We have also seen how man's unusually large brain and small muzzle have brought his head nearly into balance, and we have noted the modifications in the nuchal muscles and mastoid process that have resulted. All these changes, however, are relatively slight compared with those that have affected the locomotor apparatus.

SUGGESTIONS FOR FURTHER READING

Most of this chapter is based on the extensive investigations of Professor Adolf Schultz. A lifetime of research in primate morphology is briefly summarized in A. H. Schultz, *The Life of Primates* (New York: Universe Books, 1969); more detailed information may be obtained from the references in it. Two other publications summarize work of relevance to this chapter: A. H. Schultz, "The Specializations of Man and His Place among the Catarrhine Primates," *Cold Spring Harbor Symp. Quant. Biol.* 15 (1950): 37–53; "Age Changes, Sex Differences, and Variability as Factors in the Classification of Primates," in S. L. Washburn (ed.), *Classification and Human Evolution,* pp. 85–115 (Viking Fund Publs. Anthrop., No. 37 [Chicago: Aldine Publishing Co., 1963]).

5

Locomotion and the Hindlimb

5.1 The generalized primate limb

IT IS THE LIMBS rather than any other part of the primate body that can be described as "generalized," that is, as having varied functions and thus capable of evolution in a number of different directions. Primate limbs are not used merely to support the body on a horizontal surface like props as in most quadrupedal vertebrates. Among the primates, limbs have evolved to grasp the branches of trees and to support the body at various angles in relation to the branch. They have evolved to support the weight of the animal not only standing but also clinging and hanging. They have evolved as food-gathering organs, for cleaning the body, and as sense organs. The primates are characterized therefore by an efficient "universal" ball and socket joint, allowing movement in any direction, at the point of the attachment of the limbs at the shoulder and hip. The musculature has evolved accordingly, allowing the proximal limb bones (those nearest the trunk) to be moved in almost any direction and to support the body from a wide range of angles.

Flexibly mounted limbs are not so much unique to the primates as they are a primitive mammalian character that has been lost in other orders of mammals. Most terrestrial mammals, such as the ungulates, have undergone limb modifications in evolution that give greater stability at the expense of flexibility. Among horses, for example, not only is the direction of movement of the proximal limb bones limited to one plane, but the power of rotation of the distal bones (those farthest from the trunk) is lost too, and the digits are reduced in number, having no longer any grasping function. In the primates, however, we find the mammal limb developed without losing its generalized form, in response to the three-dimensional forest environment.

With a very flexible universal proximal joint to the limbs, the elbow and knee joints evolved not so much to give flexibility as to provide stability. Stability was achieved by retaining the primitive hinge joint with its more limited musculature, an arrangement that has the advantage of retaining the main mass of muscles near the trunk and keeping the limbs

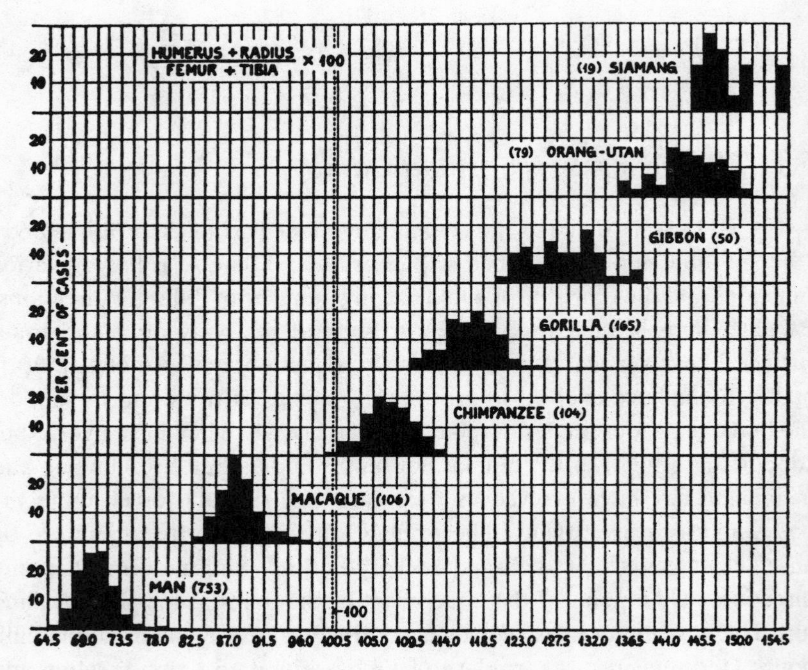

FIG. 5.1. Frequency polygons showing the distribution of variations in the intermembral index of different primates. Figures at the bottom indicate the intermembral index, those in brackets the sample size. The longer the hindlimbs in relation to the forelimbs, the lower the index; an index of 100 indicates limbs of equal length. (From Schultz 1937.)

slender—an essential mechanical arrangement for a stable center of gravity in a fast-moving animal.

The relative length of the fore- and hindlimbs is closely related to the pattern of locomotion of primates. These lengths have been studied extensively by Schultz (1937) and others, and are usually expressed by the *intermembral index*. From the data reproduced here (Fig. 5.1), it can be seen that man and the Old World quadrupedal *Macaca* have hindlimbs longer than forelimbs, while the other Hominoidea have forelimbs longer than hindlimbs. Longer limbs give greater speed of movement because they affect greater leverage: limbs primarily responsible for propulsion tend to lengthen in evolution, while those used least may tend to shorten. In this character man is closer to the Old World monkeys than to any of the other living Hominoidea; indeed, he has exceeded even the macaques in the relative length of his legs. All the forms with relatively longer arms are believed to have a history of arm-swinging or brachiation (Fig. 5.2).

The intermembral index of man has been considered suggestive of the stage of primate evolution at which he left the forest, but the considerable development of the hindlimb for bipedalism, which has almost certainly occurred since that event, together with a possible reduction of the forelimbs, has surely obscured the original limb proportions. For more reliable data we must await the discovery of sets of fossil limb bones in a good state of preservation so that the limb proportions of ancestral groups may be calculated with some accuracy.

5.2 The evolution of the pelvis

THE PELVIS links the hindlimbs to the vertebral column. It is a structure of complex shape, curved in all three planes of space, and together with the base of the vertebral column (the sacrum) it forms a rigid hollow bony structure. The pelves of three primates are illustrated in Figure 5.3, and in Figure 5.4 the relationship between vertebral column, pelvis, and legs is shown. The pelvis consists of two parts, the left and right innominate bones, which meet in the midline ventrally and are fixed to the sacrum dorsally, as is shown in the upper row of Figure 5.3. Each side of the pelvis consists of three bones, the *ilium, ischium,* and *pubis,* and from

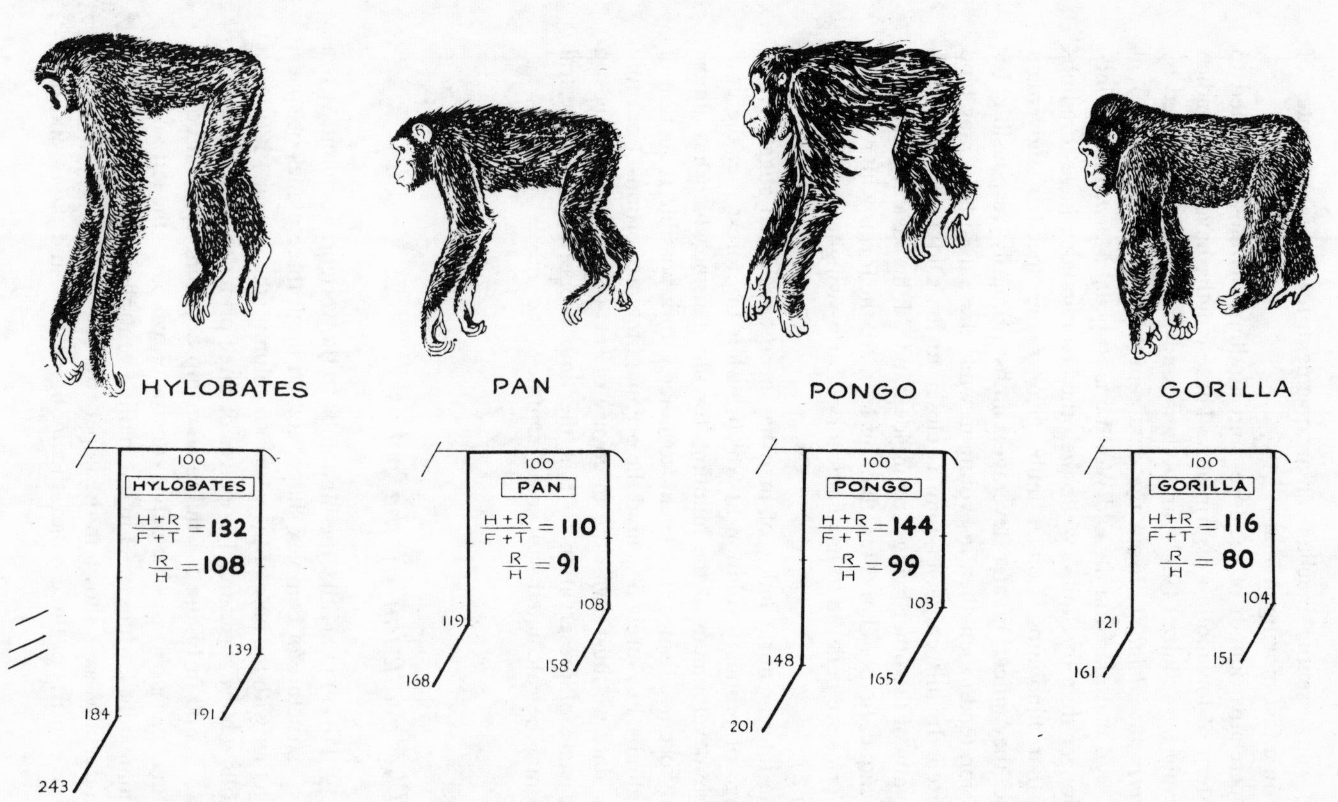

FIG. 5.2. *Above*, profile views of the four anthropoid apes; *below*, diagrams of the same, all constructed on a constant trunk length. (From Erickson 1963.)

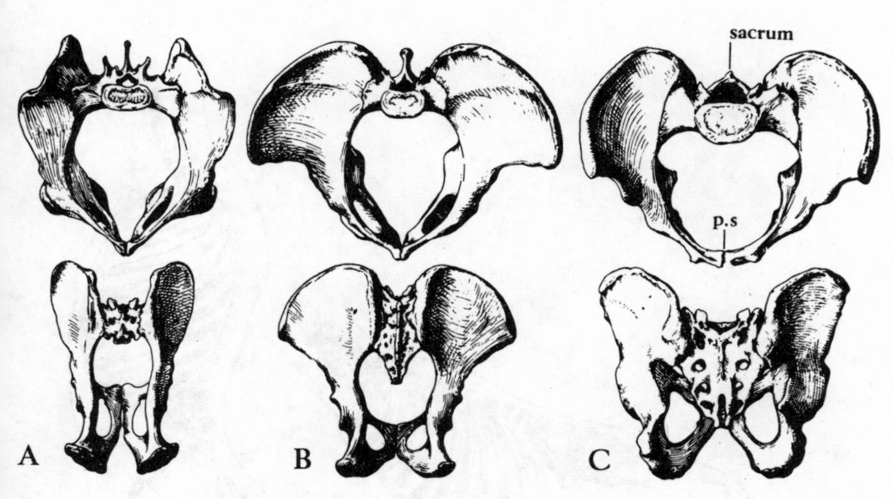

FIG. 5.3. Views from above and behind of pelves of *Macaca* (*A*), gorilla (*B*), and modern man (*C*), *p.s*, pubic symphysis. All drawn the same size. (From Schultz 1963b.)

about the time of sexual maturity these bones become completely fixed together (though there is never any movement between them). The ventral fusion of the pubic bones (the *pubic symphysis*) is finally completed in man between the ages of seventeen and twenty-five, and in other primates at an equivalent stage of development. The human pelvis is shown in Figure 5.5; the three bones and their most important landmarks are labeled.

The ilium—which becomes fused to the sacrum—is a broad flattened bone with a blade-like extension spreading in a curve to each side of the sacrum. At its lower end it fuses with both the ischium and the pubis, and at this point of fusion is formed the cup-shaped socket into which fits the head of the thighbone (the femur). This socket is called the *acetabulum,* and through the pair of acetabula is transmitted the whole weight of the body in bipedal primates. From this central point, the ischium extends backward (dorsally) and downward (to the base of the buttock in man). In some primates it is thickened where it approaches the skin in the form of two *ischial tuberosities,* which give rigid support in the sitting position. Each *pubis* extends forward (ventrally) to meet its fellow in the midline, where it holds the two sides of the pelvis together at the pubic symphysis (see Fig. 5.3).

When we examine typical primate pelves, however, we note some

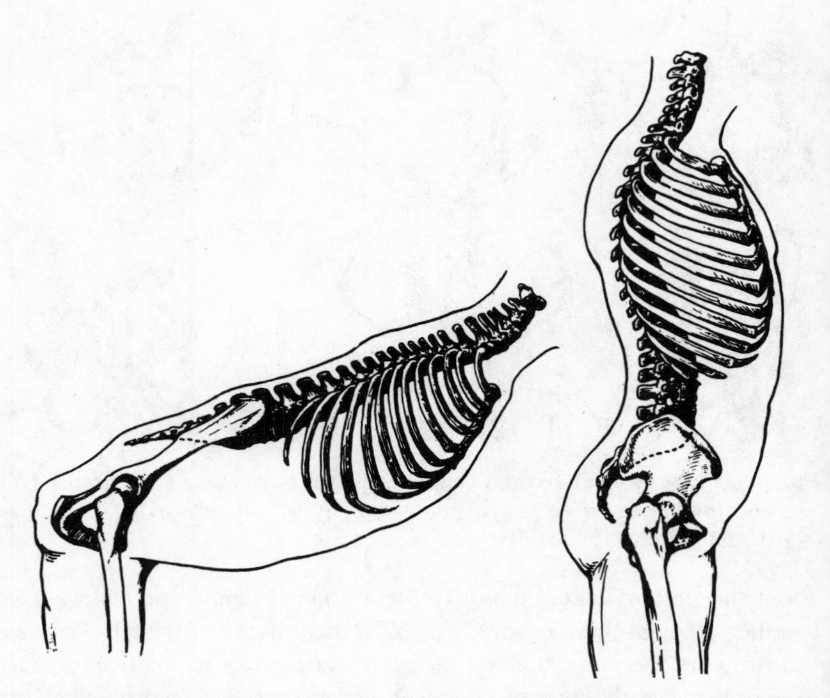

FIG. 5.4. The pelvis in chimpanzee (*A*) and man (*B*). Note the difference in shape and size and the relationship between the pelvis, veterbral column, and femur. (After Schultz 1963b.)

striking differences. Although the general function and relation of the bones is similar, a number of obvious differences are apparent in the monkey and ape pelves drawn in Figure 5.3, and these differences are even more striking in the lower primates. In the pelvis of the little tree shrew *Ptilocercus* (Fig. 5.6), it is possible to see the arrangement of the pelvic bones typical of primitive mammals and lower primates. The most strikingly different character here is the long, narrow blade of the ilium, which extends forward on each side of the sacrum. The most notable trend in the evolution of the human pelvis has been the shortening and widening of this blade from the condition seen in Figure 5.6 through Figure 5.3, *A,* to that in Figure 5.3, *C.* It is this structural change that we shall attempt to follow and understand.

If we compare the larger Old World monkeys with the lower primates, we find some enlargement of the blade of the ilium upon which

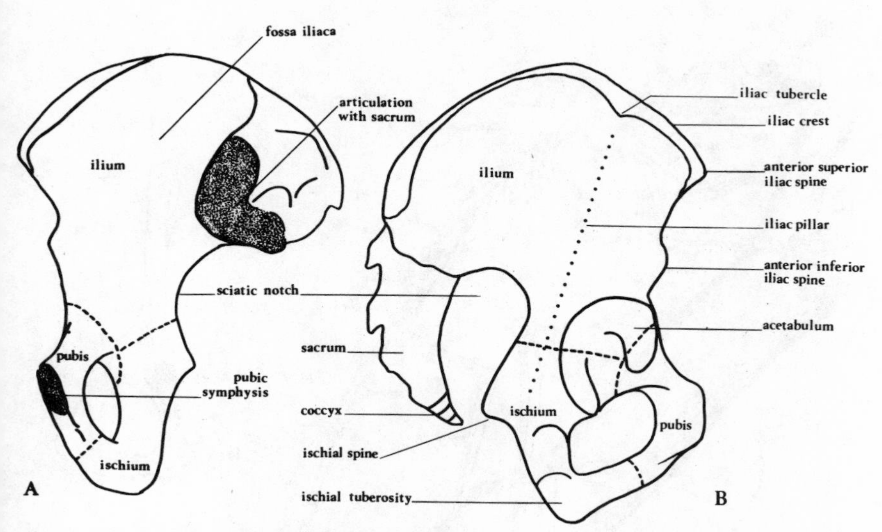

FIG. 5.5. The human right innominate bone. An internal view (*A*) and an external view from the right (*B*) showing the sacrum in position. The dashes indicate the boundaries of the fused bones; the dots indicate that thickening of the ilium which is called the "iliac pillar." The shaded areas are articular surfaces.

originate the important groups of muscles that effect movement of the femur (Fig. 4.4). This enlargement of the areas of origin results from the enlargement of the muscles as a whole in the larger animals caused by the increased power developed in the hindlimbs for jumping. The two ischial tuberosities are also farther apart to give a more stable sitting position, and they form a foundation for the *ischial callosities*—the especially hardened areas of skin characteristic of the Old World monkeys (Fig. 9.2), which commonly not only sit, but sleep sitting, often with their feet drawn up (Fig. 3.7).

Among the nonhuman Hominoid primates we see that the larger animals have still heavier musculature to support a larger body and, as a result, the blade of the ilium expands still farther (Fig. 5.3). The crest (the top ridge of the ilium) itself assumes more importance, since the abdominal muscles inserted upon it have a heavy load of viscera to carry and it must act as part of the stress member of the parabolic girder supporting the spine. But the generally elongated form of the pelvis survives, especially in the gibbon, where considerations of weight on the hindlimbs are less important.

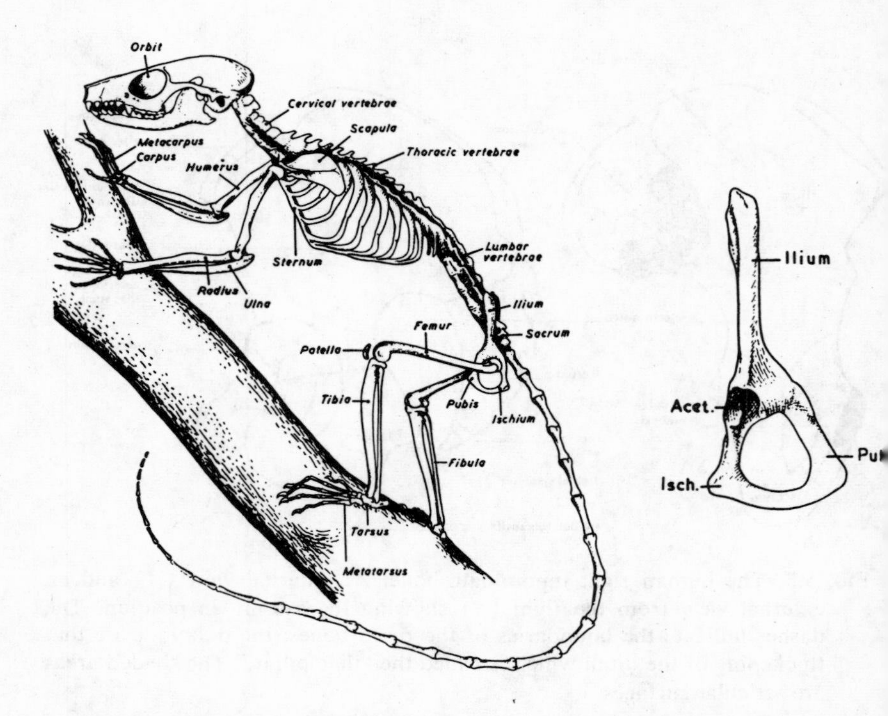

FIG. 5.6. Skeleton of *Ptilocercus*, a tree shrew, drawn about ⅔ natural size, and an enlarged drawing of the right innominate. (From Le Gros Clark 1971.)

The human pelvis shows remarkable differences compared with the pelves of other primates, differences which need detailed consideration. The Latin word *pelvis* means basin, and the basin-like shape that it assumes in man gives it that name. In the human family, the attainment of bipedal locomotion has fundamentally changed the stresses set up in the pelvis resulting in a change in shape (Fig. 5.3). Both Le Gros Clark (1971) and Schultz (1936b) have analyzed this change in shape into a number of components.

1. The muscles of the thigh move the leg forward and backward and provide the power for both quadrupedal and bipedal locomotion. The muscles that move the leg forward are called the *flexors* because they bend the leg at the hip; those that move the leg backward are called the *extensors* because they extend the leg at the hip joint. The development of these muscles is relatively greater in man than in any other primate because man alone depends entirely upon them for his

locomotion, and his leg is proportionally larger and heavier. At the same time, there are changes in man's bone structure that tend to increase the leverage of the muscles. These changes involve as a whole the movement of the areas of origin of the muscles outward from the point of pivot: a small horizontal extension here greatly increases effective muscular power (Fig. 5.7).

Probably the most important flexor muscles are the *ilio-psoas* and *rectus femoris;* the areas of origin of the latter are indicated in Figure 5.8. The human pelvis has a much more pronounced *anterior inferior iliac spine* (where a portion of the muscle originates) than has the nonhuman pelvis.

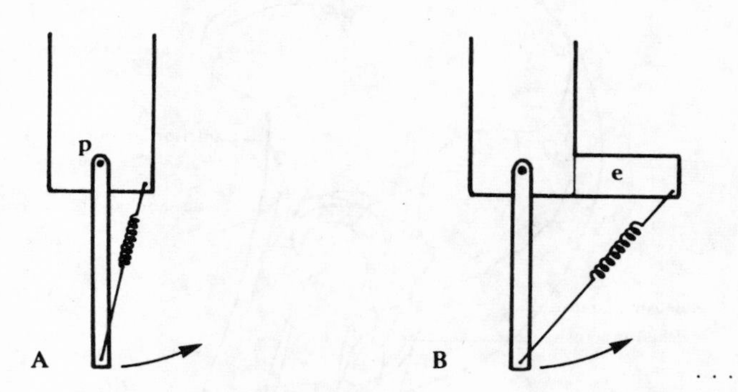

FIG. 5.7. In this diagram the effective power of the spring to move the lever to the right (*A*) is increased by moving the fixed end of the spring farther away from the pivot (*p*) as in *B*. The structural extension (*e*) is a model of the increased horizontal dimension that is typical of the human pelvis as compared with nonhuman pelves and that increases the effective power of all muscles used in locomotion as well as of those required to maintain the balance of the trunk above the legs.

The most important extensor muscles among higher primates are parts of the *adductor magnus* and *biceps femoris* (one of the hamstring muscles in man), which have their origin on the ischial tuberosity and their insertion on the femur and fibula respectively. The origin and position of the biceps is shown in Figure 5.8. However, in man the extensor action of these muscles, so important in locomotion, comes to be reinforced by the action of another muscle, the *gluteus maximus,* which is an *abductor* muscle in other primates (see p. 157). This change in function has been brought about by a very complex change in the form of the ilium and in particular in the proportion and curvature of its blade, which has brought the muscle to lie behind the acetabulum rather than to one side.

The most obvious difference between the nonhuman and the human ilium is that the bone has shortened and widened so that the articular surface, where the bone is connected to the base of the vertebral column (the sacrum), has been brought nearer to the acetabulum, the socket of the femur. This change gives greater stability, since the weight of the trunk

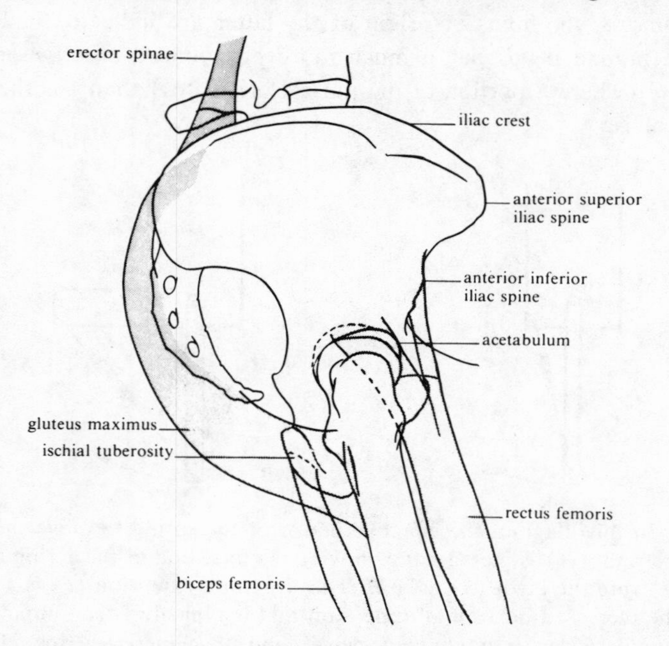

FIG. 5.8. Diagram of flexor and extensor muscles of the thigh in man.

is transmitted more directly to the leg. As Schultz (1963b) has shown, the ilium occupies only 24 per cent of trunk length in man, whereas it occupies 36–38 per cent in the great apes.

Washburn (1950) has also pointed out that, as in evolution the ilium becomes shorter, its axis—from sacrum to acetabulum—must have a greater angle with the ischium in order to keep the same diameter in the pelvic birth canal (Fig. 5.9). Shortening of the bone must therefore go with outward displacement, and this "bending" causes the change in the position of the origin of the gluteus maximus (which arises on the iliac crest) in relation to the acetabulum and femur. As Figure 5.8 shows, the gluteus maximus in man acts as a very powerful extensor of the thigh, but can get an effectively straight pull only when the thigh has already been moved back some way by the other extensors. The gluteus maximus makes it possible to transmit great power through the leg at the end of the stride and in climbing a slope or stairs. The extensors move the thigh back at the beginning of the stride; the gluteus maximus gives the final drive (Washburn 1950). It is the existence of this muscle as an extensor rather than as an abductor that differentiates the human stride from the weak bent-knee gait of the apes and that gives man his buttocks.

From Figure 5.7 and 5.8 it can be seen that the greater the extension of the pelvis bones in a ventral or dorsal direction, the greater leverage a given muscle will exert on the leg for locomotion. Although Figure 5.8 is a great oversimplification of the musculature, the mechanics of the muscle action are straightforward. The leverage of the muscles is increased by the changes that have occurred in the pelvis

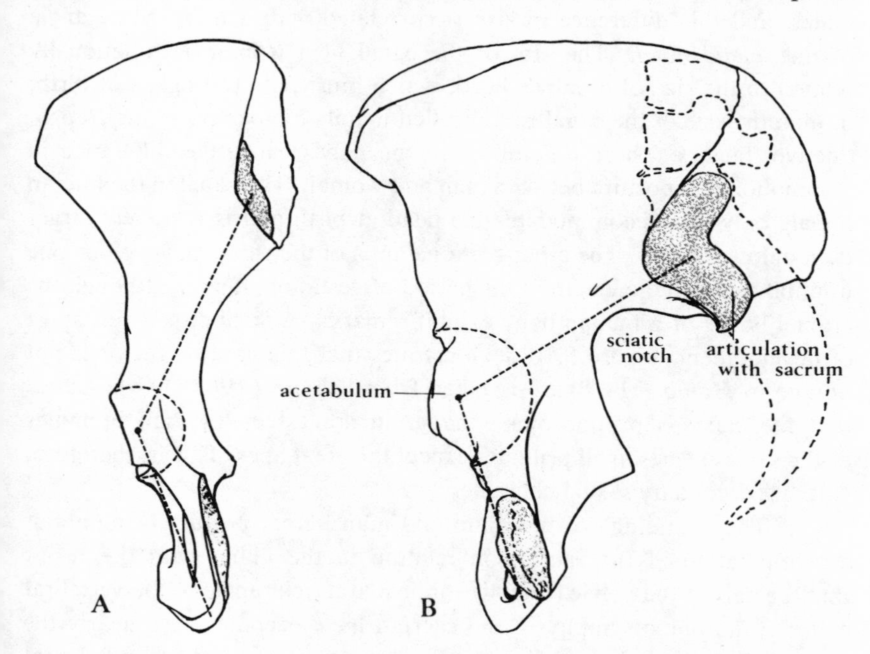

FIG. 5.9. Pelves of chimpanzee (*A*) and modern man (*B*) drawn the same size to show that as the articulation with the sacrum (*shaded*) moves down toward the feet, it must also move away from the acetabulum so that the size of the pelvic canal will be maintained. The dotted line shows the change in structure of the pelvis that results in its "bent" appearance.

during the evolution of man. The ischial tuberosity has moved outward and upward to give more leverage to the biceps muscle about the acetabulum—the pivot of the femur (Fig. 5.8). As we shall see, this evolutionary trend is found to be at an intermediate stage in *Australopithecus*. Another advantage of the increasing extension of the pelvis dorsally is the greater leverage given to the *erector spinae* muscles, which are inserted on the ilium and maintain the upright posture of the vertebral column (Fig. 5.8).

2. The "bending" of the ilium mentioned above has resulted in the formation of a relatively deep sciatic notch, which is associated with an accentuation of the ischial spine, the point upon which important muscles forming the floor of the pelvic basin are inserted (Fig. 5.5).

Since we have seen that the "bending" of the ilium affects the diameter of the pelvic birth canal, we find here a sexual character. The "bending" has proceeded further in males than in females; the canal is smaller in males, and this difference in size is correlated with a more acute angle of the *sciatic notch*. The size of the canal in a woman is functionally related to the size of a baby's head, which must pass through it at birth; in man the size of the canal is controlled mainly by locomotor, not reproductive, factors. These differences in the pelvis cause the difference in locomotion and posture between man and woman. The canal in the human female only just accommodates the head at birth and is relatively larger than in lower forms. The cross-sectional area of the canal may well be one limiting factor in the length of the period of gestation in man, although the cranial bones of a human baby exhibit remarkable flexibility at this stage of development. Figure 5.10 demonstrates that this limiting factor is not unique to *Homo*. The figure is taken from Schultz (1963b), who states that the canal is proportionally larger in adult females than in males of the same species in all primates except the great apes. This is, therefore, clearly a secondary sexual character.

3. The "bending" of the ilium and man's erect posture bring about a reorientation of the sacrum in relation to the ilium, with the result that the axis of the pelvic canal lies almost at a right angle to the vertebral column; the pubic symphysis and sacrum have become more nearly the floor and roof of the pelvic cavity rather than the ventral and dorsal walls (Fig. 5.4; see also Fig. 4.8). At the same time, because greater weight is transmitted through the area of contact between sacrum and ilium, this area has increased relative to that of the ilium as a whole (Fig. 5.9). For the same reason, the acetabulum and the head of the

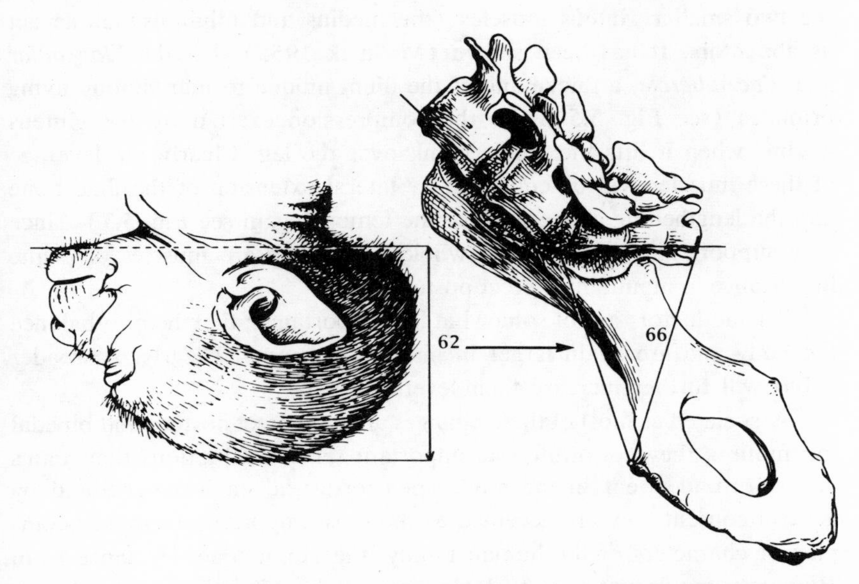

FIG. 5.10. The pelvis of the adult baboon and the head of its full-term fetus, drawn to the same scale. Mother and young died immediately after the difficult birth. (From Schultz 1963b.)

femur have also increased in relative size during the course of human evolution.

4. The *adductor* and *abductor* muscles, which move the limb from side to side in an arboreal primate according to the position of a branch or the demands of the terrain, have great importance for a bipedal creature (compare the extensors and flexors, which move the limb backward and forward). They become essential in maintaining the lateral balance of the trunk upon the legs.

During bipedal walking all the weight of the body is borne by one leg at a time, and it is the abductor muscles (together with the oblique muscles which run from thorax to ilium) which at each step raise the trunk and pull it vertically over the thigh, so that the center of gravity lies over the triangle of the foot and the weight is transmitted directly down the leg through the knee. The adductors, on the other side of the thigh, help to hold the balance (Fig. 5.11).

In the ape, all three *gluteus* muscles (*maximus, medius,* and *minimus*) lie to the side of the hip joint and act as abductors. The gluteus medius is the largest, the gluteus maximus merely half as big. In man, the two smaller gluteus muscles (the medius and minimus) alone act as abductors. It has been shown (Mednick 1955) that the *iliac pillar* and *iliac tubercle,* a thickening of the ilium unique to man among living primates (see Fig. 5.5) take the compression exerted by the gluteus medius when it lifts the whole trunk over the leg. Clearly the leverage of these muscles will be enhanced by lateral extension of the iliac bone and the lengthening of the neck of the femur (again see Fig. 5.7). Since they support the weight of the whole body, they are muscles of prime importance in maintaining erect posture.

The adductors are of somewhat less importance; they help to balance the body and are again larger in man than in apes. Clearly, a broader pelvis will further increase their leverage.

Associated as most of these changes are with erect posture and bipedal locomotion, they constitute one important set of adaptations that man's ancestors underwent in the more open terrestrial environment, and by general consent they are accepted as the most important diagnostic complex of characters of the human family, the Hominidae. Evidence from *Ramapithecus* is not yet available but will be of the greatest interest; but these characters are already present in *Australopithecus* in an evolved condition. One of the most important of the fossil discoveries that have helped to elucidate the course of human evolution is that of the pelvis of *Australopithecus*. Remains of seven fragmentary innominate bones have been reported from South Africa and are described by Robinson (1972):

Sterkfontein	almost complete mature pelvis
	right ilium and pubis
Kromdraai	left ilium
Swartkrans	imperfect right innominate
Makapansgat	juvenile left ilium & right ischium
	juvenile left ilium

These specimens are remarkably similar to those of modern man, and nearly all the characters of the human pelvis listed above are present. The differences between the *Australopithecus africanus* pelvis and that of modern man are slight and include the following features (Le Gros Clark 1955, Napier 1964):

1. The bones are smaller and lighter, being more comparable to those of the Bushman than to the heavier races of modern man. The muscle attachment areas on the blade of the ilium are less marked; the iliac pillar and iliac are slight. These differences can be seen in Figure 5.12.

2. The blade of the ilium is more flared to each side (see Fig. 5.12)

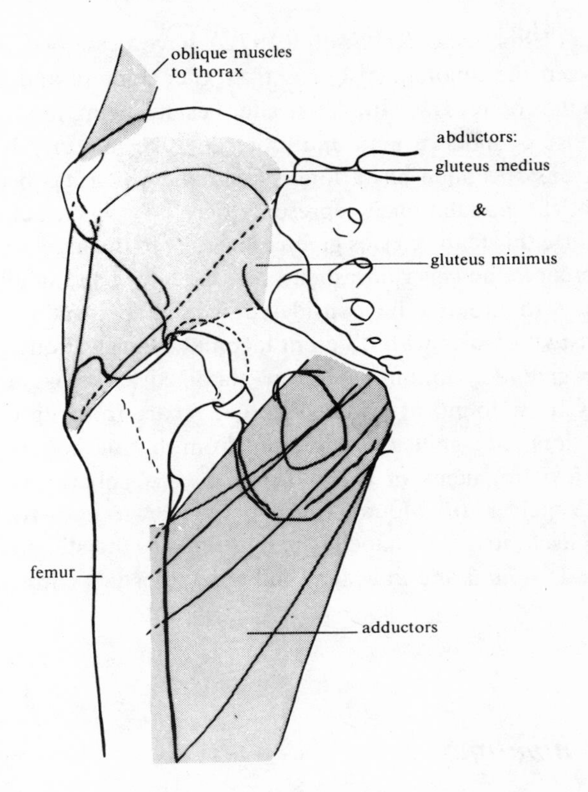

FIG. 5.11. Diagram of abductor and adductor muscles of the thigh in man.

than in modern man so that the abductors which originate on this bone have a better leverage than in man.

As a whole these differences indicate that *Australopithecus* was smaller and more lightly built than modern man. They also indicate that the form of the ilium has since changed in human evolution so that the blade does not flare out so widely. This is probably due to the fact that the birth canal has increased in size while the maximum pelvic breadth between the hips was limited for reasons to do with the balance and mechanics of walking. Thus we have less leverage available for the abductors in modern man than in *Australopithecus* and this has resulted in greater forces being transmitted through the acetabulum. This particular development is the reverse of that generally indicated in these pages for the pelvis as a whole and shown in Fig. 5.7. It is a secondary response to the special condition of large cranial size among the newborn which we associate with the genus *Homo*.

Napier (1964) and Robinson (1972) have described some differences between the innominate bones from Sterkfontein and Swartkrans. They state that bones from the latter site, besides being more robust, are less like those of modern man and more ape-like in form. In particular the Swartkrans specimen has a longer ischium, which has been taken to suggest that the population it represents may have been active arboreal climbers, since this feature gives greater leverage to the hamstring muscles. These differences, however, are slight, and in view of the small size of the sample, they should not be considered for the present to signify the presence of two species with different locomotor adaptations.

One fragmentary innominate bone classified as belonging to *Homo erectus* has been found at Olduvai gorge. Apart from its considerable robusticity, it is not significantly different from that of modern man (Day 1971). A few fragments of early *Homo sapiens* pelves have been collected. The pelves of Mount Carmel (McCown and Keith 1939), Neandertal itself, and La Chapelle are described as indistinguishable from those of modern man and indicate a fully erect posture (Straus and Cave 1957).

5.3 *The hindlimb*

THE MORPHOLOGY of the limb bones reflects the mode of locomotion of the animal. The relative lengths of the different bones and the extent of movement possible are related to the function of the limb as a whole. The thickness of the bones reflects the size of the animal and the weight transmitted by each bone; it is to be expected that the bigger animals have longer limbs and heavier bones.

The *femora* or thigh bones of the lower primates and small monkeys are, on the whole, slender and straight, while the *greater* and the *lesser trochanter,* upon which extensor and flexor muscles, respectively, are inserted, are prominent processes. The femora of the larger quadrupedal monkeys and great apes are slightly bowed in the anterior-posterior plane, but those of an arm-swinging form like the langur (*Presbytis*) have a long straight shaft. Remains of fossil femora of *Dryopithecus* have been found (Le Gros Clark and Leakey 1951) with a long straight shaft, and they have been compared and found to have much in common with femora of arm-swinging monkeys as well as with modern man (Napier 1964) (Fig. 5.13).

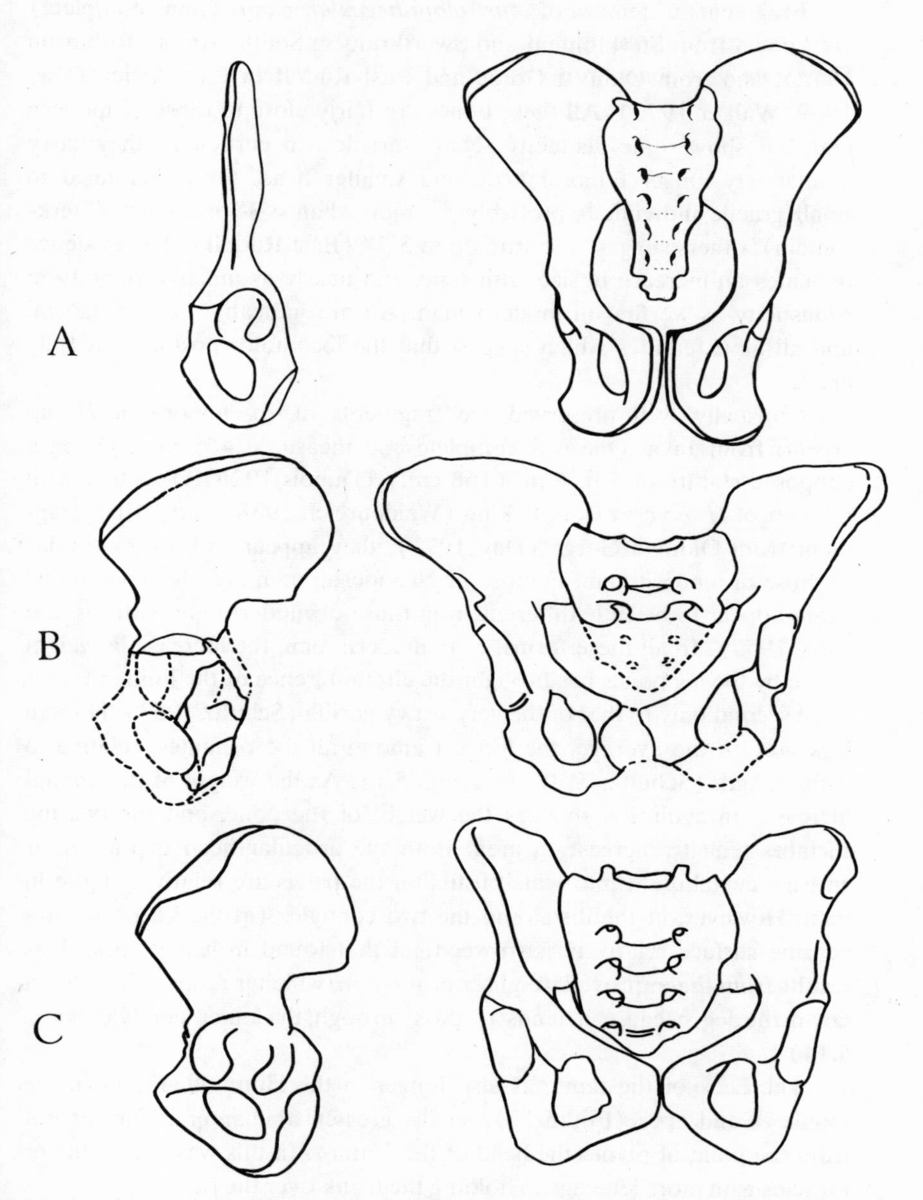

FIG. 5.12. The pelvis of a chimpanzee (*A*) compared with that of *Australopithecus africanus* (*B*) and a Bushman (*C*); *on the left*, the left lateral view, and *on the right*, the entire pelvis from the front. Note the strong similarity between the human and *Australopithecus* pelves. (Not drawn to scale.) (Redrawn from Le Gros Clark 1971 and Dart 1949.)

Fragments of femora of *Australopithecus africanus* (none complete) are known from Sterkfontein and Swartkrans in South Africa (Robinson 1972), and from Olduvai Gorge and East Rudolf in East Africa (Day 1969, Walker 1973). All these bones are fairly close to those of modern man but show some distinctive characteristics; in particular, they carry a relatively longer femoral neck and smaller head. Some belonged to small gracile individuals probably no more than 4′3″ in height (Sterkfontein), others suggest a stature up to 5′7″ (East Rudolf). The evidence indicates an increase in size with time, and nearly as much variability in robusticity as we find in modern man. All are indicative of bipedalism, and all have features which suggest that the locomotor posture was fully erect.

Unusually well preserved are fragments of six femora of *Homo erectus* from Java. One was complete and measured 455 mm. giving a supposed stature of 5 ft. 6 in. (168 cm.) (Dubois 1926). Together with a femur of *H. erectus* from Peking (Weidenreich 1938) and a shaft fragment from Olduvai Gorge (Day 1971), they appear to be very similar to those of modern man. Femora of Neandertal man are also known and again appear to be little different from those of modern man (Straus and Cave 1957). In all these forms, as in modern man, the increase in weight borne by the leg bones has brought the circumference of the human femur to be second only to that of the very heavy gorilla (Schultz 1953). Human legs are, on the average, the longest among all the primates, relative to trunk length (Schultz 1950) (see Fig. 5.1). As the weight of the animal increases in evolution so does the weight of the bone, and the bearing surfaces tend to increase in area. Both the articular head of the femur and the acetabulum into which it fits on the pelvis are relatively large in man. However, at the distal end the two condyles (at the knee) form a bearing surface relatively narrower than that found in heavy apes. This results from the improved weight transmission, which is more or less direct down the leg because it tends to pass through the outer condyle. (Fig. 5.14).

The neck of the femur is also longer in the Hominidae than in the monkeys and apes (Fig. 5.13), so the greater trochanter is farther out from the point of pivot (the head of the femur). In this way the abductor muscles gain more leverage in holding the trunk over the leg.

Full extension at the knee does not occur among non-human primates. By a slight medial rotation of the femur on the tibia, man can lock his knee joint in the fully extended position.

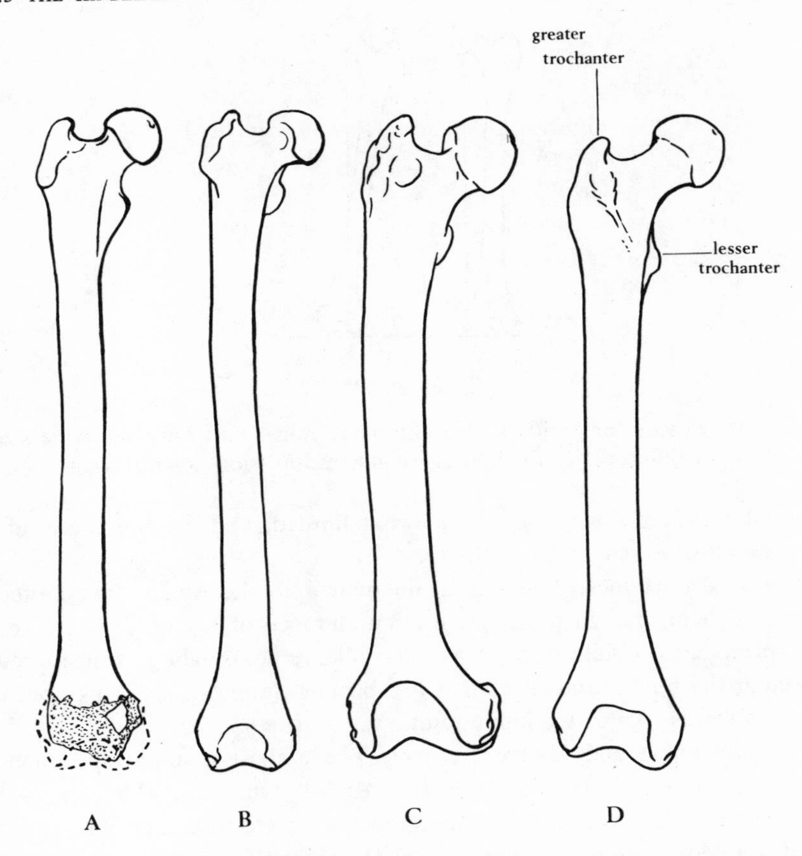

FIG. 5.13. Femora of *Dryopithecus* (*A*) and (*B*), chimpanzee (*C*), and modern man (*D*), all drawn the same size. Femur (*A*) is a reconstruction from fragments of two different bones.

The distal bones have also been modified in the course of human evolution. In the primitive mammal, the existence of two distal limb bones, rather than one, made possible the rotary movement of the hand and foot (*manus* and *pes*). The *tibia* can be considered the fixed bone that transmits the majority of the weight (the *ulna* in the forelimb), while the distal end of the *fibula* (and *radius*) can move in a rotary manner round the base of the tibia, and this movement in turn revolves the pes (or manus) (Fig. 5.6). Such rotation is essential to an arboreal creature for grasping the trunk and branches of trees at all angles. The retention of this generalized feature of primitive limb structure is characteristic of the primates, though modified in *Tarsius* and some lemurs. In the ter-

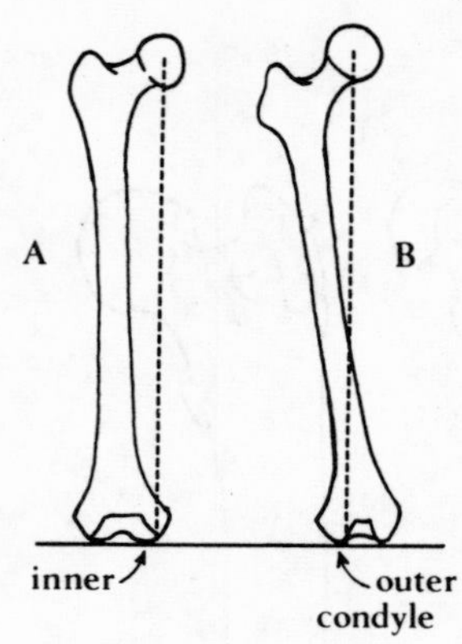

FIG. 5.14. Femora of gorilla (*A*) and modern man (*B*) drawn the same size to show difference in the line of weight transmission down the leg.

restrial forms, the rotation is somewhat limited, and in man the fibula plays almost no part in this activity.

The size of these bones is again related to the weight transmitted through them, and in the relative circumference of his tibia man leads the primates (Schultz 1953); as a corollary, no weight is transmitted through the fibula, and in man it has become reduced, as is his ability to revolve his foot at the ankle joint.

Fossil tibiae and fibulae are rare. The earliest hominid specimens we have are tibial fragments from East Rudolf, which are older than 2.6 million years BP. We also have an almost complete tibia and fibula from Bed I Olduvai Gorge belonging to *Australopithecus africanus*. Davis (1964) has described the bones and shown them to belong to a creature habitually bipedal, but whose gait may well have differed from that of modern man. The bones are already quite long, and the fibula is reduced in mobility and indistinguishable from that of modern man. Fragments of these bones are also known from Neandertal and Rhodesian man; they appear to fall within the range of variation of modern man.

All these changes, which have clearly occurred in human evolution, are directly related to the development of erect posture, to the position of the femur in relation to the pelvis, to its position in relation to the tibia, and to the transmission of weight directly through the bone rather than through a system of stressed and compressed members of muscle and bone. The reduction of the fibula in hominids is related to their reduced power to rotate the foot—an adaptation no longer necessary in an animal that walks and runs on the ground.

5.4 *The evolution of the foot*

THE HIND FOOT or *pes* has a number of fairly distinct functions in primates, and it is the changing balance between these different functions that characterizes its evolution. The functions may be considered in turn.

1. The foot acts as a lever that adds to the propulsive force of the leg, a function clearly seen in modern man. By contraction of the *gastrocnemius* and *soleus* muscles of the calf of the leg, the body is raised upon the ball of the foot. The point of pivot is the *crural joint* of the *talus* (or *astragalus*). The power arm of the lever is the *calcaneus, tarsals* and *metatarsals;* the contraction of the gastrocnemius is transmitted through the *Achilles tendon* to the heel. The load arm is the tarsal section of the foot anterior to the pivot, together with the metatarsals (Fig. 5.15).

The relative lengths of the load arm and power arm have changed in primate evolution according to the mode of locomotion. Extensive movement in the foot lever is characteristic of the leaping prosimians, where power is less essential. The load arm of *Tarsius* represents $8/9$ of the power arm. In the heavier Hominoidea, power is more important, and, in man, the load arm is less than $3/4$ of the power arm of the foot lever.

Within the load arm itself, the proportions of the tarsal and metatarsal segments have varied, but only in a few prosimians (e.g., *Tarsius*) has the tarsal segment assumed great importance. Among the Anthropoidea, the metatarsals are longer, the tarsal bones remaining short and compressed. Slight lengthening of the tarsal segment is, however, characteristic of human evolution (Fig. 5.16).

When we examine the mechanism for the transmission of weight from above, we see some significant variations in the load line. Among the monkeys that have retained a quadrupedal mode of locomotion, weight is transmitted mainly through the middle digit of the foot, which is the longest. These monkeys retain the primitive phalangeal projection index of decreasing length common to quadrupedal mammals, so from the longest to the shortest the order of digits is $3 > 4 > 2 > 5 > 1$ (Fig. 5.17). Later in evolution, as body weight increased and the grasping functions of the foot increased to some extent at the expense of the propulsive function, the *hallux* (great toe) became increasingly opposable to the other digits. In the feet of apes, we find the weight is transmitted to the branch not through digit 3 but through the webbing between digits

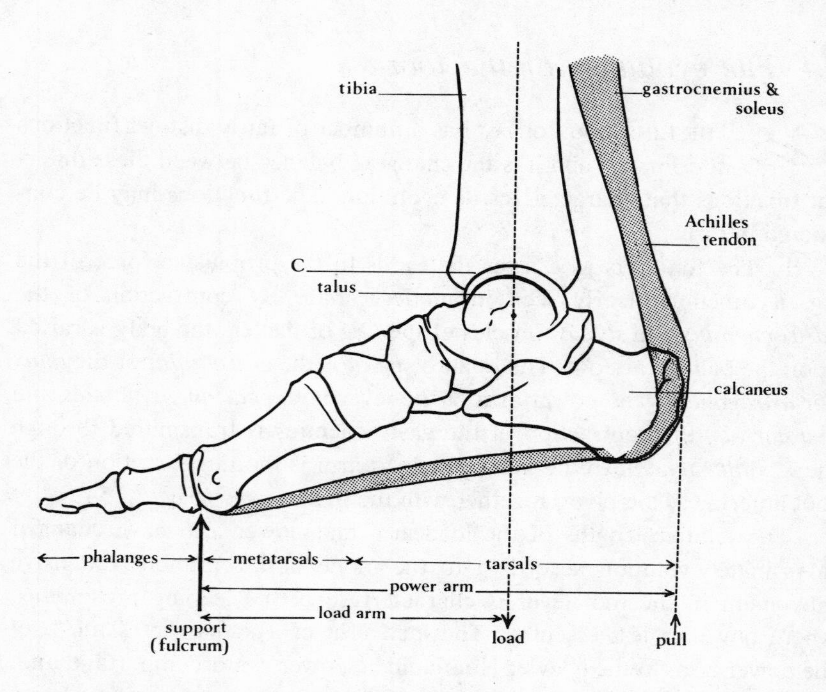

FIG. 5.15. Diagram of the structure and mechanics of the human foot. *C*, crural joint.

1 and 2 (Figs. 5.17 and 5.18). This shift in the load line has been extended in the human foot almost to digit 1 (the hallux), and the existence and origin of the shift is of the utmost importance in human evolution. The fact of this shift of load line suggests very strongly that the Hominidae evolved from a primate that had increased somewhat in weight, developed arm-swinging and a partially opposable hallux. In contrast, the baboons have retained the primitive weight line through the third digit, suggesting their origin from quadrupedal monkeys.

The human foot has the formula of phalangeal projection $2 > 1 > 3 > 4 > 5$, and Schultz (1950) has shown that this change has been brought about by a reduction of digits 2–5 rather than by an elongation of the hallux. A significant comparison can be made with two subspecies of gorilla. The mountain gorilla (*Pan gorilla berengei*) is almost entirely terrestrial, and its foot is much closer to that of man, since the opposability of the hallux is partially lost. It should be compared with the lowland gorilla (*P. gorilla gorilla*), which is still to a considerable extent arboreal

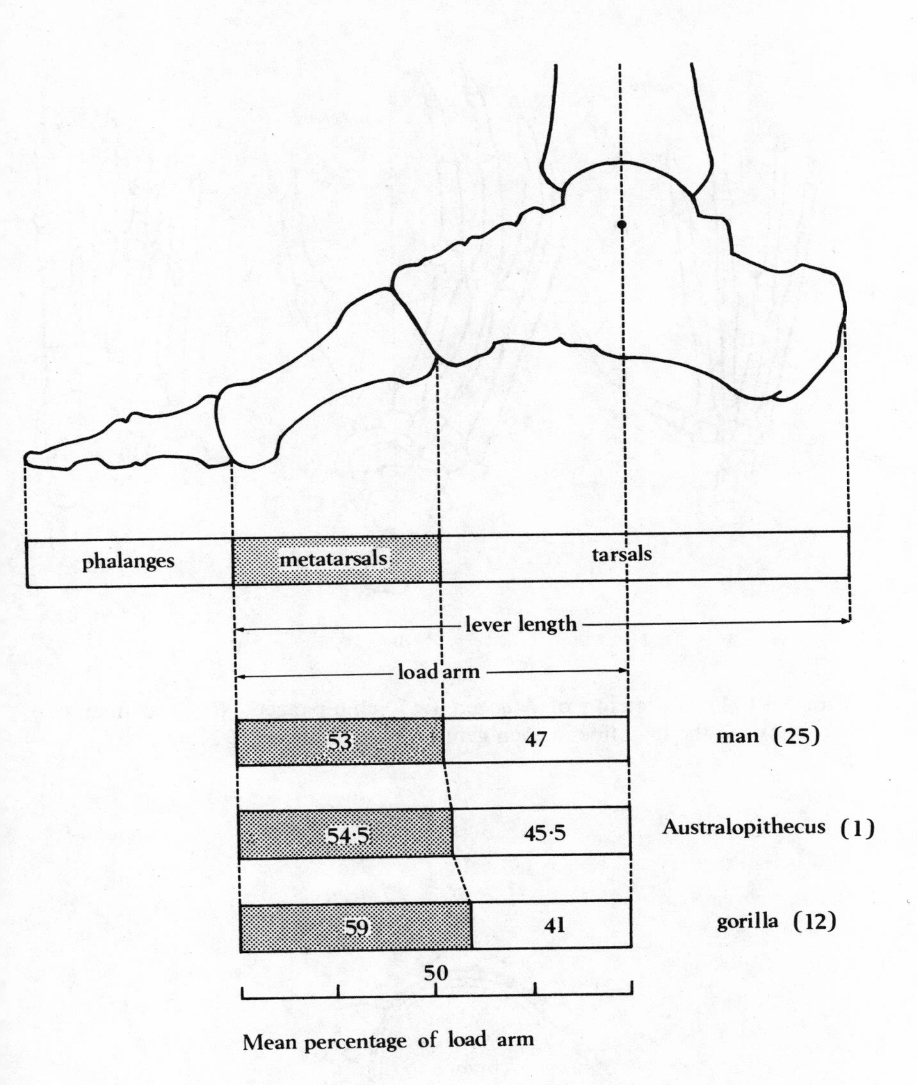

FIG. 5.16. Mean proportions of different parts of the feet of three primates suggest that the tarsal section of the load arm has lengthened in human evolution. The number following each name shows the sample size (n) used for this calculation. The proportion in modern man is of course variable and overlaps the figures for *Australopithecus*. As always, the significance to be attached to a sample of $n = 1$ is questionable, but in the absence of further evidence there is no reason to discard these figures as a basis for discussion. (Data from Day and Napier 1964.)

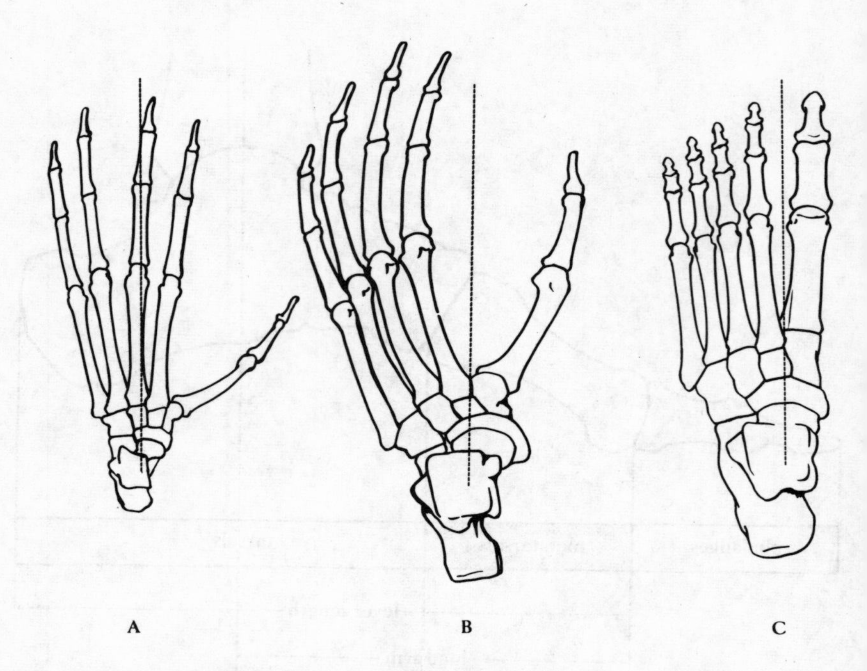

FIG. 5.17. Foot skeleton of *Macaca* (*A*), chimpanzee (*B*), and man (*C*), showing the load line in each genus. (After Morton 1927.)

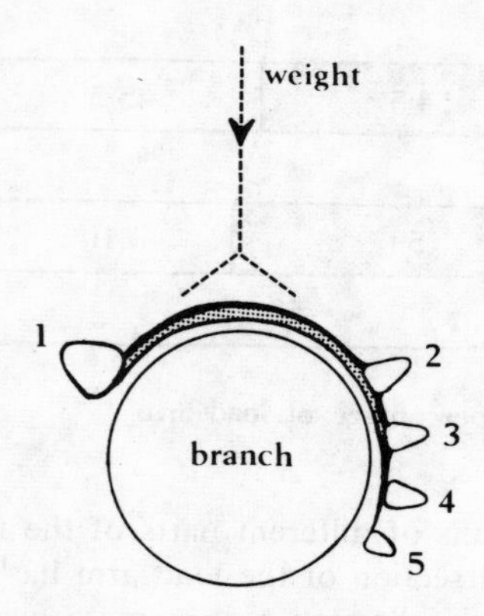

FIG. 5.18. The transmission of weight to the branch in apes is not through a digit but through the musculature between digits 1 and 2. This figure shows a diagrammatic transverse section through foot and branch. (After Morton 1927.)

(Fig. 5.19). The pattern of robusticity of the metatarsals is also significant in this respect: in the fully terrestrial forms alone, more weight is transmitted through digit 5, which is more robust than digits 2 and 3 (Day and Napier 1964) and which, by making the weight distribution like a tripod, helps lateral balance (Fig. 5.20).

2. The foot may also act as a means of gripping the branches and trunks of trees. This is primarily the property of the phalanges. In an arboreal primate, the phalanges must be of such a length as to enable the animal to grasp the branches among which it moves. As has been mentioned, claws alone are not altogether satisfactory in a fully arboreal animal.

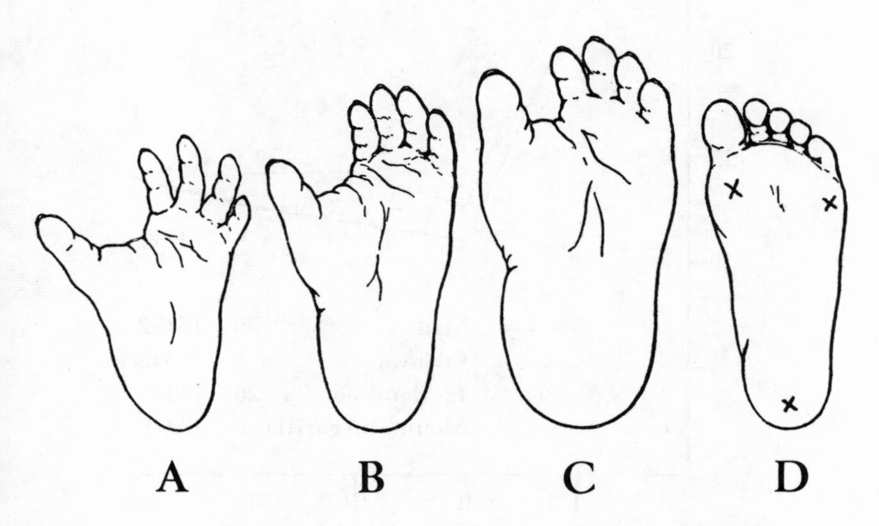

FIG. 5.19. Feet of chimpanzee (*A*), lowland gorilla (*B*), mountain gorilla (*C*) and modern man (*D*). The three centers of weight in the human foot are marked with crosses. (From Morton 1964.)

The effective size of grasp of the phalanges can be increased by the evolution of an opposable hallux (or *pollex* in the forelimb), a common adaptation among the prosimians, the small size of which poses special problems in gripping the larger branches. Among the relatively large quadrupedal monkeys, opposability of the hallux has not evolved to any great extent; instead, long phalanges are the rule. Only among the slower-moving great apes do we find full opposability of the hallux, and, as has been described, this opposability gives rise to a change in the load line, an important feature in human evolution.

A second development of great importance has effected the grasping power of the primate hand and foot (together called the *chiridia*). The function of the hooked claws of the insectivores and lower prosimians

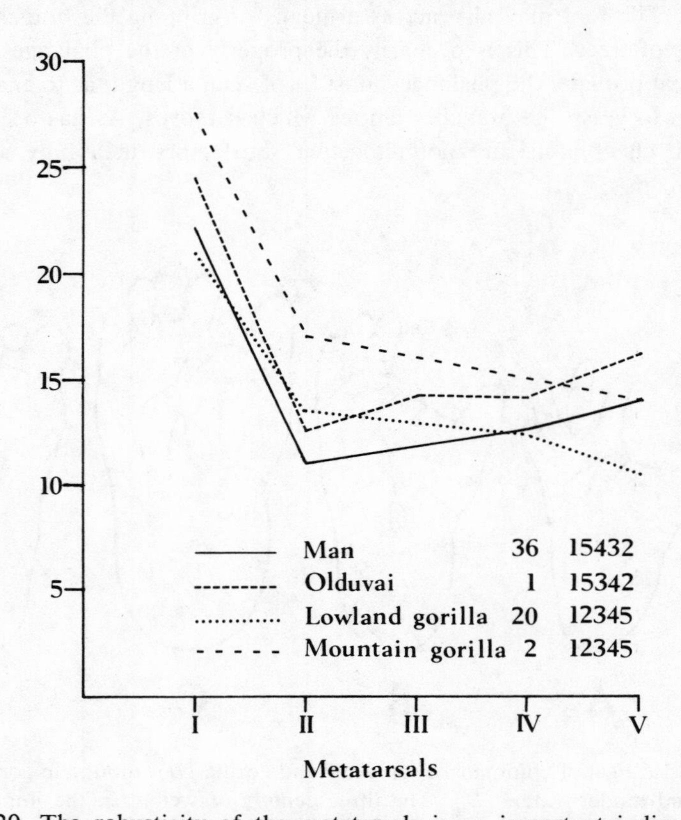

FIG. 5.20. The robusticity of the metatarsals is an important indicator of weight distribution through the foot. The robusticity index is defined as $\frac{\text{mean diameter x 100}}{\text{mean length}}$ of the different metatarsals. In this diagram, this index is shown for the metatarsals of modern man, the Olduvai foot, and the two gorillas. Both in modern man and in the Olduvai foot, the fifth metatarsal is the second most robust after the first. In the gorillas, it is the least robust, though in the more terrestrial mountain gorilla the index of all metatarsals is greater, especially that of the fifth metatarsal. The five-figure numbers in the key show the order of robusticity (the robusticity formula) of the different samples; the smaller numbers indicate the sample size (n). (After Day and Napier 1964.)

has been taken over in the primates, not only by the long phalanges, but also by the evolution of *volar pads* (friction pads) on the palms, which were enlarged and eventually fused. Volar pads are characterized by a special ridged friction skin with a microscopic folding of the outer layer. This skin lacks hair but is well supplied with sweat glands. The claws themselves were reduced in evolution and modified to flat plates, the nails, which gave support to the terminal pads (nail pads) that are so important in maintaining a grip on a branch.

3. The foot may also act as a tactile organ, and there was an immense evolutionary proliferation of the tactile receptors in the ridged skin. In this way, the chiridia came to replace, as prime organ of touch, the tactile rhinarium and whiskers on the snout that we associate with most other mammals. The importance of this development, particularly as it applies to the hand, is immense in human evolution; it is further discussed in chapters 6 and 7. In the human foot we find an organ perhaps unnecessarily well-endowed with tactile receptors, since during human evolution its functions have been reduced from three to one, that of locomotion.

Unfortunately, we have little knowledge of early hominid feet. There is one talus bone from Kromdraai and one from East Rudolf, together with a nearly complete foot from Bed I, Olduvai Gorge. All three specimens suggest that the line of weight transmission through the foot was different from that in modern man, though the Olduvai specimen is more human than the others. The presence of a facet between the first and second metatarsals of the Olduvai foot suggests the hallux was much less divergent than in apes and close to that of modern man. In other respects the foot suggests the upright stance and bipedal gait of modern man, to which it lies morphologically close, though it is probably somewhat broader and the relative shortening of the metatarsals has not gone quite so far as in man (see Figs. 5.16 and 5.20) (Day and Napier 1964).

No fossil foot bones are known of *Homo erectus*, but of Neandertal man we have both bones and a footprint. They indicate a broad, short foot, but they can be matched in modern man, and, although the proportions are unusual, they are not significantly simian, as has been claimed. The best preserved specimens are those of La Ferrassie.

In summary, it is possible to distinguish six different but related trends in the evolution of the human foot:

1. There was probably a reduction of the load arm of the foot lever compared with the power arm in the evolution of the anthropoid foot.

This reduction would have occurred as the monkeys became heavier animals. In the Hominoidea, we see a shortening of the phalanges themselves.

2. The volar (friction pad) area of the foot has increased to cover the whole contact surface.

3. With the hairless volar skin, we find sweat glands and a great density of tactile sensory nerve endings; both these features were perhaps a little reduced in the most recent stages of human evolution.

4. There was a change of load line from metatarsal 3 to between metatarsals 1 and 2; this change was a correlate of the freely opposable hominoid hallux. In human evolution the load line moved on towards the hallux, which became relatively large—a significant pointer to human origins.

5. The high proportion of load arm made up by the metatarsals (as compared with tarsals) is somewhat reduced in man by a shortening of metatarsals 2–5.

6. The foot is narrowed, yet at the same time a new secondary load line evolves along phalanx 5, which becomes more robust and thus enables the weight of the bipedal animal to be distributed through three centers in each foot—the heel, the hallux, and the small toe—giving the human foot the stable character of a tripod.

These changes, which summarize the evolution of the human foot, involve not so much the evolution of a new function as a reduction in the original primate functions. We see the foot changing from a tactile grasping organ to a locomotor prop. The phalanges are shortened, and the musculature is reduced. Although some nonlocomotor function is still possible in the foot (especially if the full use of the arms and hands is restricted or inhibited over an extended period of time) we see that it has clearly been evolving from a generalized to a specialized organ.

5.5 Summary

FROM A CONSIDERATION of the lower limbs it is possible to conclude that the evolution of erect posture and bipedalism has involved fundamental changes in the function of two structures, the pelvis and the foot, together with a host of minor changes elsewhere.

The pelvis, structurally very complex already, has changed shape markedly in accordance with the different characteristic angle of the femur in relation to the vertebral column and the changed function of its associated musculature. In particular, the pelvis, in its role as anchor for the important muscles of forward propulsion and lateral balance, has broadened to give these muscles greater leverage about the hip joint. This broadening has been accompanied by a shortening of the ilium and a closer approximation of the points through which the body weight is transmitted—the sacral articulation and acetabulum. The pelvic basin has undergone re-alignment with the spine and gives essential support to the viscera.

The leg is straight, and in man we see the whole body weight transmitted directly through the knee joint from acetabulum to ankle. The leg has also lengthened in relation to the trunk and arms, so that the proportions of the body characteristic of other hominoids have been completely reversed, as can be seen from the chart of intermembral indices (Fig. 5.1).

The foot itself has undergone evolution from a generalized to a specialized condition. It has been modified to carry the whole weight of the body in walking, one foot at a time. The points of weight transmission have become precise and threefold rather than general, and the primary line of weight has moved close to the hallux. Digits 2–5 have shortened, and the outer metatarsal has thickened.

The immense power of the human stride comes from three muscles in turn. The biceps brings the body forward over the knee; the gluteus maximus completes extension at the hip; the gastrocnemius gives a final lift by raising the foot lever on the distal end of the metatarsals. The bipedal stride of modern man is unique in the animal world and, as we have seen, was already advanced in *Australopithecus,* although there are still a few minor features in the bones of this genus that have not evolved as far as in modern man. *Australopithecus* was a biped at least two million years ago; by the time of *Homo erectus*, the evolution of bipedalism was evidently complete.

It seems clear that the evolution of bipedalism was one of the early character complexes to appear in human evolution. If it was almost complete two million years ago, it must have begun some millions of years earlier; we shall return to this matter when we come to discuss the origin of the Hominidae in chapter 11.

SUGGESTIONS FOR FURTHER READING

The evolution of the primate and in particular the hominid pelvis has been treated in some detail by W. E. le Gros Clark in two books: *The Antecedents of Man* (3rd ed.; Chicago: Quadrangle Books, 1971) and *The Fossil Evidence for Human Evolution* (2d ed.; Chicago: University of Chicago Press, 1964). The evolution of the human foot has been described by D. J. Morton, in *The Human Foot* (New York: Haffner Publishing Co., 1964). The evolution of bipedalism is discussed by J. R. Napier, "The Antiquity of Human Walking," *Scientific American,* 216 (1967): 56–66. The hindlimb of *Australopithecus* has been discussed in detail (but with a different taxonomy) by J. T. Robinson, *Early Hominid Posture and Locomotion* (Chicago: University of Chicago Press, 1972). Its mechanics have recently been reviewed in an important paper by C. O. Lovejoy, G. Kingsbury, G. Heiple, and A. H. Burstein, "The Gait of *Australopithecus*," *American Journal of Physical Anthropology*, 38 (1973): 757–780.

6

Manipulation and the Forelimb

6.1 The mammalian forelimb

In the locomotion of the terrestrial quadrupedal mammals, including the quadrupedal primates, the main driving force is derived from the hindlimbs. The body is propelled forward from the rear (not pulled), and mechanical efficiency requires a rigid bone-to-bone connection between the pelvis and the vertebral column. The function of the forelegs, even in a generalized mammal, is different. In such movements as running and jumping, the forelegs usually hit the ground first and take a great deal of shock. This shock would be transmitted throughout the body by a rigid bone-to-bone connection. Such a connection does not exist; the shock is absorbed by the mass of muscle that suspends the spine and thorax from the shoulder girdle (Fig. 6.1).

Although the shoulder or *pectoral* girdle consisted in its primitive condition of three bones, as does the pelvis, it is reduced in mammals to two—the *scapula* and *clavicle*. The scapula is the dorsal bone and is equivalent to the ilium; the clavicle is ventral and equivalent to the pubis. Unlike the pelvis, however, they are not fused but are connected by a movable joint (the *acromioclavicular joint*). The clavicle makes contact with the rest of the skeleton through a second movable joint, the *sternoclavicular,* where it articulates with the *manubrium* or upper part of the *sternum*

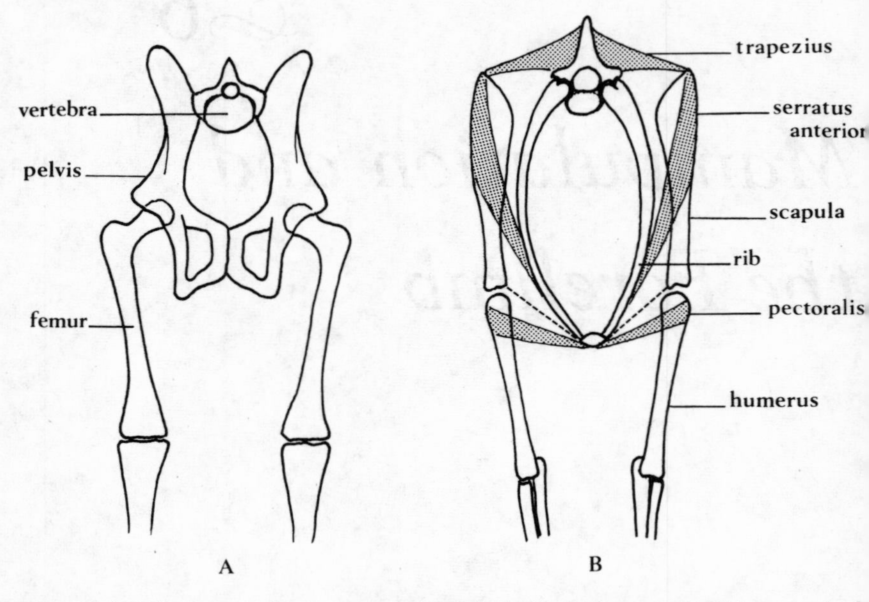

FIG. 6.1. Diagrammatic representation of the limb "suspension" in a typical quadrupedal mammal (such as a horse). The pelvic girdle, supported by the hindlimbs, is fixed to the backbone by a rigid bone-to-bone connection (*A*). Upon the pectoral girdle (*B*), the body is suspended mainly by two muscles, the *trapezius* and *serratus anterior,* which transmit the animal's weight from the backbone and ribs. The lower end of the scapula is steadied by the *pectoralis,* which holds the head of the humerus near the trunk. The dotted line indicates the position of the clavicle if it is present. It is lost in ungulates, since it serves no function in a strictly quadrupedal animal.

(Fig. 6.2). The clavicle maintains the distance of the scapula from the sternum and, acting as a strut, allows movement of the scapula at a constant radius from the manubrium. The weight of the animal is suspended on the *serratus anterior* muscle. The clavicle merely keeps the *glenoid cavity* (the pectoral equivalent of the acetabulum), into which the *humerus* fits, at a fixed distance from the midline of the body. Figure 6.1 shows, however, that when the body weight alone is transmitted by the pectoral girdle (as among ungulates, which do not leap or climb), the clavicle is not an essential component, because movement occurs only in the anterior-posterior plane, and there is no lateral swing of the forelimbs. The clavicle has been lost in these groups.

The muscle suspension of the forelimbs gives them greater flexibility than the hindlimbs, and this flexibility has been exploited, as we shall see, by the primates.

6.2 *The pectoral girdle*

UNLIKE MANY quadrupedal mammals, the primates have retained the two-boned pectoral girdle, and both bones are flexibly mounted. They can move some 40 degrees about the sternoclavicular joint in each of two planes. At the same time, the scapula can move freely in relation to the clavicle about the acromioclavicular joint. These two movable joints allow the scapula to move over the surface of the thorax in both the vertical and the horizontal planes. The acromioclavicular joint allows the scapula to twist in relation to the clavicle and so remain flat on

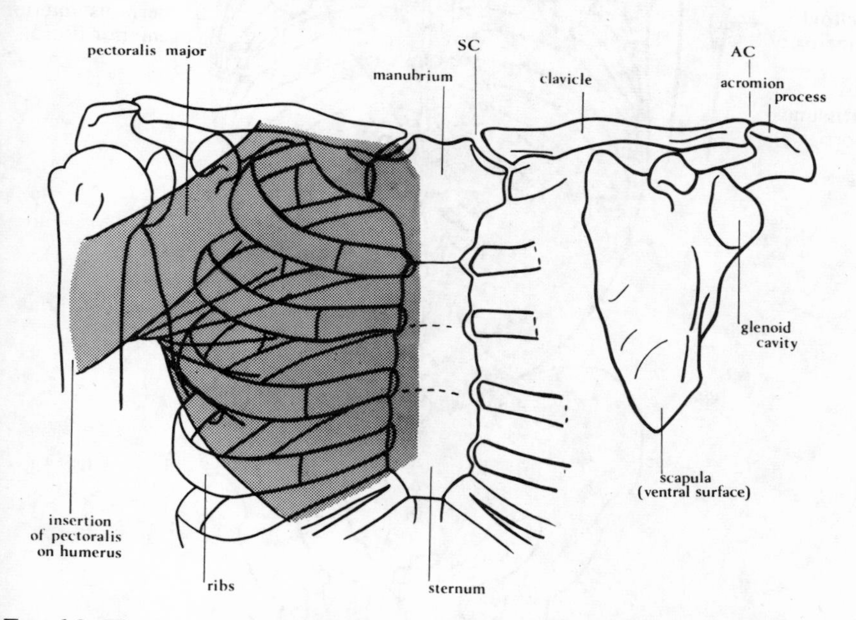

FIG. 6.2. The human thoracic region is here shown cut away on the right so that the clavicle can be seen from the front. On the left the thorax is intact, and upon it is shown the *pectoralis major* muscle on its ventral side, which in quadrupeds propels the body forward and in brachiating primates lifts the body in brachiation. *SC* = sternoclavicular joint; *AC* = acromioclavicular joint.

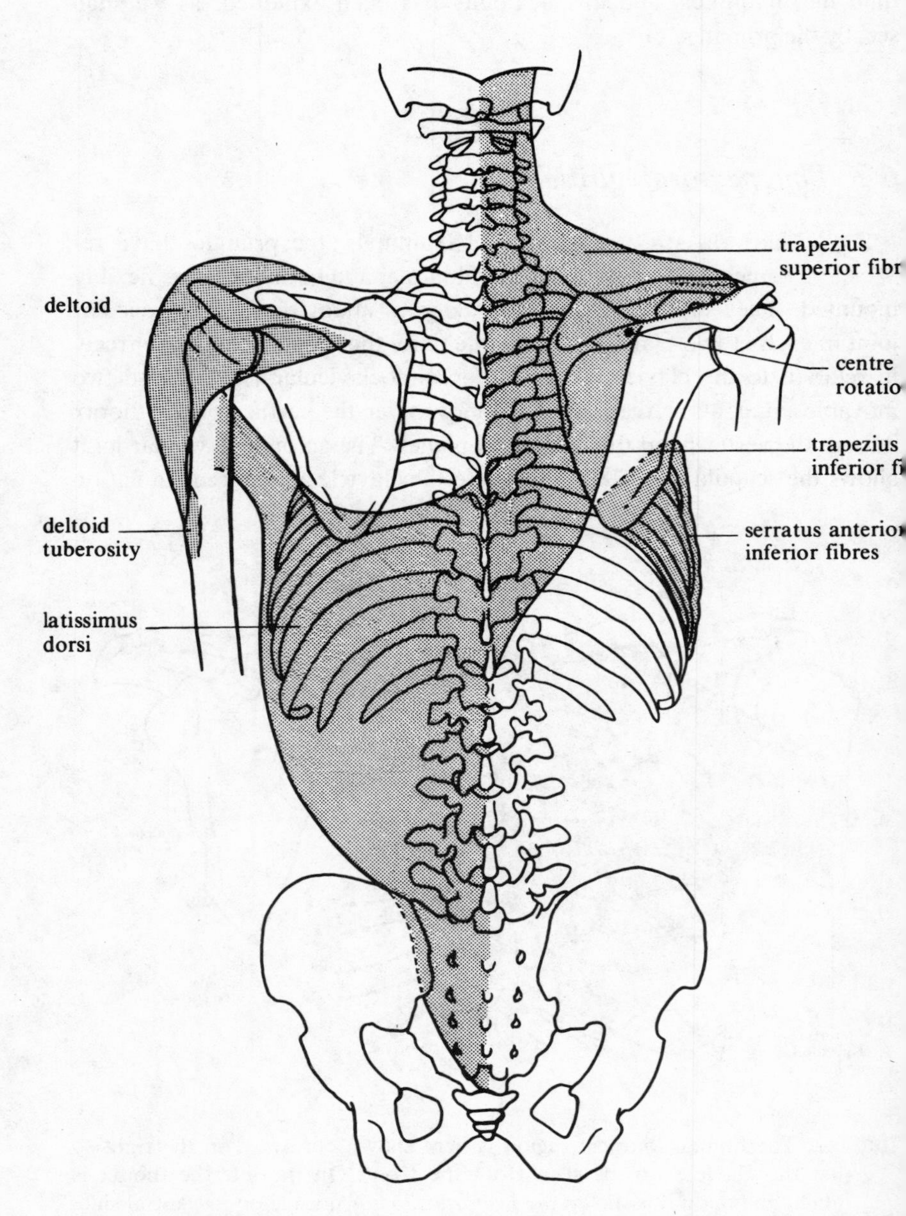

FIG. 6.3. Dorsal view of the human trunk showing (*on the right*) the two muscles that can effect rotation of the scapula about the center of rotation and (*on the left*) muscles that raise and lower the humerus.

the surface of the thorax, in spite of the fact that the latter is not spherical. The scapula moves forward round the thorax in pushing or thrusting and to the back in pulling the body forward or upward in climbing. In addition, by the action of the *trapezius* and serratus anterior muscles, the scapula can be rotated (Fig. 6.3, right side) so that the glenoid cavity (equivalent to the acetabulum in the pelvis) can face in different directions. This rotation nearly doubles the range of movement of the limb in a vertical plane. It may also be noted that the glenoid cavity is more open than the acetabulum and therefore allows a greater arc of movement in the forelimb than in the hindlimb. Here, stability has been sacrificed to mobility.

From this short account of the pectoral girdle, it can be seen that, in accordance with its different function, its structure is different from that of the pelvis. In fact, the structure is more generalized—that is, capable of more kinds of movement—and we might accordingly predict a greater variety of adaptations in the course of primate evolution. Such is indeed the case. While differences in the clavicles are slight, a glance at Figure 6.4 will show the differences between the scapula of the tree shrew, gibbon, gorilla, and man. Figure 4.10 shows rather more striking differences in the position of the shoulder girdle upon the thorax, differences related to the change of function of the forelimbs in the evolution of the primates.

We can trace three distinct functions of the forelimbs and follow the changing proportion of each. These functions are as follows: (1) to support the weight of the body in quadrupedal locomotion; (2) to suspend the weight of the body in brachiation; and (3) to manipulate objects.

The change from quadrupedalism to brachiation is reflected in the form of the scapula and its musculature. As mentioned above, the shoulder girdle can be rotated to allow the arm to rise vertically, a movement that is clearly important in brachiation (Fig. 6.3). Oxnard (1963) has shown how the importance of the two rotatory muscles, the serratus anterior and the trapezius, varies among the main locomotor groups of primates (Fig. 6.5). The data show that, as might be expected, the muscles are better developed in brachiators than in quadrupeds and that in this character man is close to the brachiators. At the same time, there is a relative increase in length in the bony lever to which these muscles are attached (as we saw in our consideration of leverage about the hip joint). That is to say, the insertion area of the muscles on the scapula tends to move away from the central point around which the scapula rotates. Therefore in brachiating forms we find that the scapula is lengthened; it is longer in the monkeys, especially the arm-swinging species, and longest

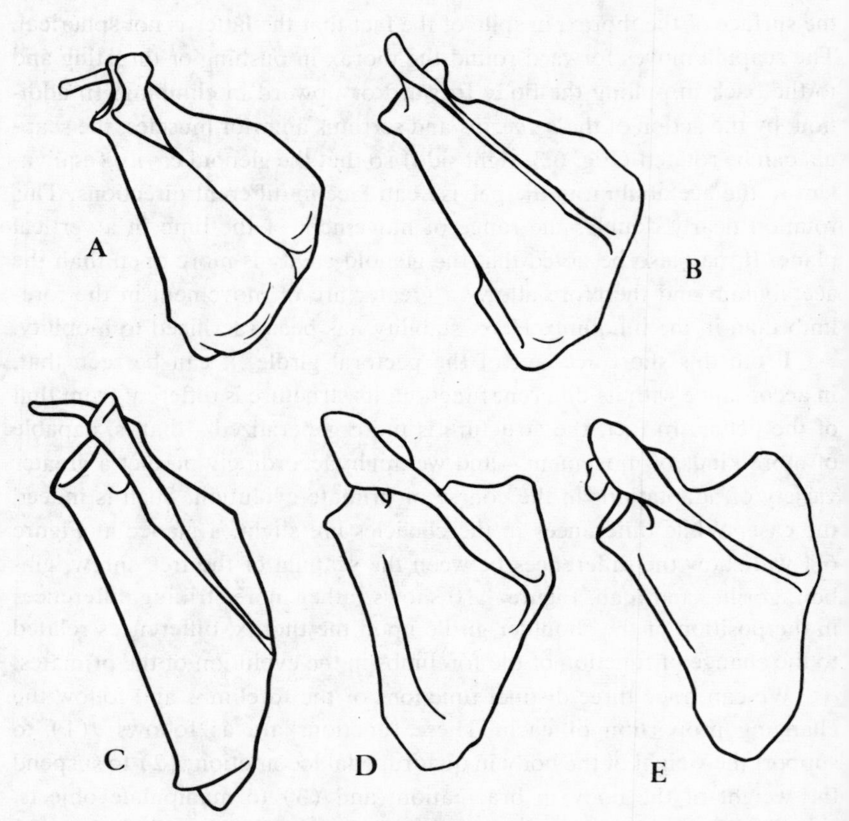

FIG. 6.4. Scapulae of *Tupaia,* a tree shrew (*A*), *Macaca* (*B*), *Hylobates* (*C*), modern man (*D*), and gorilla (*E*), all drawn the same size.

in the Hominoidea, in which the *acromion process* (which, with the spine of the scapula, carries the trapezius and *deltoid* muscles) is also well developed, for the same reason (Figs. 6.3, 6.4).

The actual plane of the glenoid cavity has similarly changed in relation to the rest of the scapula, reflecting the "characteristic" position of the arms in relation to the body, and the greater mobility of the brachiators' shoulder joint. Quadrupeds therefore tend to have the glenoid cavity pointing more or less horizontally; brachiators have the cavity pointing upward. The way the angle is measured and the figures for it are shown in Figure 6.6.

We see that in this angle man is aligned with the quadrupeds, since the plane of man's glenoid cavity is more nearly parallel to the lateral

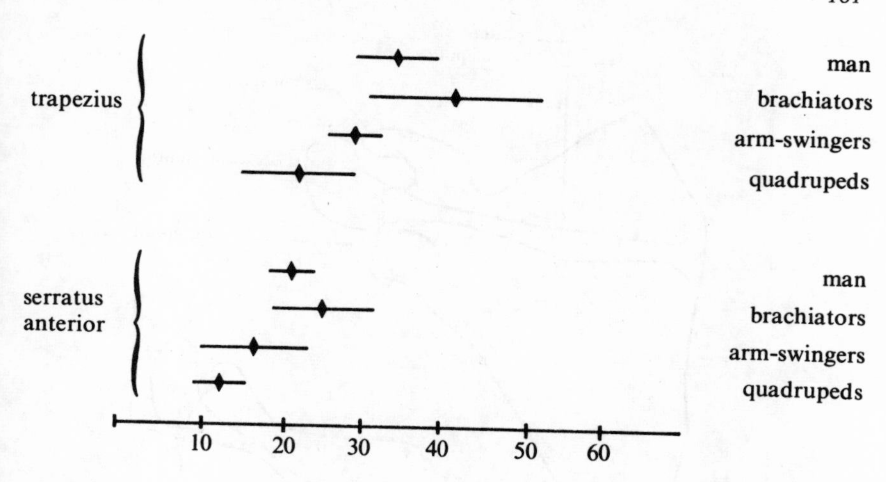

FIG. 6.5. Diagram illustrating the relative mass of the trapezius and serratus anterior muscles in the three main locomotor groups of higher primates and in man. Note that in these two features man lies near the brachiators. The numbers are units of relative mass. (Data from Oxnard 1963.)

margin of the scapula than it is in brachiators. The reason is that man does not brachiate but, on the contrary, carries his arms hanging downward from the shoulder. His bipedal stance has clearly changed the shape of the scapula in this respect; we need not necessarily conclude that the larger angle indicates a history of quadrupedal locomotion.

On the other hand, when we turn to fossil hominid clavicles and scapulae we find a different situation. Up to the present, fossil remains of pectoral girdles are rare. The lightness of the scapula and its position within the surrounding musculature probably render it liable to destruction by predators and scavengers. Clavicles are rather more common. There are two fragments belonging to the Miocene *Dryopithecus*, but, apart from their obviously hominoid character, they are too incomplete for further diagnosis (Le Gros Clark and Leakey 1951). Of *Australopithecus* we have fragments of three clavicles, one from Makapansgat, which is very fragmentary, and two from Olduvai. Of the latter, which are quite well preserved, one is described as only slightly different from that of modern man (Napier, in Tobias *et al.* 1965) but suggesting a history of suspensory locomotion (Oxnard 1968). One clavicle is known of Peking man from Choukoutien (*Homo erectus*), and it appears to be quite similar to the clavicle of modern man (Weidenreich 1938).

There is one fragmentary scapula from Sterkfontein, described by

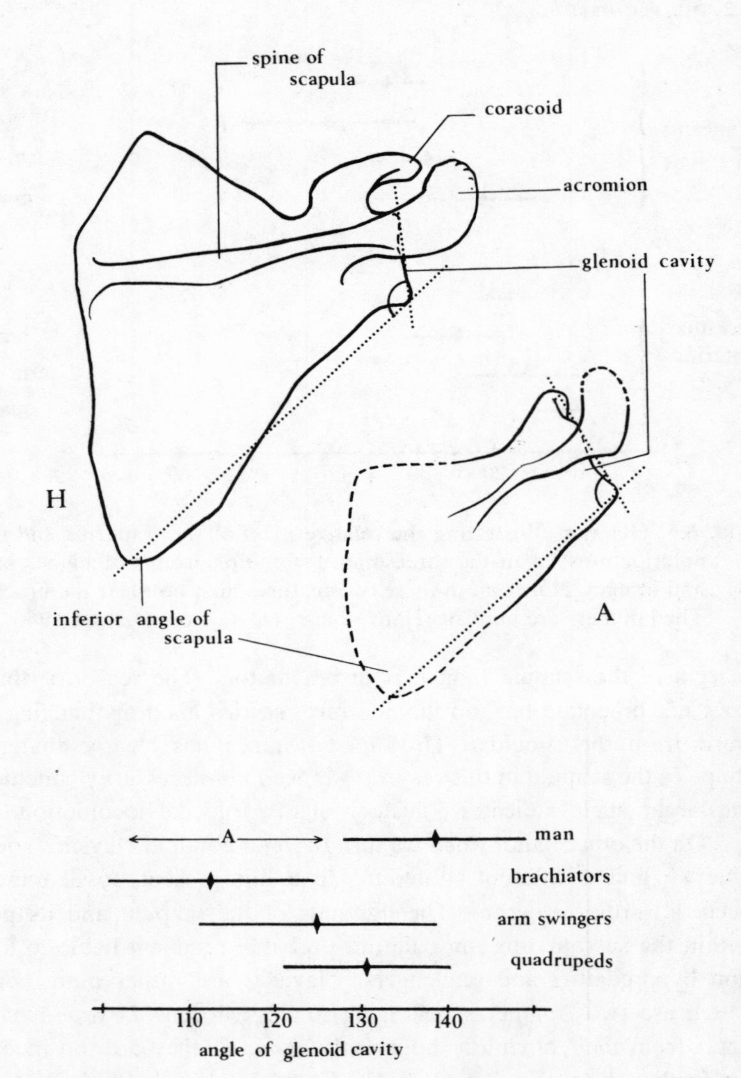

FIG. 6.6. The angle of the plane of the glenoid cavity can be measured in relation to the lateral margin of the scapula, as shown in the human scapula (*H*). This angle has been measured by Oxnard in many different primates, and his results are recorded in diagrammatic form below the drawings. The smaller, incomplete scapula on the right (*A*) belongs to *Australopithecus.* Although observation of the specimen (which is still in its rocky matrix) suggests the angle of the glenoid cavity to be greater than that indicated by Broom (103°), even the highest estimate places it among the arm-swinging monkeys (about 126°), and it may have been much less. It is shown on the chart as *A*. (After Broom *et al.* 1950.)

Broom *et al.* (1950), attributed to *Australopithecus africanus*. Fortunately, it consists of the anterior part, including the glenoid cavity, the *coracoid process,* and part of the spine (Fig. 6.6, *A*). The coracoid is curved strongly and carries a large attachment area for the biceps muscle—a brachiating character. The angle between the spine and the lateral margin of the bone (the *axilospinal angle*) is such that they are nearly parallel. Oxnard (1963) has shown that these two characters are associated with brachiation, and Figures 6.4 and 6.6, *A* suggest this association. Also, the small angle that the plane of the glenoid cavity makes with the lateral margin of the scapula again puts the bone into the brachiating class (chart in Fig. 6.6). Broom summarizes his description by stating that the bone falls between that of the orang and modern man in its overall form; or, to put it another way, it has a more brachiating form than has the scapula of modern man.

Although it is only a small fragment, this scapula may provide an important clue to the locomotor history of *Australopithecus africanus* from Sterkfontein. A single bone is a small sample, but it does suggest either that *Australopithecus* used his arms to brachiate, at least to some extent, or that his ancestors did so. Which of these hypotheses is correct

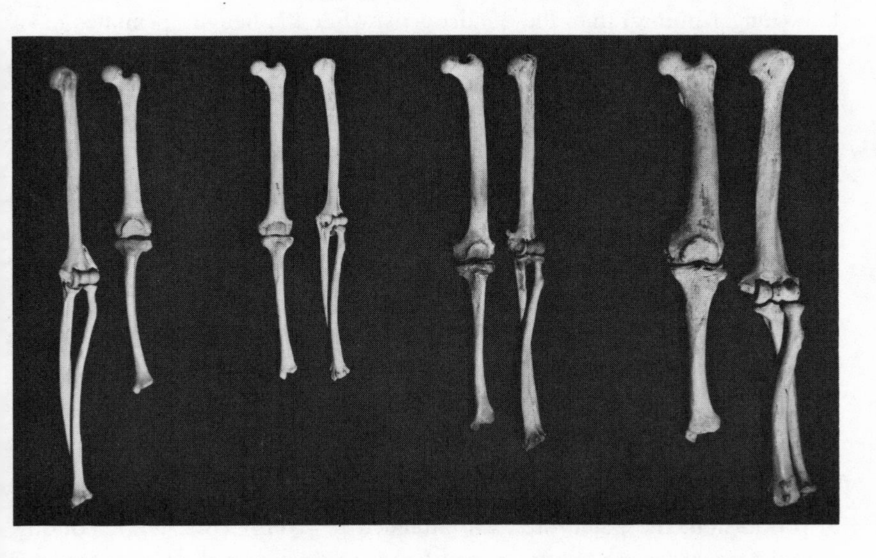

FIG. 6.7. Forelimbs and hindlimbs of—from left to right—orang-utan, pygmy chimpanzee, chimpanzee, and gorilla. The arms are relatively long in the orang-utan, while the legs become relatively short in the knuckle-walking apes as body size increases. (From Pilbeam 1972.)

will be clarified by the evidence of the skull and our discussion of the ecology of the group (chapter 7).

We can distinguish, therefore, three types of scapula, associated with three locomotor patterns: the quadrupedal scapula, shorter and squarer in proportion, with a downward-pointing glenoid cavity; the brachiating scapula, elongated, with an upward-pointing glenoid cavity; and the human scapula, still elongated, but with a downward-pointing glenoid cavity. The anatomy of the australopithecine scapula suggests some degree of arm-swinging in the ancestry of hominids, and in the final evolution of man we see a change in the angle of the glenoid cavity, associated with bipedalism.

6.3 *The evolution of the upper arm*

THE MOST striking differences among the primate humeri are in length and thickness, the functional significance of which has already been broadly discussed. As might be expected, the brachiators have longer humeri than the quadrupeds, while the heavier primates have thicker bones. Figure 5.1 shows the intermembral index, and Table 6.1 shows arm length in relation to trunk length.

TABLE 6.1 MEAN ARM LENGTH (HUMEROUS AND RADIUS) AS A PERCENTAGE OF TRUNK LENGTH (DATA FROM SCHULTZ 1953, P. 287)

Primate	*Index*
Macaca	83
Hylobates	184
Pongo	148
Pan gorilla	121
Pan troglodytes	113
Homo	113

The humerus is a powerful bone, which carries a considerable proportion of the body weight in quadrupedal mammals; the forward muscles of propulsion, the *pectorales*, are attached to it (Fig. 6.2, *left*). Among brachiators, these muscles are used to lift and support the body, and the *latissimus dorsi* muscles, which are dorsal to the thorax, also help to lift the body in climbing (Fig. 6.3, *left*). Oxnard (1963) has shown that they are relatively larger in brachiators than in quadrupeds, for only in the former do they lift the total body weight. The arm itself, on the other hand, is lifted by the deltoid muscle (Fig. 6.3), which again is larger in brachiators. At the same time, the point of insertion of the muscle on the humerus moves down the bone to give greater leverage (Fig. 6.8).

Less striking differences include the plane of the articular head of the humerus in relation to the elbow joint. This relationship has changed in evolution according to the position of the scapula on the thorax (Fig. 4.10). Thus in quadrupedal forms, the center line of the ball joint of the humerus, which fits into the glenoid cavity, is just about at right angles to the elbow (Fig. 6.9, *C*) facing directly backward, while in a species such as the chimpanzee or *Presbytis,* the humerus has twisted so that the angle between elbow and head is near 30 degrees, and the head points inward (Fig. 6.9, *A* and *D*). Modern man shares this character with the brachiating groups: in both arm-swinging species and bipeds the arm is able to move laterally as well as backward and forward.

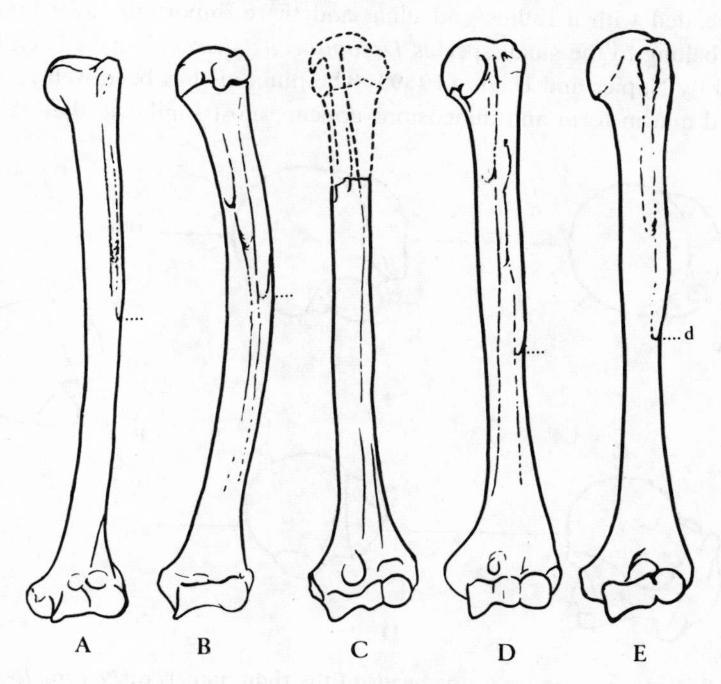

FIG. 6.8. Humeri of *Presbytis* (*A*), *Cercopithecus* (*B*), *Dryopithecus* (*C*), *Pan* (*D*), and modern man (*E*), all drawn the same size; *d* marks the lowest extent of the insertion of the deltoid muscle. (After Napier & Davis 1959.)

Finally, it is characteristic of brachiators as well as of man, that the hinge joint at the elbow, the *trochlea,* allows full extension of the forearm and is relatively broad, giving much greater lateral stability at that point (Fig. 6.8).

In the upper arm, therefore, we find that man shares four characters with the brachiators: (1) a relatively long and stout humerus, (2) relatively large arm-raising muscles and a low point of insertion of the deltoid, (3) the head of the humerus pointing inward rather than straight backward, and (4) a fairly broad trochlea at the elbow. In one character, however, man's upright stance has led to a reduction in musculature, compared with all other primates and especially the brachiators—that is, in the propulsive muscles, the pectorales and latissimus dorsi.

Let us now turn to the fossil evidence for the evolution of the hominoid forelimb. We have fragments of two humeri of *Dryopithecus,* four of *Australopithecus*, and one of *Homo erectus*. One *Dryopithecus* humerus is associated with a radius and ulna, and these important fossil bones, which belong to the small species *Dryopithecus africanus,* have been described by Napier and Davis (1959). The humerus has been well reconstructed and in form and dimensions appears most similar to that of the

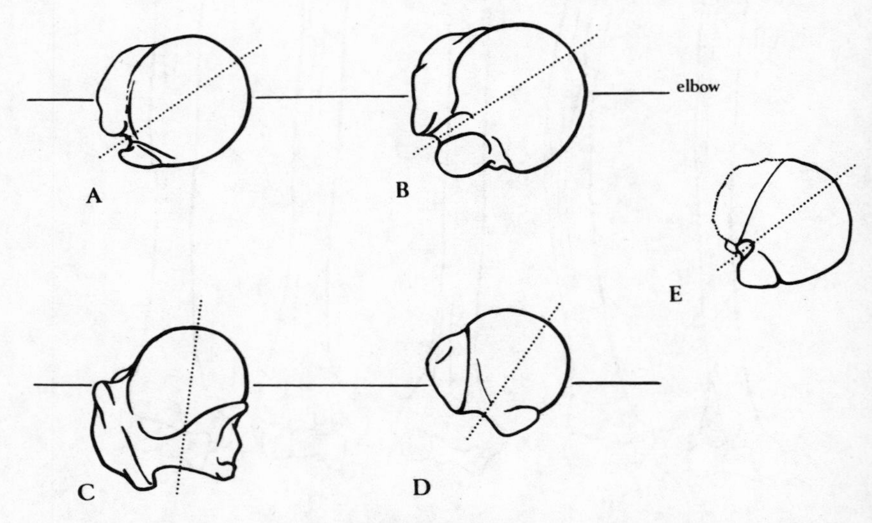

FIG. 6.9. View from above of the heads of the right humeri of *Pan troglodytes* (*A*), modern man (*B*), *Macaca* (*C*), and *Presbytis* (*D*). They are aligned according to the plane of the elbow joint. *E* is the humerus head of *Australopithecus*; it is not correctly aligned since the elbow joint is missing. It is, however, added for comparison of the form of the head and groove.

arm-swinging monkey *Presbytis*. The point of insertion of the deltoid muscle is well down in the brachiator position, and the trochlea closely approaches the condition seen in *Pan*, allowing great extension at the elbow joint and (by its broadness) great stability (Fig. 6.8). This in turn suggests that arm-swinging was combined with the kind of intermittent quadrupedalism we find in the chimpanzee (Conroy & Fleagle 1972).

Small portions of a femur and humeral shaft have also been reported from the chimpanzee-sized *Dryopithecus nyanzae* (Le Gros Clark and Leakey 1951). These fragments do not bear clear characters of brachiators, but this may perhaps be accounted for by the fact that larger forms of living primates (apart from *Pongo*) do not show such extreme specialization in this respect as do the smaller forms (such as the gibbon).

It is not possible to draw more than very limited conclusions from these few fragments of *Dryopithecus,* but it is fair to admit evidence of an arm-swinging and perhaps knuckle-walking primate in the Miocene—*Dryopithecus africanus*. It is of great interest that the dentition of this genus shows close affinities to the great apes (see 7.7). While we can never know for sure what place this Miocene species had in the evolution of the Hominoidea, at least the limbs do not preclude its close association with a hypothetical Miocene pongid stock, as well as a hominid stock.

Humeri of *Australopithecus* were found at the following sites:

Kanapoi	Distal extremity (Patterson and Howells 1967)
Sterkfontein	Proximal extremity (Broom *et al.* 1950)
Makapansgat	Small piece of shaft
Kromdraai	Distal shaft and trochlea with ulna fragment (Broom and Schepers 1946)
Olduvai	Distal shaft & trochlea
East Rudolf	Complete except for head (McHenry 1973)

The Kanapoi fragment, dated at about 4.0 million BP, is the most ancient, and with the Kromdraai specimen, has been statistically determined to show similarities to modern man. The Makapansgat fragment is too small to be of great interest, and the Olduvai fragments have not yet been described. The Sterkfontein humerus consists of the head and greater part of the shaft and has been described by Broom (Broom *et al.* 1950). In general, he finds the head of the humerus almost typically human, but the lesser tuberosity is very well developed and suggests a powerful *subscapularis* muscle—the muscle that draws the humerus forward and downward—an important muscle to a brachiator. In other characters, the bone appears almost human, but, according to Broom, it has characters that recall those of the brachiating orang. All these fragments may be classified as *Australopithecus africanus*. Only the East Rudolf specimen stands out as peculiar and distinct from existing hominoidea (McHenry 1973); its status is in doubt, but it probably belongs to the heavily built *A. boisei.*

A humerus shaft has been described from Choukoutien, associated with the fossils of Peking man. Though unusually thick-walled, this bone appears to fall within the range of that of modern man. The specimens of femora and humeri are few, yet it seems clear that *Homo erectus* had limbs not dissimilar to our own. If it were not for the evidence of the skull, we would certainly assess these limb bones as belonging to *Homo sapiens.*

From the data presented here, it seems clear not only that the upper arm of modern man carries characters associated with the arm-swinging primates but also that his fossil relatives, in particular *Australopithecus,* show quite strongly the characteristics of brachiation. Broom's comment on both the scapula and the humerus of *Australopithecus* was that in his opinion these bones were intermediate between the bones of modern man and the orang, the most advanced of the great apes in the direction of full brachiation. The evidence for suspensory locomotion in man's ancestry accumulates.

6.4 The forearm

THE BRACHIAL INDEX has been devised as an indicator of the relative lengths of the humerus and radius. Fig. 6.7 shows that an elongation of the forearm relative to the upper arm is a correlate of brachiation. The spider monkey (*Ateles*) is an extreme case in the New World, and the gibbon (*Hylobates*) is an extreme in this respect in the Old World (Brachial Index = 113). Some secondary shortening may have occurred in the gorilla (B.I. = 80); if so, it is possibly associated with that heavy animal's more quadrupedal mode of locomotion. The brachial index of man varies a great deal but is always well under 100 (usually 65–85), which places him with the quadrupedal primates. The same general proportions of length and thickness are shown in the radius and ulna as in the humerus.

These bones of the forearm, the radius and ulna, are flexed and extended about the hinge-like elbow joint by muscles that originate on the humerus and lie along it. The *triceps* is the extensor of the forearm; the *biceps* is the flexor (Fig. 6.10). As might be expected, the extensor is the important muscle in a quadrupedal mammal because it supports the weight of the animal at the elbow, and the flexor is important in a brachiating form because it flexes the arm and lifts the animal. Oxnard (1963) has again shown that the relative mass of the triceps is, as we might expect, diminished in brachiators. The length of the power arm

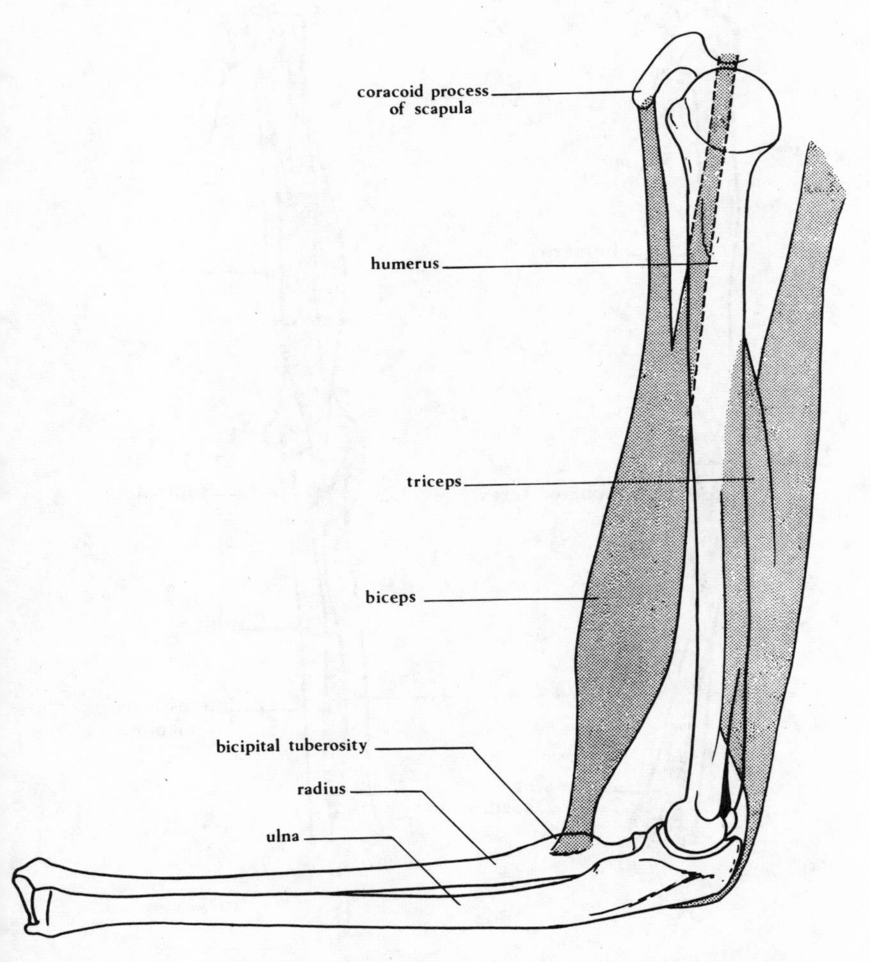

FIG. 6.10. Diagram showing the flexor and extensor muscles of the human forearm.

of the forearm lever (that is, the distance of the *bicipital tuberosity* from the elbow joint) also varies with locomotor function (see Fig. 6.10). In this character, modern man shows a shift from the quadrupedal condition toward the brachiating condition.

The retention of the two bones in the forelimb is of course due to the need for effective rotation of the hand—*pronation* and *supination* (in pronation the thumb lies nearer the body—in supination digit 5 does so;

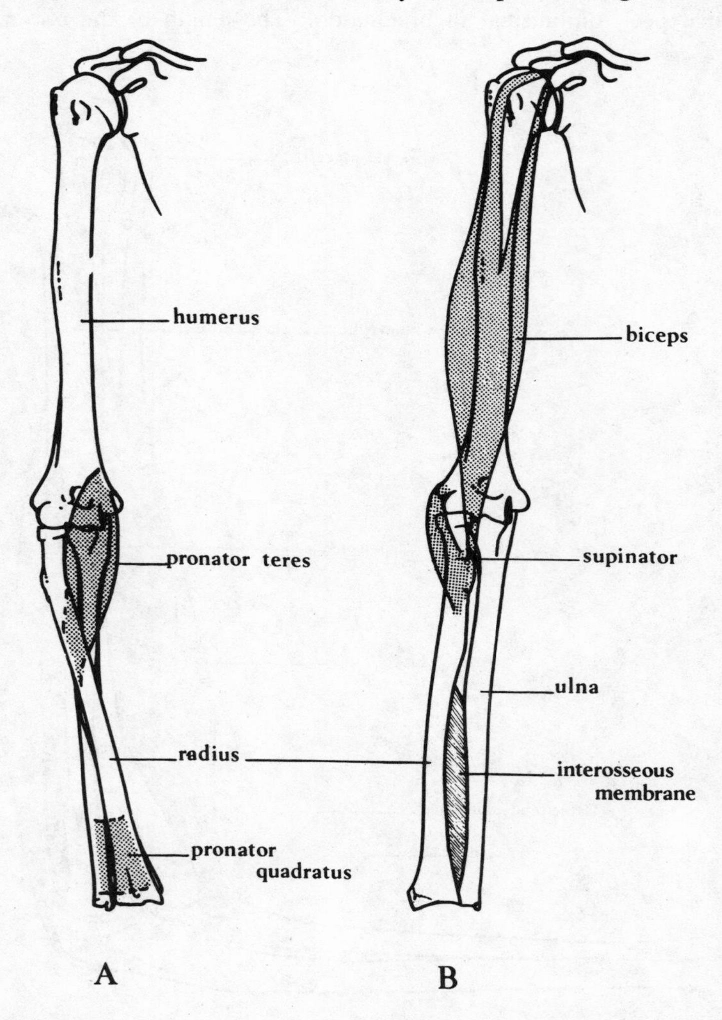

FIG. 6.11. Pronation (*A*) and supination (*B*) in man, shown from the front. The muscles used are shown in each drawing.

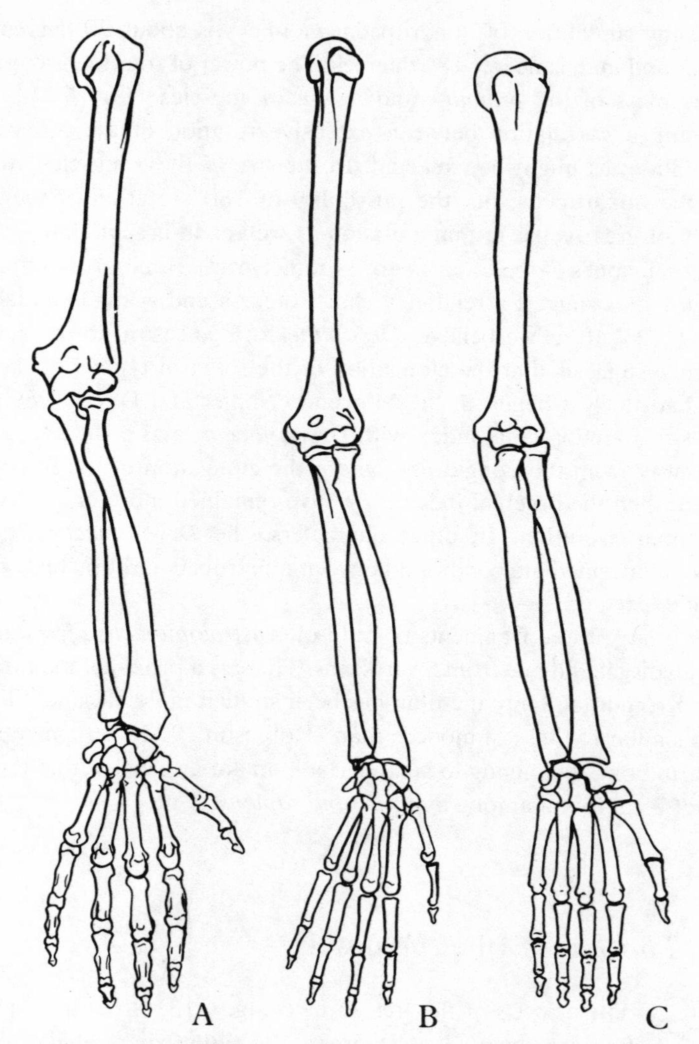

FIG. 6.12. Forelimb skeleton of chimpanzee (*A*), *Dryopithecus africanus* (*B*), and modern man (*C*), (reduced to the same total length). (After Le Gros Clark 1971.)

Fig. 6.11). The articulation at the elbow joint is primarily between humerus and ulna, and at the wrist between radius and the *carpal bones* (Fig. 6.12). Forces of compression and tension are transmitted between the two forearm bones by the *interosseous membrane*. The extent of rotation is limited by the extent to which the radius and ulna are bowed.

While the supination of quadrupedal monkeys is about 90 degrees, that of apes and man is nearer 180 degrees. The power of rotation is dependent on the mass of the *pronator* and *supinator* muscles (Fig. 6.11). There is, again, a correlation between extensive rotation of the forearm and brachiation, as might be expected. In the size of these muscles, man lies with the quadrupeds; but the possibility of 180° rotation of the arm is clearly of great value in manipulation as well as in brachiation.

Fossil bones of the forearm are extremely rare. Bones from *Dryopithecus africanus* suggest a relatively short forearm and a low brachial index (Fig. 6.12). If, as we believe, *Dryopithecus* is ancestral to the apes, this evidence suggests that the elongation of the forearm typical of the living apes had not yet begun in the Miocene (Napier and Davis 1959). Man shares this low brachial index with *Dryopithecus*, and if the hominid line split away from the pongid line before the elongation of the forearm occurred, then the brachial index may have remained more or less constant in human evolution. In other characters, the *Dryopithecus* fragments show an intermediate condition between quadruped and brachiator, as we might expect.

We have three fragments of radii of *Australopithecus africanus* from Makapansgat and one from Swartkrans. There is a proximal ulna fragment from Kromdraai. Only the ulna has been studied in detail, and all appear to be similar to those of modern man (Robinson 1972). Fragments of the forearm bones belonging to Neandertal man are as usual within the variation of form found among living *Homo sapiens*.

6.5 The hands of primates

THE MANUS of the tree shrew (Fig. 6.13, *A*) is a simple, five-fingered (*pentadactyl*) organ, the rather short phalanges making it appear little more than a paw. Each *phalanx* carries a claw, and there are six volar pads of friction skin on the palm and one terminal pad on each phalanx. Just above the wrist lies a small skin projection (*papilla*) from which grow long hairs, the *carpal vibrissae*. This papilla is a tactile sense organ common to most mammals, similar to those that bear whiskers on the face in dogs and cats. The vibrissae are implanted in richly innervated skin and form an extremely sensitive but quite undiscriminating tactile sense organ.

Studies of comparative anatomy make it clear that, as the volar pads increase in area and sensitivity in primate evolution, the importance of the carpal vibrissae decreases. They are poorly developed in the New World and absent in the Old World Anthropoidea. In the human fetus, a small transient cutaneous papilla can sometimes be seen in the carpal region (Schultz 1924). During evolution, the innervation of the volar pads, and particularly in the terminal digital pads, increases in density so that they become tactile sense organs of great importance (see sec. 6.6 and Fig. 6.13).

While it happens that in man the tactile function of the hand completely replaces the locomotor function, the manus remains an important locomotor organ in all other primates. In the Old World monkeys the forelimb may not be as important as the hindlimb in the propulsion of the quadrupedal groups, yet as a grasping organ it is very important. This grasping function is enhanced among the brachiators. There are two means of achieving a hold on a branch with the manus. The first, which is typical of the prosimians, involves a (*pollex*) thumb, which diverges from the phalanges and can be opposed to them round the branch (Fig. 6.13, *B*). The second, typical of brachiators, involves the lengthening of the phalanges and, in many cases, a reduction of the thumb (see Fig. 6.13, *E* and Table 6.2). In the Old World monkeys, however, the opposable thumb of the prosimians is not lost, but is retained as a result of the evolutionary development of the hand for use in other functions, such as the procurement of food, hygiene, defense, care of offspring, improved three-dimensional tactile sense and so forth (Biegert 1963). In fact, the Old World monkeys use their hands not only in locomotion but in a wide variety of basic functions, including grooming, carrying offspring and other objects, and throwing sticks and stones. A baboon has a precision grip between thumb and fingers sufficiently well evolved to extract the sting from a scorpion (Schultz 1961).

In certain arm-swinging monkeys and some Hominoidea, however, we find the thumb much reduced, since the more the hand is involved in locomotion as a brachiating organ, the less it is available as a hand. This specialization reaches its most extreme form, not in the gibbon where the thumb is retained for manipulation, but in *Ateles*, where it is vestigal or absent.

A more general characteristic of brachiators is the hook-like manus with its long phalanges. The proportions of the hands of apes have been investigated by Schultz (1956), and some figures are reproduced in Table

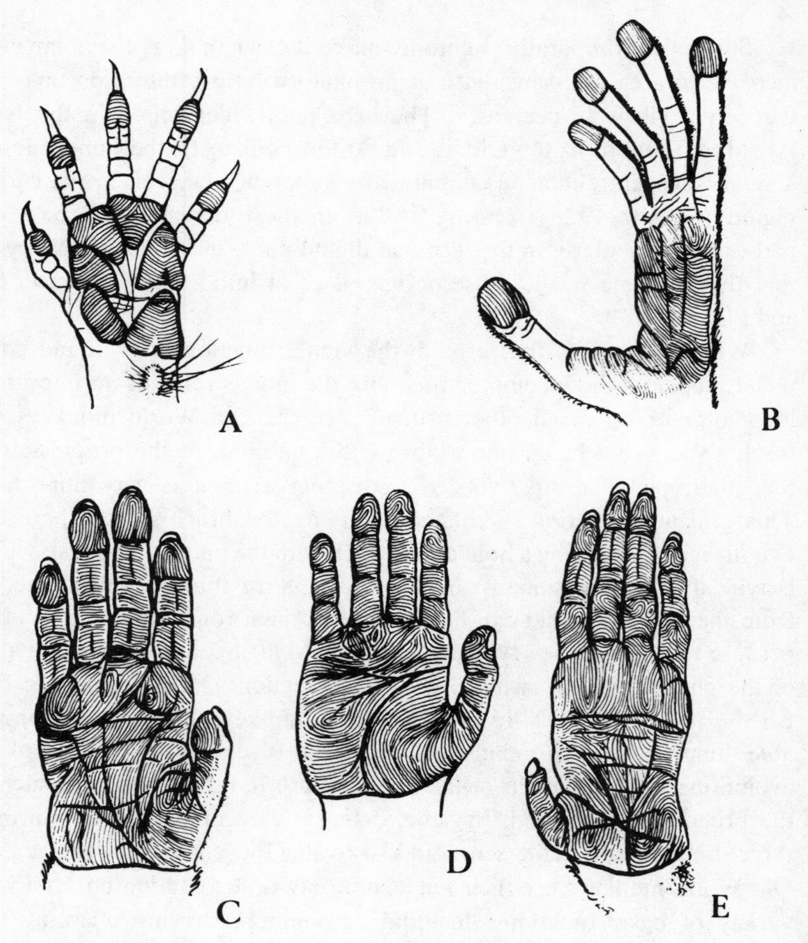

FIG. 6.13. The hands of primates: *Tupaia* (*A*), *Indri* (*B*), chimpanzee (*C*), *Homo* (*D*), and *Pongo* (*E*). Note the carpal papilla on the wrist of *Tupaia*, the tree shrew, and the increase in the area of the friction skin in the higher primates. Note also the elongated pollex of *Indri* and the short pollex of *Pongo*. (After Biegert 1963.)

6.2. When hand length was compared to body trunk length, it was found that while in monkeys hand length is usually less than one-third of trunk length, it rises to an average of 47 per cent in New World monkeys and as much as 59 per cent in the gibbon.

We see, therefore, that except among prosimians, such as *Indri* (Fig. 6.13, *B*), the thumb has not developed as an opposable digit to increase

TABLE 6.2 MEAN PROPORTIONS OF THE HAND AND THUMB IN PRIMATES (DATA FROM SCHULTZ 1956)

Primate	*Hand Breadth × 100* / *Hand Length*	*Pollex × 100* / *Hand Length*
Presbytis	29	40
Macaca	35	56
Pan troglodytes	34	47
Pongo	34	43
Pan gorilla	51	54
Hylobates	20	52
Homo	43	67

the effective grasp of branches. On the contrary, among the brachiators the thumb is relatively reduced from a longer primitive condition. Napier (1960) has shown that the great apes *Pan* and *Pongo* have such short thumbs that they cannot achieve the powerful "pulp-to-pulp" contact of thumb and digits (Fig. 6.14) that forms the anatomical basis of man's precision grip. While the gorilla thumb is somewhat longer, it is still short of the human condition (see Fig. 6.16). Apart from the prosimians, therefore, a fairly long thumb can be found only among the quadrupedal Old World monkeys and man, and, as we have seen, it is used in these groups for the manipulation of objects.

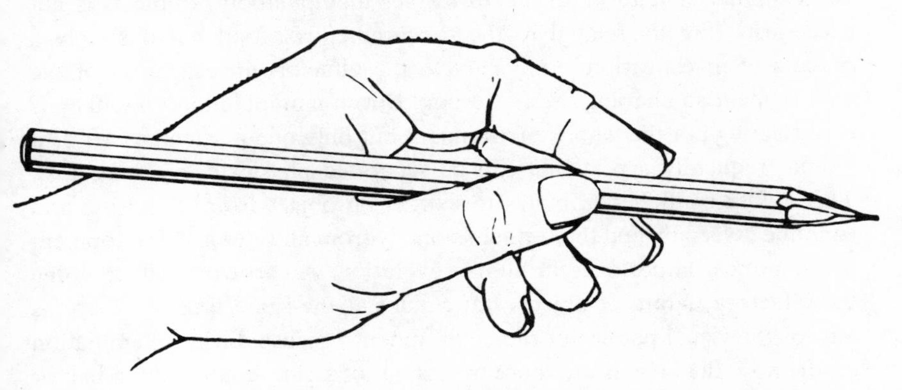

FIG. 6.14. Man's precision grip, because of the full opposition of thumb and finger, is precise and delicate, yet strong.

A fairly complete fossil hand has been described belonging to the Miocene *Dryopithecus africanus* (Napier and Davis 1959). It appears to be basically the hand of a quadrupedal Old World monkey with certain signs of brachiating adaptations. The hand is unfortunately not complete, and in particular it is not known how long the thumb was; nevertheless, its assessment as showing similarity to an arm-swinging species seems justified. A recent study of the wrist bones suggests a close affinity with *Pan* (Lewis 1973). This conclusion accords with that drawn on the basis of the forelimb bones. It seems clear that in *D. africanus* we have an incipient brachiator, and one that, on the basis of the limb bones, could conceivably be ancestral to both the living Pongidae and the living Hominidae. But this conclusion must be subject to the examination of the features of the skulls of all these forms.

6.6 *The evolution of the human hand*

WE HAVE SEEN that monkeys use their hands for a wide variety of purposes, and one of the most important, which we must now consider, is their contribution to the satisfaction of the exploratory drive (see 2.6): the hands make possible a detailed examination of parts of the environment that can be manipulated. Monkeys, apes, and man are almost the only animals that fiddle about with things, that turn them over and examine their form and texture. This manipulation of objects is not necessarily directly related to the procurement of food but is simply a process of investigation, equivalent to the olfactory investigation of the environment so characteristic of a dog. But in manipulation a monkey is investigating not the whole environment but only one particular part of it —and frequently a part that can be separated physically from the rest. This ability of higher primates to extract an object from its setting and examine it visually and three-dimensionally from all sides is a development of the utmost importance in human evolution. A carnivore will examine the olfactory nature of objects but cannot at the same time see them as part of the visual pattern of the environment because during examination by the nose the objects are more or less out of sight. Admittedly, a ball or bone can be manipulated with the paws and tossed with the mouth, but the range of objects examined is limited, and, compared with the primate hand, the paws and mouth give only a rough indication of shape and texture.

As will become apparent in chapter 10, the detachment of objects from the environment appears to be a most important prerequisite for the evolution of primate perception. This examination of things as objects we owe to the evolution of the primate hand and the opposable thumb. The recognition of different kinds of objects we owe to primate visual and tactile examination. Only a primate can, as it were, extract an object from the environment, examine it by smell, touch, and sight, and then return it to its place in its surroundings. In this way the higher primates came to see the environment not as a continuum of events in a world of pattern but as an encounter with objects that proved to make up these events and this pattern. We shall return to this development in considering perception and conceptual thought (10.2).

This all-important development may have come quite early in primate evolution, with the appearance of the opposable thumb. In man, a further advance in opposability has perfected this remarkable organ of manipulation. Napier (1961) has examined the thumb of the primate in order to analyze the development of opposability in hominid evolution. It reaches its most evolved condition in modern man and, as Napier says (p. 119), involves "a compound movement of abduction, flexion, and medial rotation at the carpo-metacarpal articulation of the pollex." That

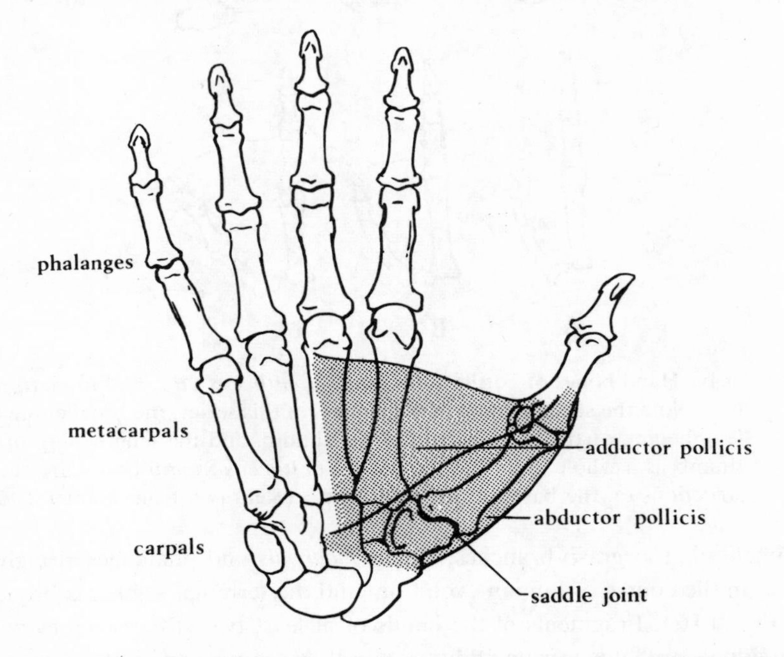

FIG. 6.15. The musculature of the thumb is well developed on the palm of the human hand. The two powerful muscles shown move the thumb toward and away from the palm. (After Napier 1962.)

is to say, a saddle joint evolves at the base of the thumb that allows a 45-degree rotation as well as movement in two planes (Fig. 6.15).

Though monkeys have an opposable thumb, it is only in man that the thumb is long enough and divergent enough to carry a heavy musculature (Fig. 6.15). This length and strength make possible a precision grip, strong yet delicate (Fig. 6.14). At the same time, with the hand no longer

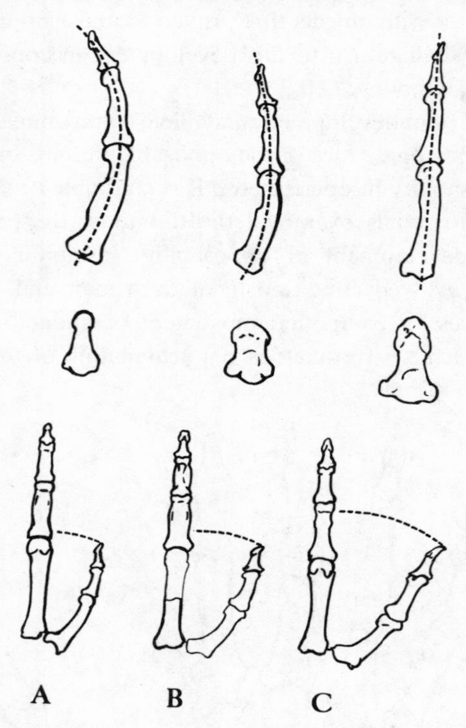

FIG. 6.16. Hand bones of gorilla (*A*), *Australopithecus* (*B*), and modern man (*C*). Note the straightening of the fingers in this series, the broadening and lengthening of the last phalanx of the thumb, and the lengthening of the thumb as a whole. Some of the *Australopithecus* thumb bones are reconstructions on the basis of other evidence. (Redrawn from Napier 1962.)

functioning to grasp branches, the *metacarpals* and phalanges straighten out in the course of human evolution, and the terminal segments broaden (Fig. 6.16). Fragments of the hands of at least two *Australopithecus* individuals are known from Olduvai, and they are reported to be somewhat less than fully human (Napier 1962) (Fig. 6.16). Recent studies of a wrist bone (the *capitate*) show that the Olduvai hand bears the characteristics associated with suspensory locomotion. This feature is also found in a capitate from Sterkfontein (Lewis 1973). Nothing is known of the hand of *Homo erectus,* but it is of interest that although Neandertal man had a brain as large as that of modern man, his hand was different and probably less dextrous (Musgrave 1971).

When apes and monkeys carry objects about, which they often do, their locomotor efficiency is impaired. Tool-carrying, food-carrying, and food-sharing (which is possible only if the food is first carried) all have been reported to a limited extent among chimpanzees (Goodall 1968), but they cannot move fast with a handful of objects and do so only for very limited distances. In early man the survival value of being able to carry objects without any disadvantageous effect on locomotion is of the greatest importance. Not only is it most advantageous to be able to carry food when surprised by a predator, but it can lead eventually to food sharing, food gathering and in particular, to hunting. A weapon at the ready is also of great survival value. The survival value of carrying also serves in turn to promote a physical structure adapted to bipedal locomotion.

Tool-using is not so rare as might be supposed in the animal kingdom. The use of sticks and rocks in agonistic display by chimpanzees and baboons is familiar (Hall 1963a). Tool modification—the next stage in the evolution of material culture—has been reported among chimpanzees, which prepare and collect sticks of a certain length for extracting termites from their nests (Goodall 1968). At Koobi Fora, East of Lake Rudolf, stone tools have been found dated at about 2.6 million years BP; they consist of battered cobbles, choppers, and flakes, and show an assured technique in their production (M. Leakey 1970). At another site nearby, of the same age, stone tools are found in association with the bones of a hippopotamus (Isaac *et al.* 1971). These are the oldest undoubted tools in the fossil record. From Olduvai Gorge, Bed I, we have a succession of increasingly varied tool-kits, just under 1.8 million years of age, often in association with one or more species of *Australopithecus,* which have been described in great detail (M. Leakey 1971) (Fig. 6.17). Here we see pebbles probably used for pounding roots and smashing bones, preserved, after flaking, as cutting tools for the preparation of meat. From South Africa we have evidence of bones modified as tools from a site and stratum inhabited by *Australopithecus africanus* (Dart 1957). The tran-

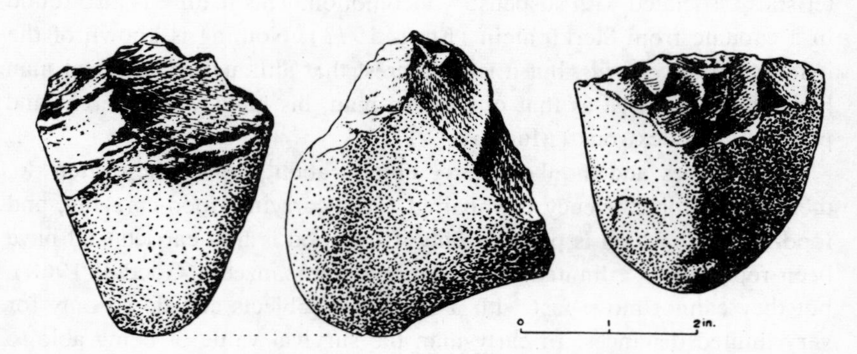

FIG. 6.17. Primitive pebble tools of the kind associated with *Australopithecus* at Olduvai, Tanzania. (By permission of the trustees of the British Museum [Natural History].)

sition from tool-modifying to tool-making is slow and continuous, but there is no doubt that a cutting tool was a great asset and is particularly necessary in the evolution of a carnivorous diet (see 7.2, 7.10).

But the tools we have found associated with the man-ape *Australopithecus* are relatively simple, and it is relevant that this creature had a brain not really much bigger in absolute size than that of a gorilla (see Table 8.2). Finely flaked stone tools are found—to date—only in association with the bones of hominids that had larger brains. Fine Acheulian hand-axes are known from Ternifine as well as Olduvai, where they are associated with remains of *Homo erectus*; in this hominid the brain may well have amounted to nearly 1,000 cc. Clearly, the final perfection of the hand was a perfection not only in anatomy but in sensory perception and motor control, with a highly developed brain. The most clear-cut neural correlate of the evolution of the human hand is the increase in representation that this organ carries on the motor and somatic sensory cortices of the brain (see Fig. 8.10). We have also noted significant improvement in the direct nervous pathways to the hand of the ascending and descending fibers (see 3.6).

We can summarize the evolution of the human hand by marking two stages in its development. First, the primate arboreal environment selects the opposable thumb, for grasping branches, and this thumb appears among the prosimians and in turn makes possible manipulation among Old World monkeys. Second, the ecological move to the open plains correlated with bipedalism frees the hand from all locomotor function and allows the perfection of the precision grip by some minor modifications of the thumb. The hand has contributed as much as the eye to the making of man; together they gave him a new perception of his environment and, with his technology, a new control of it.

6.7 Man's ancestors: arboreal, arm-swinging

FROM THE EVIDENCE presented in this chapter, it seems incontrovertible that modern man's ancestors were either arm-swinging monkeys or apes. Not only does living *Homo sapiens* share many characters with living arm-swinging forms but so does the genus *Australopithecus.* In this connection we must recall the much earlier form *Dryopithecus,* which also carries a variety of arm-swinging characters. Oxnard has reviewed the evidence relating to the evolution of the human shoulder (1969) and concluded that the most likely origin for it lies in a fully arboreal ape, genetically related to *Pan,* but functionally similar to *Pongo.* The character of the *Australopithecus* scapula and humerus is important evidence in this hypothesis.

Modern man does, of course, bear certain characters in the forelimbs that are adaptations to erect posture, and in particular these characters result from the downward-hanging characteristic position of the arms. This result is to be expected, and, although some of the characters have something in common with quadrupedal monkeys, they need not lead us to suppose that man's ancestors were quadrupedal monkeys.

Finally, we need not be surprised to find that the muscles that carry the weight of the body in brachiators are, with their areas of attachment, reduced in man (Ziegler 1964). Although man uses the same muscles in carrying as do the brachiators in swinging, these muscles do very much less work.

Man is, of course, a much more recent arrival on the earth than the kangaroo or the dinosaur, so it is not surprising that his forearms show in comparison little or no sign of reduction (see Fig. 4.5), especially as swinging arms are an important component of man's upright walking gait. It is not impossible that the arms have been slightly reduced, though as yet we have no evidence on this matter from such an intermediate form as *Australopithecus*, since the long bones known at present are fragmentary. On the whole, though, it seems likely that the value of man's arms is such as not to result in the selection of a much reduced length. For throwing stones and spears, a long arm is invaluable; for writing, it is perhaps less necessary.

SUGGESTIONS FOR FURTHER READING

Few authors have given much attention to the evolution of the forelimb. As a general introduction that runs parallel to this discussion see the appropriate chapter in W. E. Le Gros Clark, *The Antecedents of Man* (3rd ed.; Chicago: Quadrangle Books, 1971). An important study of the forearm is to be found in J. R. Napier and P. R. Davis, *The Forelimb Skeleton and Associated Remains of* Proconsul africanus (London: British Museum [Nat. Hist.], 1959). For reference to the primate hand see J. Biegert, "Volarhaut der Hände und Füsse," *Handbuch der Primatenkunde,* II/1 Lieferung 3, pp. 1–326 (Basel and New York: S. Karger, 1961), and R. H. Tuttle, "Quantitative and Functional Studies on the Hands of the Anthropoidea," *J. Morphology* 128 (1969): 309–364. The forelimb of *Australopithecus* has been discussed in some detail (but with a different taxonomy) by J. T. Robinson, *Early Hominid Posture and Locomotion* (Chicago: University of Chicago Press, 1972).

7

Feeding, Ecology, and Behavior

7.1 Functions of the head

IN APPROACHING the subject of this chapter we cannot avoid consideration of the head as a whole, as an organ that has come to assume an increasing number of functions in animal evolution. To understand the evolution of the human head, we must examine these different functions carefully. In order to reveal their phylogenetic origin, let us turn for a moment to consider the structure of a primitive living creature, the lancelet (*Amphioxus*), which is believed to be very similar to the ancestor of all vertebrates. This small marine animal looks like a minute fish, about an inch long (Fig. 7.1).

In this creature (as well as in more primitive invertebrate forms) the head can be recognized by three characteristics. First, it may be defined as the part of the body of the animal that precedes the rest in locomotion; second, it is near to, or incorporates, the mouth; and, third, it contains at least one sense organ. In preceding the body, the head is able to investigate the environment that the organism is approaching, to direct the organism toward food (in many instances by moving toward light or up a "desirable" chemical gradient), and to receive food by the mouth. Since movement toward food is a simple kind of behavior, we find the nervous system developing between the head receptors and the effectors (the swimming muscles), which lie behind it along the animal's body. In the course of the evolution of both invertebrate and vertebrate animals, a small knot or *ganglion* of nerves appears near the receptors and the mouth. Within this small ganglion of nervous tissue, the primitive brain, information about the environment (or about the food) is analyzed and an appropriate pattern of behavior is effected. In the evolution of vertebrates, the extent of sensory investigation of the environment has vastly increased, for increased sensory input relating to environmental conditions has characterized the appearance of more complex individuals. The evolution of multiple sense organs has been accompanied by the evolution of a relatively massive central nervous system, which analyzes the vast input from the receptors and mediates a wide range of behavioral activity.

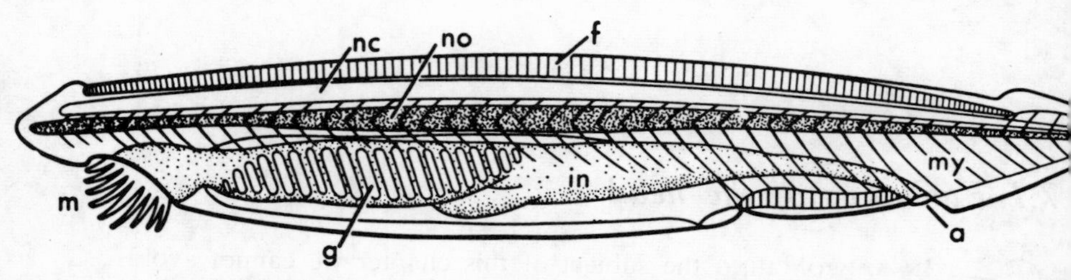

FIG. 7.1. The lancelet, a primitive animal still found in shallow sea water, believed to be very like the ancestor of vertebrates; *m* = mouth, *in* = intestine, *a* = anus, *g* = gills, *no* = notocord (the primitive spinal column), *nc* = nerve cord, *my* = myotomes (the lateral swimming muscles), and *f* = fin. (From Colbert 1955.)

In vertebrates, therefore, with the brain evolving near the sense receptors in the head, we find that the head serves the animal in two quite separate ways, first as a complex organ of interaction with the environment and second as a protective box for the brain. The functions of the head may be summarized as follows:

I. Relations with environment
   A. Metabolic
      1. Food intake through mouth and masticatory apparatus
      2. Inspiration and expiration of air or water-born oxygen through the mouth, with which the lungs or gills are connected (for they arose as outgrowths from the alimentary canal)
   B. Sensory and motor
      1. Deployment of sense receptors concerned with input from en-

vironment, both distant and contact (especially food recognition, mate recognition, and enemy recognition)
2. Communication (a secondary function of metabolic structures): movement of facial muscles, and sound production through trachea and mouth
3. Defense and offense: a secondary function of teeth of masticatory apparatus, and evolution of horns on the skull in some groups
4. Grooming and gripping: a secondary function of the masticatory apparatus in many animals

II. Protective case for the brain

The evolution of these functions of the head will be considered in the following chapters. Here, we are concerned with the part played by the head in feeding and digestion, including the sensory receptors concerned with food-finding and -testing (I, A, 1 above, together with B, 1). Other functions of the masticatory apparatus will also be considered (B, 3). Chapter 8 refers to further aspects of the evolution of the head (I, B, 1 and II), and chapter 10 refers to the evolution of communication (I, B, 2).

Before investigating the evolution of feeding in primates, it is convenient to classify the sense organs functionally into two kinds: the *contact receptors,* which supply data about the immediate environment (taste and touch), and the *distance receptors*, which measure intensity and direction (or extent of change) of external conditions (see Fig. 10.4). We recognize three kinds of distance receptor: the *chemo-receptors* measure the density of certain vaporized substances in the air (smell), the *photo-receptors* measure the intensity and wavelength of certain electro-magnetic waves (vision), and the *mechanoreceptors* measure vibration, sound waves, and gravity (hearing and balance) (see Table 8.1). The evolution of these senses will be discussed in this and the following chapters.

7.2 *Food-finding and diet*

TOGETHER WITH mating behavior, food-finding is a basic activity of all animals. An animal may spend most of its waking lifetime searching for food, and only an increase in food availability will allow for the development of other behavior patterns less immediately essential. It is an important fact of primate biology that in their adaptation to an arboreal environment primates were able to exploit a rich source of food, which had not been tapped before to any great extent. An enormous proportion of the energy received from the sun is converted into trees, as distinct from herbs, and the primates were able to consume not only the foliage but also the fruit and seeds of the forest. Fruit and seeds have high food value, and arboreal animals obtained access to food resources of quite a different order of richness from those available to the grazing mammals of the plains. Plants have evolved a highly concentrated food store in seeds which nourishes the developing seedling, and a very attractive food source in fruit. Edible fruits have evolved in parallel with their consumption by animals, which in turn serve the plant by distributing the seeds in their feces. Thus, though the primates are broadly herbivorous, it is clear that the type of food they live on is different in texture and quality from that available to the terrestrial quadrupedal mammals which browse or graze.

Primates identify their food by sight, smell, touch, and finally taste. The pre-eminent importance of primate vision has already been mentioned, and it is certain that such an acute sense is of value in recognizing fruits and seeds, especially the former, which are often brightly colored. Birds and primates are the two groups of vertebrates known to have a well-developed color sense, and the colored fruits of the forest have evolved, no doubt, in conjunction with their developed color sense. The development of the visual receptors and visual cortex has been of very great survival value to primates, in this as well as in other ways.

Although the visual sense is highly evolved in the carnivores and some ungulates, the sense of smell is by far the most important distance receptor in most mammals other than primates, and even in primates its importance should not be underestimated. The early mammals (which were probably nocturnal) and most recent mammals depend almost entirely on distance chemo-reception in food-finding and mate-finding. For that reason, the olfactory sense is an essential determinant of behavior. While sexual behavior is probably predominantly innate (see chapter 9), a great deal of food-finding behavior is known to be learned, and the olfactory lobes are closely connected with the cerebral cortex (see 8.6). It is significant that the cortex itself, with its vast memory store, evolved from the part of the brain originally concerned with smell (the *rhinencephalon*). It is not surprising, therefore, that, although in the evolution of the primates the visual sense overtakes the olfactory sense in overall importance, the latter plays some part in the deeply rooted behavior patterns of feeding and mating, and this importance is not altogether lost in modern man.

When food has been located and approached, the evolving primate hand plays an important part. The evolution of an organ of manipulation has not greatly affected the diet, but it has enabled the animal to carry food to the mouth, rather than having to pluck food with the mouth alone. Therefore the neck of primates has never lengthened as an adaptation to food-gathering; the prehensile and sensory hand has evolved instead.

When food has been identified by the eyes and tactile senses of the hand it is passed to the lips, where again we find a tactile sense organ of great importance. The sensory and motor areas of the cerebral cortex that relate to the lips (see Fig. 8.10) are very large; the primate lips are a highly innervated organ of investigation, and this character is not lost in human evolution. As the food is brought to the mouth, it is held close to the nostrils, which can thus obtain reliable olfactory information about it. Innate physiological responses follow olfactory stimulation, such as the secretion of saliva and other digestive substances. Any food that has passed the lips and been subjected to so much investigation is unlikely to be falsely identified.

Though it is probable that primates were originally insectivorous, today different primates have different feeding habits and diets according to the nature of their various environments. Flora varies with climate, altitude, and continent, and diet varies accordingly. The diet of species that remain almost permanently in the highest tree-tops will be confined mainly to vegetable matter, leaves, fruit, young shoots, and new bark. Primates that visit the ground from time to time will eat roots and even some meat, besides termites, ants, and tree frogs. The mountain gorilla, though predominantly ground-living, is, however, fully vegetarian (Schaller 1963). The chimpanzee, on the other hand, is not averse to eating termites and even small mammals (Goodall 1968). The fully terrestrial baboons also eat meat when they can catch a young animal, but all are predominantly herbivorous (DeVore and Washburn 1963). The baboon lives mainly on grass, shoots, and other succulent vegetation.

Schultz (1961) claims that some animal food is consumed by all primates and that some New World monkeys and most prosimians eat more worms, insects, frogs, lizards, and bird eggs than plant food. The important conclusion for our studies is that primates have at all times been able to consume and digest animal food—indeed, they appear to enjoy it. The limiting factor in the evolution of a more carnivorous diet among higher primates was, first, probably a lack of any selection pressure to do so, since the forest contained a nearly unlimited supply of plant food of good quality, and, second, the fact that primates are not in any way pre-adapted to catch animals larger than frogs, lizards, and so forth.

Modern man is an omnivore, and his diet varies a great deal; it is probably more variable than that of any other animal species. Meat represents a large portion of the diet only in north-temperate peoples who live in an environment short of vegetables during the winter. For that reason, Eskimos are almost completely carnivorous. Tropical man eats meat, and some will go to much trouble to obtain it by hunting, but most tribes, even the pastoralists, have a primarily vegetarian diet. Most hunter-gatherers consume only about 35% meat to 65% vegetable food (Lee, in Lee and Devore 1968). (Exceptions include the Masai tribe of eastern Africa whose members subsist primarily on blood and milk obtained from their cattle; they live in an arid region, where the supply of vegetable matter is limited.)

It therefore appears that there was no really fundamental change of diet between the primate ancestors of man and *Homo sapiens*. An analysis of animal bones on the living floors at Olduvai in Bed I times (about 1.7 m. yrs BP) shows that at that stage the meat diet of *Australopithecus* included not only a variety of small animals, but that some large mammals were also eaten, even an elephant (M. Leakey 1971). The change that occurred, therefore, in the diet of evolving man, was to the consumption of larger animals—the plains mammals—which could be caught only with the technology and social organization associated with hunting.

The evolution of the social hunt brought varying proportions of mammal meat into man's diet. Clearly, this new and virtually unlimited food source was of immense survival value to groups of primitive men living on the open plains, where good vegetables were neither common nor succulent, especially during the dry seasons. But man remained to some extent dependent on insects, reptiles, and birds, as he does to this day.

It is an important fact that every new evolutionary radiation has involved the exploitation of a new major source of food. Of course, the herbivorous mammals of the wide grass plains had already been exploited by the carnivores, such as the lion and leopard, but the natural balance between those two groups appears to be about one carnivore to every hundred herbivores. The effectiveness of social hunting among humans enabled them to tap this huge reserve more effectively; there was enough food for both men and lions.

It seems possible that the availability of an immense supply of mammal meat made the synthesis of certain proteins a much simpler matter for *Homo*; they were already synthesized by the herbivorous mammals and, though broken down in the process of human digestion to their component amino acids, were available in the right proportions for re-synthesis. The availability of mammal meat may have been very important in the final evolution of *Homo*. It is of interest that the baboon, another plains-living primate, has been observed to kill and eat mammals regularly in both east and south Africa.

Meat is a concentrated form of food, comparable to seeds, which have been exploited by the rodents and are no doubt a contributory factor to their great success. Meat contains a high percentage of protein and, when digested, will release the whole range of amino acids necessary for the synthesis of body tissues. It also contains vitamins (in particular in the liver), which are not readily available in a vegetable diet. There is little doubt that the final stage in human evolution (since the Lower Pleistocene) was correlated with the exploitation by man of the large terrestrial mammals. Man's immensely successful evolutionary radiation must be associated, then, not with a fundamental change in diet, but with an important change in emphasis from a diet that was mainly vegetarian to one that was increasingly omnivorous, if not distinctly carnivorous. The change was not so much one of internal evolution of the alimentary canal and masticatory apparatus, as a change in ecology, in man's whole environment. In leaving the forest for the plains, man's ancestors changed not only their diet but their whole way of life. The change in diet was a reflection of the new environment, not the reason for it.

7.3 Taste and the tongue

THE TESTING and identification of food does not end when it is placed in the mouth. The inside of the mouth is rich in tactile sensory nerve endings, as is the tongue. In addition, the tongue carries numerous *papillae*, containing nerve endings sensitive to taste (the taste buds), which in man (and presumably other primates) can distinguish the flavors sweet, bitter, acid, and salt. Such particular characteristics of food are not readily detectable by the nose (one cannot smell the sweetness or saltiness of a dish), for the relevant molecules are not easily vaporized. But the identification of these flavors is important: sugar is a valuable source of food, and salt is an essential mineral. Our marine origin has left us with saline body fluids, the salinity of which must be exactly maintained. Primates have been observed licking rocks containing salt and obtaining it from the salty sweat deposits in each other's fur. Bitterness and acidity are the characteristics of some natural poisons.

The human sense of "taste" is of course for the most part due to the activity of the olfactory organ, the nose. The smell of food in the mouth enters the nose by the *pharynx*, the internal passage from the mouth to the nose. There is no doubt that, although the sense of smell is now less important to man than it was to some of his ancestors, he still has the power of great chemical discrimination (in a "trained palate"); while the total volume of the brain involved with this sense may be relatively reduced, it is probably not absolutely reduced and is still of considerable magnitude. In man, the importance of smell has been more strikingly reduced as a determinant of sexual behavior than in feeding behavior.

Since the tongue lies between the teeth, it is able to sample food during mastication. Stimuli interpreted as undesirable can still result in rejection of the food. It is interesting that the nerve fibers from the taste buds lead to the brainstem and are strongly tied to innate reflexes, such as salivation or rejection. The significance of taste is primarily innate, that of smell is to a great extent learned.

The tongue has other functions besides bearing the taste buds. In prosimians as well as in many other animals it has a roughened horny surface and aids in grasping the food in the mouth. The tongue initiates the process of swallowing and is important as a means of removing food particles from the teeth and in keeping them clean. In the Hominoidea this function is still important, ranking after tasting, swallowing, and speech, though the rough surface has been sacrificed to increased sensory discrimination.

The musculature of the tongue is an important consideration in a study of human evolution because the tongue makes possible the act of speech. This musculature is well developed in most mammals for the various functions described above, and little anatomical change was needed to turn it into an organ of speech. The tongue is anchored to the skeleton at four points, which include the inner surface of the *mandible* or jawbone at the *genial tubercle* (Fig. 7.9) and the *hyoid* bone (Fig.

10.6). The tongue also carries its own internal muscle, which makes possible its complex changes of shape. However, the important changes that occurred in the evolution of speech were not so much in the tongue's anatomy as in the motor control of the tongue by the brain (see 10.6).

7.4 *The masticatory apparatus*

MASTICATION is an essential process in the realization of the total value of foods. The enzymes and other substances that effect digestion cannot operate on large masses of food except very slowly. It is remarkable that some reptiles (such as snakes) are able to digest entire animals without any form of mastication, but the resulting rate of output of digested food substances is low and the animals undergo a period of postprandial sluggishness, or even coma. The warm-blooded mammal and the bird require a constant and high rate of digestion, and each has developed its own masticatory apparatus; among mammals, it evolved from the jaws, and, among birds, the gizzard evolved from the alimentary canal. Besides breaking down large pieces of food, mastication also destroys plant cellular structure and frees the porteins, fats, and sugars from the insoluble cellulose cell walls that enclose the living cell contents, thus releasing an immense quantity of food for immediate digestion—a development clearly essential to the mammal way of life.

The evolution of primate dentition is indeed of great interest, not least insofar as the structure of the teeth reflects different diets. But there is another reason for its interest to us: of all parts of the body, the teeth have been most successfully preserved as fossils, and these fossil teeth have enabled us to understand something of the evolution of the living primates. Jawbones, too, are very strong structures and for that reason have also been preserved in considerable numbers. The evolution of the human masticatory apparatus is probably more completely documented than that of any other part of the body, and it deserves a detailed description.

In accord with changes in tooth function, the mammal jaw itself has evolved. As we have seen, the reptile jaw was primarily a food trap, so it needed to be large, quick-operating, and escape-proof. It was not necessary for the jaws to develop large compressive forces between the teeth. In the python, for example, all the bones of the forepart of the skull and mandible are loosely connected, to allow the swallowing of large food animals without mastication (Fig. 2.3). The most primitive primate jaw is to be found among the tree shrews, where little more than grasping action is required. As grinding action evolved, more power was required, and the jaw changed shape. Before consideration of these important shape changes, however, it is necessary to review briefly the musculature of the primate jaw.

The jaw is operated by paired muscles that move the mandible about its pivots, the *mandibular condyles*. The maxillary bones (or *maxillae*) that carry the upper teeth are firmly fixed to the skull, and the compressive force used in crushing and grinding is achieved by raising the mandible against the maxilla. The muscles that raise the mandible and close the jaw are the *temporalis* muscles, assisted by the *masseter* and *medial pterygoids* (also called *internal pterygoids*). The mandible is moved forward and sideways by the *lateral* (or *external*) *pterygoids*. Figures 7.2, 7.3, and 7.4 show the arrangement of these muscles. The architecture of the skull is

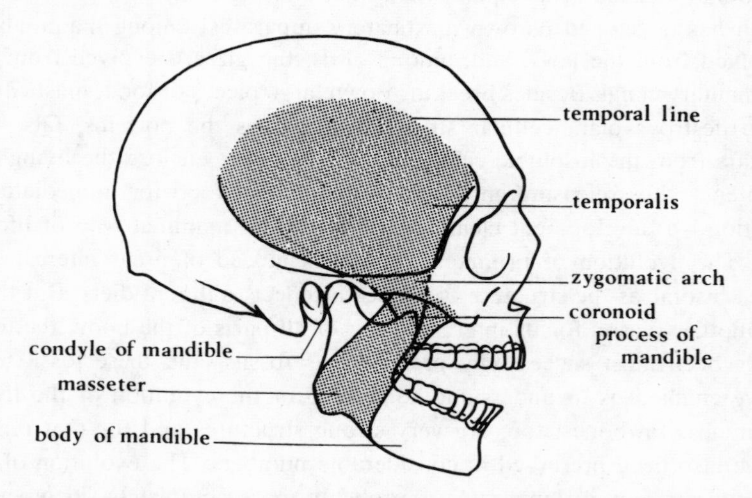

FIG. 7.2. The human skull, showing the masseter and temporalis muscles, which raise the mandible to bite. The medial pterygoids (which also raise the mandible) are not shown, since they lie within the skull.

such as to form a framework around the nasal passages and orbits to transmit these forces of mastication (imposed upon the maxillary bone) to the top and sides of the skull. The heavy browridges—the *supraorbital torus*—have evolved partly in order to spread this force and may be considered as part of the masticatory apparatus. Their form and function is discussed in greater detail in chapter 8.

In the evolution of chewing among primates, two features of the jaw itself have changed:

1. The *occlusal plane* is the line on which the teeth meet when the jaw is closed. In primitive primates, the point of pivot—the mandibular condyle—lies more or less on this plane; in higher primates, it has moved up well above this plane (Fig. 7.5). The result is that, while in the lower primates the molar teeth meet first and the incisors later (like a pair of scissors), in the higher forms the whole dentition occludes simultaneously, which means that all the molars are equally effective as crushers and grinders. At the same time, the increased distance between pivot and teeth

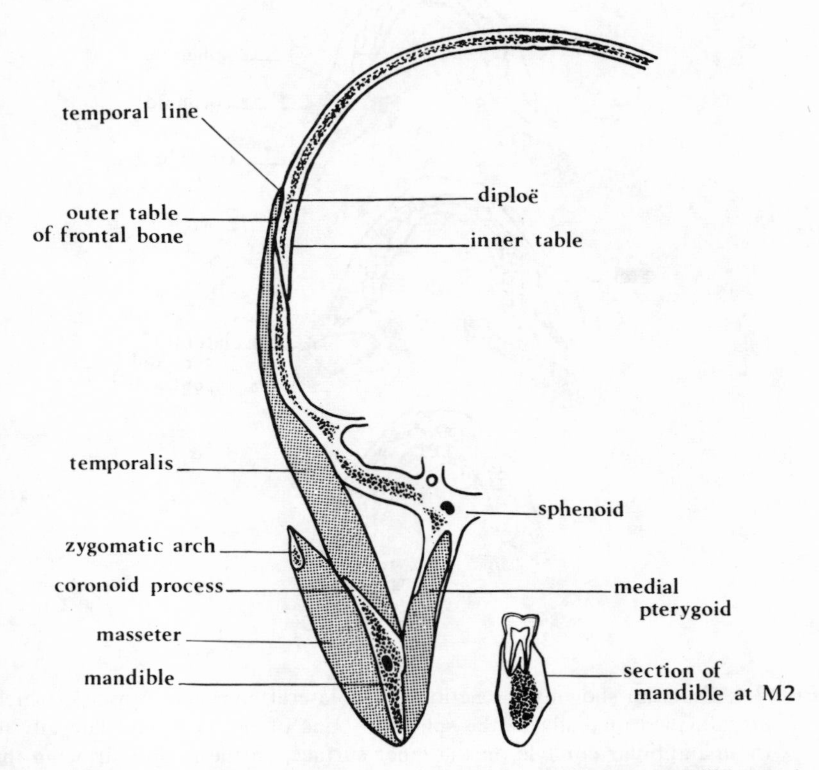

FIG. 7.3. A section through the right side of the skull and mandible at the coronoid process showing the arrangement of the three biting muscles. *Bottom right,* a section of the mandible at the second molar showing the bone structure into which the tooth is embedded.

allows the lateral movement necessary for grinding. The overall form and musculature of the jaw is modified accordingly. Thus the jaw becomes an angled bone instead of a more or less straight one, the *ascending ramus* (the part bearing no teeth) being ascending rather than horizontal as its name implies (Fig. 7.5). The displacement has been achieved in the skull itself by lowering the floor of the nasal chamber and the dental arcade in relation to the braincase and orbits, and in some species the palate is vaulted. The changes in the general form of the skull resulting from this modification are discussed in chapter 8.

2. The second primate development relates to the power developed between the molar teeth. This power is proportional to the relationship

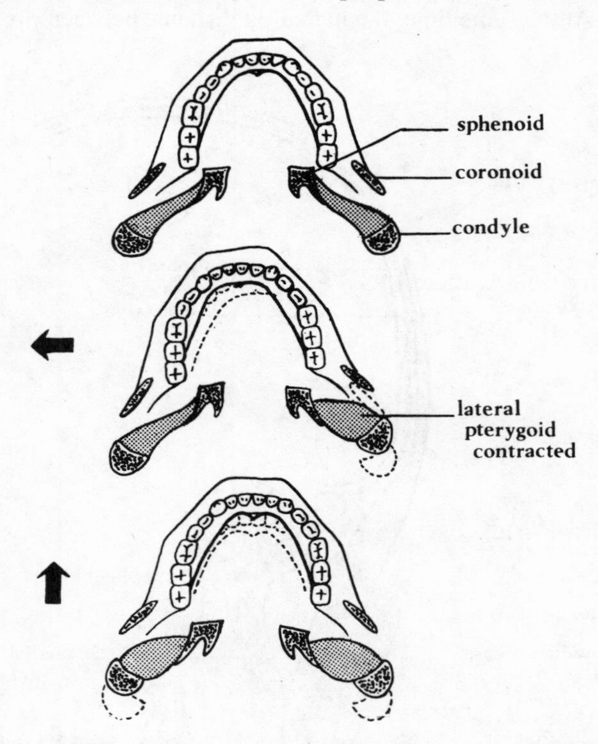

FIG. 7.4. Diagram showing the action of the lateral pterygoid muscles, which are attached medially to the sphenoid bone of the skull and laterally to the mandibular condyle on the inner surface. In the middle drawing the mandible is pulled to the left by the action of one muscle only—the right lateral pterygoid. In the lower drawing the mandible is pulled forward by the contraction of both lateral pterygoid muscles. (Redrawn from Testut 1928.)

between the length of the power arm and the load arm of the mandible lever, for a given volume of muscle. The shorter the load arm (that is, the distance between the teeth and pivot point projected onto the occlusal plane), the greater the power of compression developed between the molars for grinding food (Fig. 7.6). In the evolution of the primates

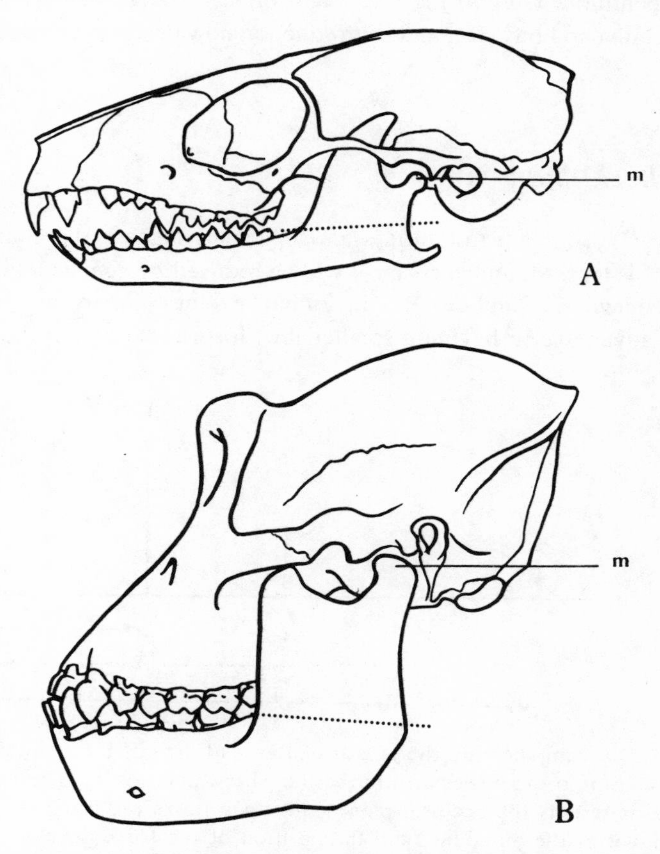

FIG. 7.5. Skulls of the tree shrew *Ptilocercus* (*A*), and a female gorilla (*B*). Note the relative sizes of the jaw and neurocranium and their relative positions. In particular, note the height of the mandibular condyle (*m*) above the occlusal plane of the teeth (*dotted line*). (From drawings in Le Gros Clark 1971.)

the relative shortening of the load arm has been achieved to a great extent by bringing the dental arcades backward and under the braincase, which has had the added advantage that the maxillary bone can transmit the forces of compression more directly to the skull vault from which arise the temporal muscles. Development of a short and powerful jaw was made possible when the need for snapping action by the mouth became of secondary importance in the evolution of a herbivorous diet.

These two trends have operated simultaneously in the evolution of the human dentition. Only in the last stages of the process was jaw size reduced relative to body size, and this reduction now deserves consideration.

7.5 The human jaw

THE CONCEPT OF BIOLOGICAL efficiency suggests that if an organ is larger than necessary it will be reduced by genetic adjustment in the course of evolution. But in human evolution there was another distinct advantage in having a smaller jaw, for a reduction in the size of

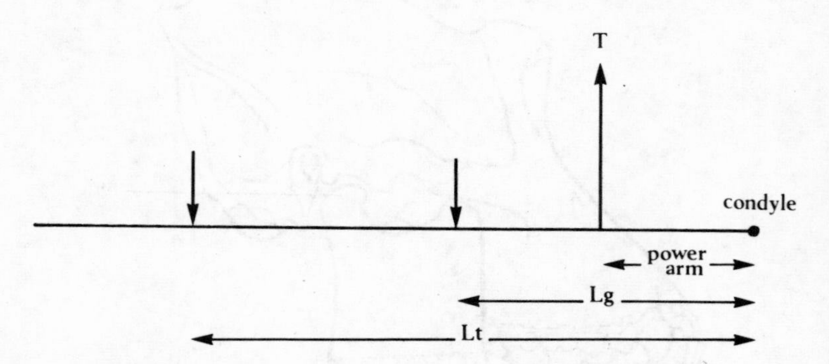

FIG. 7.6. Diagram showing the length of the load arm of the mandible lever in relation to the power arm in the tree shrew and gorilla. The horizontal line represents the occlusal plane, and upon it are projected the condyle or pivot of the mandible, and the position of the force developed by the temporalis muscle and transmitted through the coronoid process (*T*). The relative length of the load arm is calculated on the basis of a constant power-arm length and is defined as the distance of the first molar tooth from the condyles. This length is indicated by *Lt* for the tree shrew and by *Lg* for the gorilla.

the masticatory apparatus in an erect animal will help toward achieving a better balance of the head upon the spine. A balance has been almost achieved in modern man, and the reduction in the size of the masticatory apparatus (a reduction of subnasal prognathism) is a factor of prime importance in this evolutionary development (4.4). We see this development as early as the early Pliocene in the mandible of *Ramapithecus,* which in Figure 7.7 is shown alongside that of a typical fossil pongid (of the subfamily *Dryopithecinae*) of the same antiquity. The reduction of the teeth has allowed the mandible to curve inward at the level of the first molar in *Ramapithecus*; in the dryopithecine animal it curves much farther out, level with the premolar teeth (see also Figure 7.14).

Experiments on animals have demonstrated conclusively that in ontogeny the final size of the mandible and masticatory muscles depends on the amount of use the masticatory apparatus is put to during growth—an example of developmental homeostasis. On the other hand, the size of the teeth is not affected in this way by environmental influence but is under more or less direct genetic control. The crowns of the teeth are of course subject to wear, but the length and breadth of the crown is not modified unless the tooth is damaged.

Two other more detailed features must concern us in a study of the later stages of human evolution: the form of the *mandibular body* and the evolution of the chin.

The mandibular body or *corpus* is the part of the mandible that carries the teeth. Its function, besides carrying teeth, is to transmit to the teeth themselves the forces put upon the ramal part of the mandible by the contraction of the masticatory muscles. Large teeth, like canines, have large roots as anchors, and of course it follows that a deep mandibular body is necessary to carry large teeth. In man, with much-reduced teeth, especially canines, the body may be quite shallow, but the bone structure has to transmit various forces and so must be of a certain cross-section beyond that necessary to house the roots of the teeth. The depth of the mandible allows it to transmit the vertical forces involved in closing the jaw when the temporalis, masseter, and medial pterygoid muscles contract; the body of the mandible can be considered as a girder, and its depth is directly related to the vertical bending stress developed at each point along its length. The teeth make it, in effect, a U-girder with internal webbing (see Fig. 7.3).

Forces in the vertical plane are not, however, the only forces acting on the body of the mandible. The lateral grinding movement of the mandible is an important masticatory movement and is brought about by the lateral pterygoid muscles (the action of these muscles is shown in Figure 7.4). Lateral movement is caused by the action of each muscle alter-

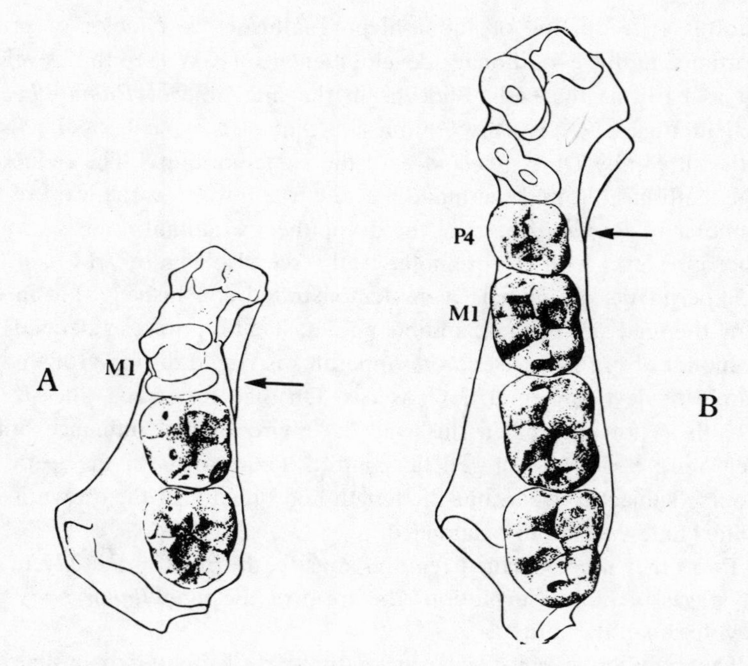

FIG. 7.7. Comparison of Mio-Pliocence hominid mandible fragment of *Ramapithecus* (*A*), with contemporary ape *Dryopithecus* (*B*). In the hominid, the mandible begins to turn inward toward the midline at the level of the first molar (*M1*), while in *Dryopithecus* the curve does not begin until the second premolar (*P4*). This feature shows that the reduction typical of the hominid jaw and dentition was already apparent at this early date. (From Simons 1964.)

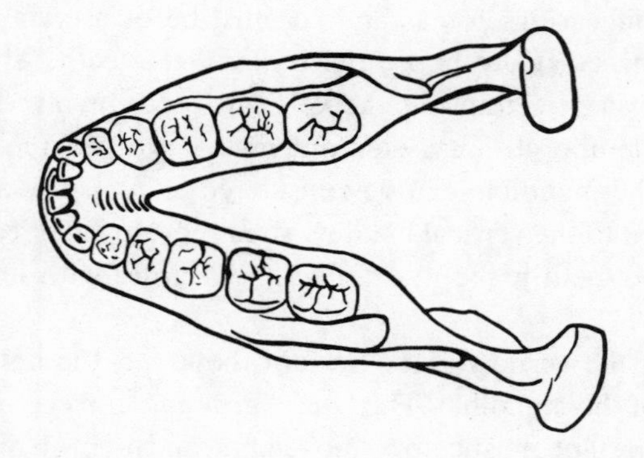

FIG. 7.8. The heavily built *Australopithecus* mandible. Note the parabolic dental arcade without projecting canines and the small lower incisors.

nately: the right lateral pterygoid pulls the mandible to the left; the left lateral pterygoid pulls it to the right. The power for lateral grinding action therefore comes from one side only and is transmitted through the body of the mandible to the molar teeth on the other side. The bending stress on the mandible in the horizontal plane can therefore be very great, especially when the jaw muscles are strongly contracted and the friction between the molars is at its maximum.

In species that have evolved a powerful lateral grinding action of the molar teeth, therefore, we find a thickening of the body of the mandible in the horizontal plane. Thickening is most apparent at the point at which the mandible is most curved, at the *symphyseal region* (the midline between the first incisor teeth). In the mandibles of higher primates, this point is strengthened by the development of internal buttressing, which may occur either at the lower margin, when it is called a *simian shelf,* or halfway up the body of the mandible, when it is called a *mandibular torus.* In some forms both kinds of buttressing occur together (Fig. 7.9).

In the evolution of man both kinds of buttressing have been lost, although their traces can be identified in some mandibles. The internal buttress has been replaced by a weakly developed external buttress, the chin. Clearly the stresses set up in the mandible of modern man are not great, but, as can be seen in the wear of human teeth, grinding action has by no means been entirely lost and some strengthening is still necessary. The reason for this change of structure in the buttressing of the mandible of *Homo* has been revealed by a functional analysis by DuBrul and Sicher (1954).

Both the recession of the dental arcade and the movement of the head backward on top of the spine (with the forward movement of the occipital condyles) to achieve a better balance have brought the mandible into very close proximity to the neck. If the form of the lower margin of the mandible had not changed, it would certainly have constricted the windpipe, larynx, and soft viscera of the neck, including the vital veins and arteries to the brain, which lie just behind the angle of the mandible. DuBrul and Sicher have shown how the lower margin of the mandible has been everted to avoid this state of affairs (Fig. 7.10). The chin is the result of the eversion of the lightly buttressed *symphysis*; it is a necessary result of the reduced masticatory apparatus of modern man.

It is of interest to study the skull topography of the orang in this context. Biegert (1963) has pointed out that in order to accommodate the laryngeal sacs, in the laryngeal space under the jaws, the occipital con-

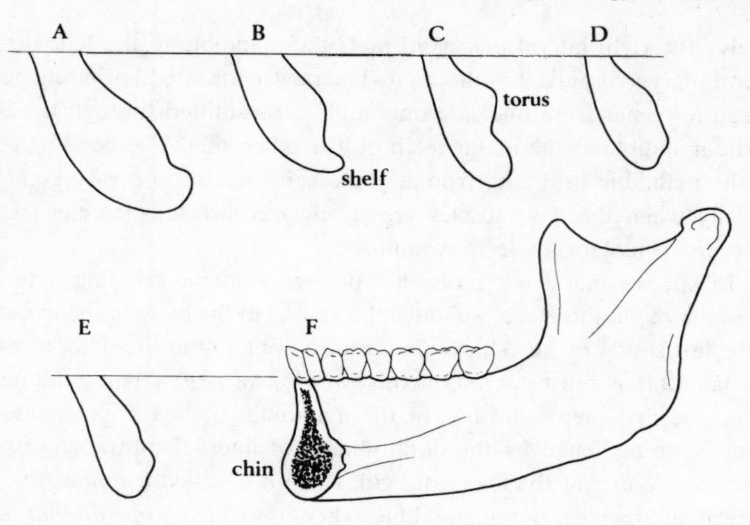

FIG. 7.9. Section of the symphyseal region of different jaws: gorilla (*A*), chimpanzee (*B*), *Australopithecus africanus* (*C*), *Homo erectus* (Heidelberg) (*D*), *Homo sapiens* (early form from Krapina) (*E*), and modern man (*F*). Note the different means of horizontal stiffening: the simian shelf, mandibular torus, and chin. On the inner symphyseal surface of (*F*) are to be seen two small protuberances, the genial tubercles, to which are attached the muscles supporting the tongue.

dyles have moved back under the skull; the jaws have come forward and a more "primitive" alignment of the skull has resulted. This development appears to be an alternative means of protecting the neck viscera.

The changes in the primate jaw that led to that of modern man can therefore be summarized as follows: (1) evolution of the ascending ramus, at right angles to the corpus, (2) retraction of the dental arcade under the skull, (3) reduction of the jaws and dentition, and (4) eversion of the lower border of the mandible.

7.6 Dentition: the incisors

BEFORE WE CONSIDER the different kinds of teeth, it is necessary to review their arrangement in the jawbones (maxilla and mandible). This arrangement is termed the *dental arcade*, since in man it is seen to take the form of a parabolic arc. In other primates, however, the teeth lie either in two converging rows or as three sides of a rectangle (Fig. 7.11). Fossil evidence suggests that human evolution involved a change in the dental arcade from one with straight rows to a continuous curved row of teeth, already seen in *Ramapithecus* (Fig. 7.7). At the same time, the jaws shortened and became smaller in relation to the rest of the head. This reduction in the size of the masticatory apparatus has already been discussed (7.5).

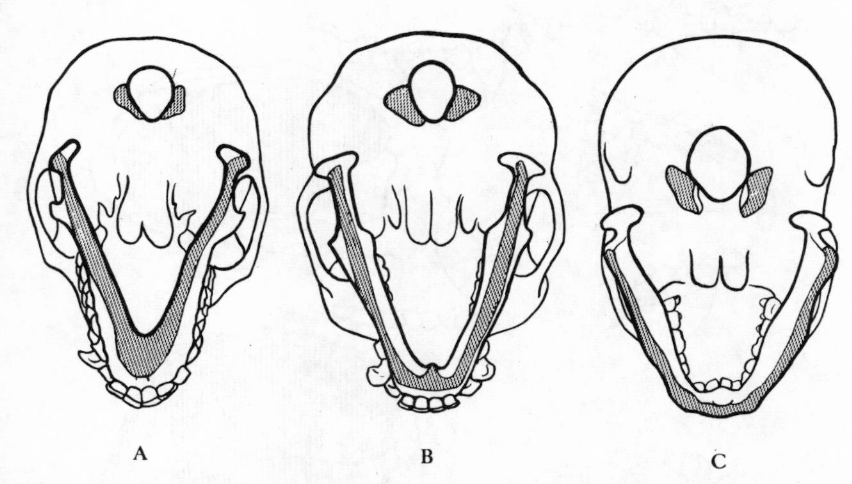

FIG. 7.10. Basal views of the skulls of a monkey (*Cercopithecus*) (*A*), the gibbon (*Hylobates*) (*B*), and modern man (*C*) drawn the same size. Note that the lower margin of the mandible (*shaded*) remains a more or less constant distance from the occipital condyles (*also shaded*), which support the head upon the neck. The buttressing of the jaw in man takes the form of a chin in place of a simian shelf, visible in the monkey. (After DuBrul 1958.)

Since the different kinds of heterodont mammalian teeth have evolved from the similar teeth of a homodont reptile, there are no very fundamental historical differences among them, and they are in some species indistinguishable in form or function. In most primates, however, the four kinds of teeth remain distinct, though they are often reduced in number. They are the incisors (abbreviated as I), the canines (C), premolars (P), and molars (M). The teeth are conventionally numbered from the front backward, so the primitive hypothetical mammal shown in Figure 7.12 had four rows of teeth as follows: I1, I2, I3, C, P1, P2, P3, P4, M1, M2, M3, making a total of twenty-two in each jaw.

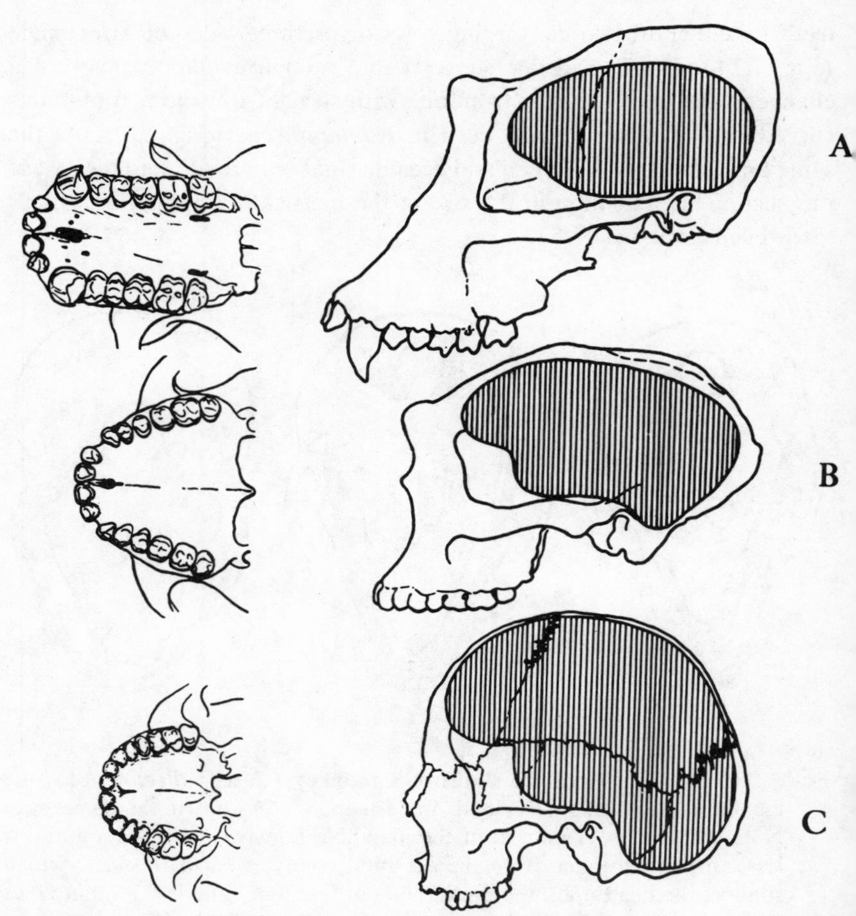

FIG. 7.11. Upper dentition (*left*) and complete skulls (*right*) of gorilla (*A*), *Homo erectus* (*B*), and modern man (*C*). Note size and form of dental arcade and compare the size of the jaws and that of the cranial cavities (*shaded*). (From Weidenreich 1939–41.)

The incisors, the most anterior of the whole dentition, are generally cutting teeth, as their name implies (though they grow into the tusks of elephants). Exceptions do occur among the primates, and in the lemurs we find a total change in function: the incisors of the upper jaw are very small, and those of the mandible are procumbent and modified into pointed pegs to form a comb for cleaning the fur. They no longer serve any function related to feeding. In the higher primates, however, the

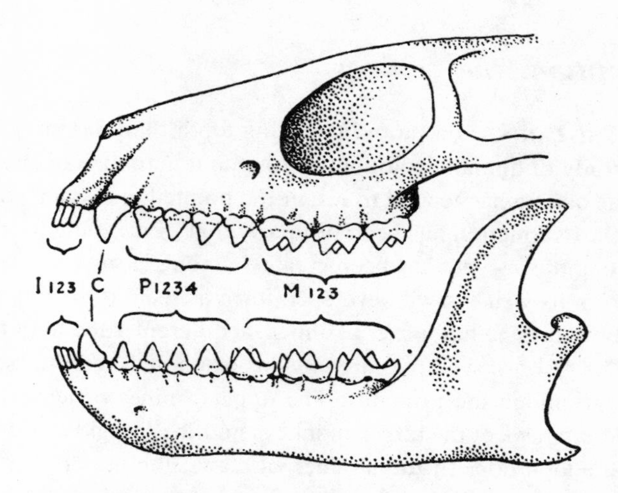

FIG. 7.12. Diagram of the dentition of a hypothetical early placental mammal showing the arrangement and grouping of teeth from which we presume the various primate dentitions evolved. (From Le Gros Clark 1971.)

incisors are spatulate (chisel-shaped), and serve to cut food. They are used for biting into fruit, nuts, and shoots, for chopping leaves and stripping bark. They have enabled many groups of mammals as well as the higher primates to utilize food that is too big to swallow whole and thus have made available a vastly increased food supply.

Among higher primates there are only four incisors in each jaw, two on each side (I1, I2). In most primates they tend to lie in a nearly straight line (Fig. 7.11 *A*) and together form an efficient cutting edge. The incisors of the Pongidae are the largest and broadest and tend to make the dental arcade rectangular, though the lateral upper incisors do not share this enlargement.

In the evolution of man no great changes have occurred in the function of the incisor teeth, so their form has remained more or less unchanged. Their size has decreased in accordance with the reduction in size of the whole masticatory apparatus. They lie in a curve instead of in a straight line (Fig. 7.11, *C*) and lose their procumbency.

We have incisors of *Australopithecus* showing no very significant differences from human incisors (Fig. 7.14). The incisors themselves tell us less about hominid evolution than do the other teeth, which we shall now consider.

7.7 Dentition: the canines

THE CANINE is a more interesting tooth than the incisor for the study of human evolution. Throughout mammalian evolution as a whole, the canine has tended to retain the pointed and rounded form of reptile teeth. Its function has been primarily that of grasping food, and its importance in this respect has been greatest among carnivores. It receives its name from its well-known developed form as seen in the dog (*Canis*). Among herbivores it has often assumed a different function: that of a weapon. The tusks of wild pigs are obvious examples, and this adaptation is not absent among the primates. The upper canines of lemurs and the canines of both jaws of the larger monkeys and the Pongidae are a case of this special adaptation. In the absence of claws, the importance of these large canines is probably enhanced. Since the animals are primarily herbivorous, the function of the canines in feeding is clearly not primarily responsible for this unusual enlargement.

In the Pongidae, and especially in the gorilla, the canines are larger in the male animal than in the female, as is the supporting bone structure and skull. This is an instance of *sexual dimorphism,*—of the form of a character varying between the sexes. The enlargement of the teeth is due to some behavior unique to the male animal—fighting for defense of the social group and sexual competition by threat or display.* The development of sexual dimorphism is a character of social significance, since it affects the status of different individuals within the social group; it will be discussed in chapter 9.

Animals with large canines generally have a *diastema,* a gap in the opposing tooth row into which the canine fits so that the jaws may be closed. The lower canines always fall into a diastema in front of the upper canines, and the upper canines fall behind the lower (Fig. 7.13). Le Gros Clark (1971) points out that the diastema is absent until the canines have fully erupted. It appears that the diastema is merely the result of active tooth occlusion (the way the teeth of the two jaws in fact interlock) rather than a genetically predetermined character.

The *hypertrophy* (enlargement) of the canines has evolved as a

* Among species in which the masticatory apparatus is used for fighting and killing, we can expect to find extra powerful development in the neck complex of cervical vertebrae, skull, and their musculature. Since it seems clear that with their small canines the Hominidae did not fight or kill with their teeth, this factor has probably not been significant in human evolution.

striking feature of the living Pongidae, and the condition of modern pongids appears to be an instance of steady evolution from a *Dryopithecus* ancestor whose canines resembled those of modern monkeys (Fig. 7.14).

In the evolution of man, the opposite trend has occurred. The exact form of the canine varies among individuals (as the reader may easily verify) from a spatulate tooth, indistinguishable from an incisor, to a rather pointed tooth. It is significant that the root of the modern canine is longer than that of the neighboring teeth, suggesting that it may have been

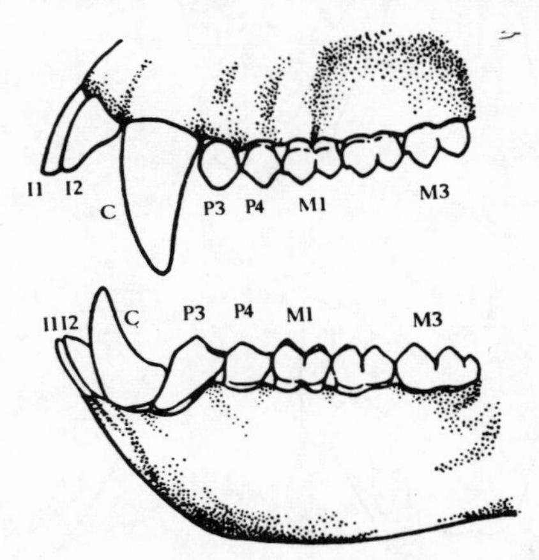

FIG. 7.13. Dentition of the catarrhine monkey *Macaca*. Note the canine teeth (*C*), diastemata, sectorial lower premolar (*P3*), and bilophodont molars (*M1–3*) (see 7.9). (From Le Gros Clark 1971.)

reduced from a larger tooth in the course of human evolution. The fossil evidence gives no certain indication of great reduction; the earliest hominids may not have had canine teeth proportionately any larger than those of the specimen of *Ramapithecus wickeri* from Fort Ternan (Fig. 3.15). However, if the first hominids evolved from an earlier *Dryopithecus* such as *D. africanus*, then some considerable reduction of the canine may have occurred. In either case we could account for the relatively large canine roots in modern man as having the function of anchoring teeth which are still used to tear food. Kinsey (1971) believes that no good evidence of canine reduction is to be found in man's evolution, though the very small canines in some groups of *Australopithecus* do show that canine reduction occurred among some if not all hominids. To summarize, the evidence suggests that: 1) the canine tooth of man may have been reduced in the course of his evolution, 2) it was certainly never as large as it is found in the living Pongidae today, and 3) the evidence of some groups of *Australopithecus* which may be ancestral, and of *A. boisei* which is not,

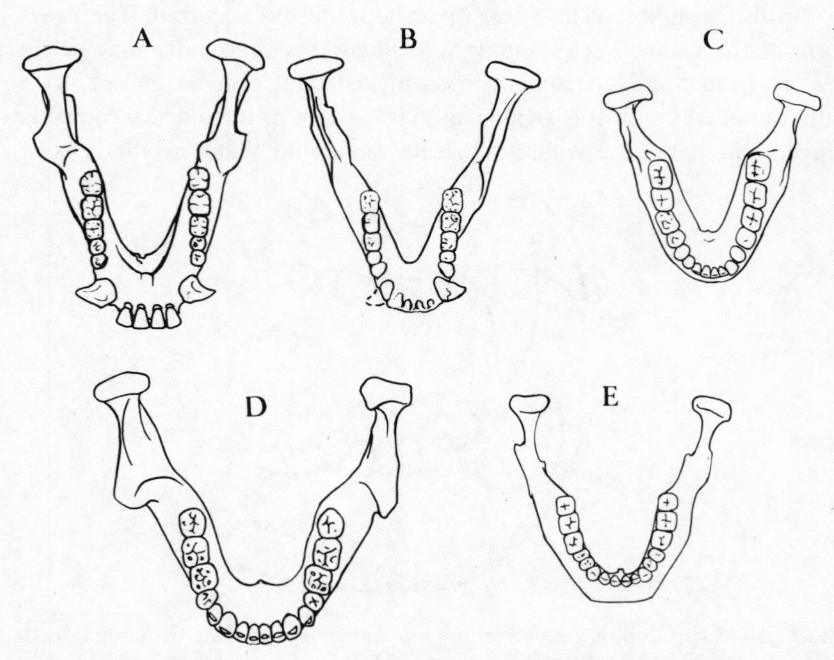

FIG. 7.14. Lower dentition of various primates: chimpanzee (*A*), *Dryopithecus nyanzae* (*B*), *Australopithecus* (*C*), *Homo erectus* (*D*), modern man (*E*). Note the different forms of the dental arcade and canine teeth. (Not drawn to scale.)

suggests that canine reduction was a common trend among Pliocene Hominidae. The fact that the human canine was not so reduced as the canine of *Australopithecus boisei* suggests that it did retain some special functional value to man's ancestors.

A. boisei is a species closely related to *A. africanus* and approximately contemporary with it. It is remarkable in having a canine tooth that does not project at all beyond the level of the incisors and premolars and together with the incisors is greatly reduced. For that reason (among others) we can conclude that it was ancestral to neither *A. africanus* nor *Homo erectus,* both of which have relatively larger canines. It seems probable that *boisei* maintained a wholly herbivorous diet and had lost the need for a

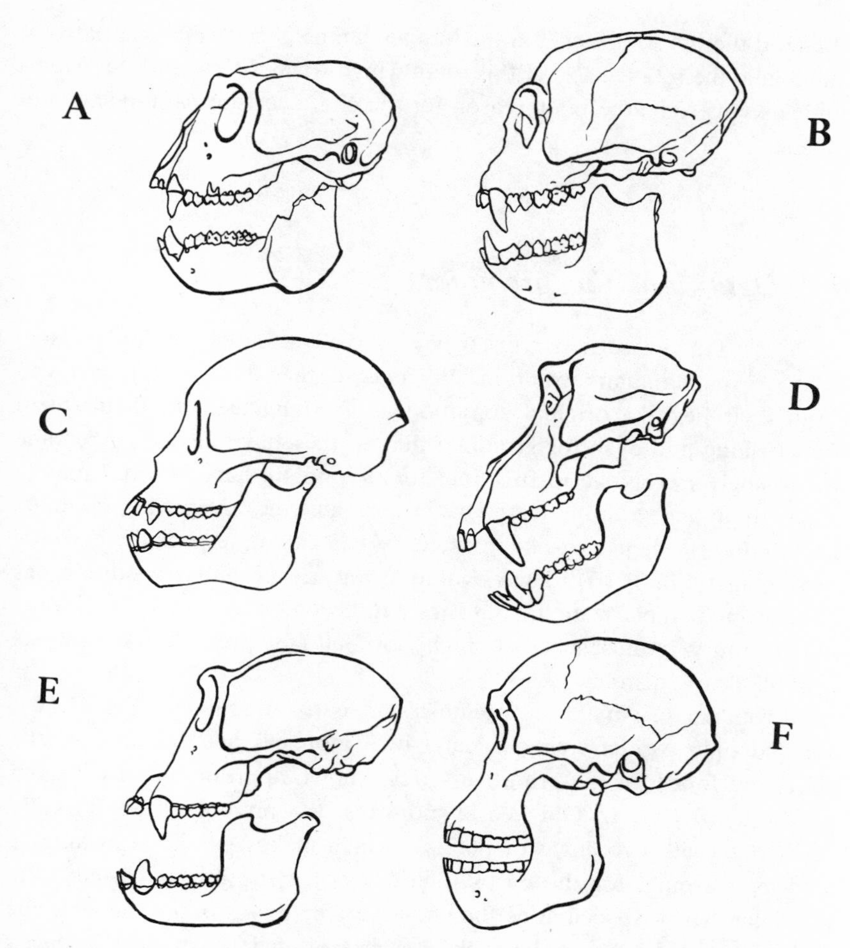

FIG. 7.15. Skulls of *Mesopithecus* (*A*), *Presbytis* (*B*), *Dryopithecus* (*C*), *Papio* (*D*), *Pan* (*E*), and *Australopithecus africanus* (*F*). (Redrawn from Le Gros Clark 1971.) (Not drawn to scale.)

grasping tooth. It had apparently also lost the need for such a tooth in fighting or display, which suggests that *A. boisei* had possibly perfected the use of weapons (such as stones and bones) for defense and offense and did not use dental display for threat as do many other higher primates.

In the human lineage, we can assume that there was at least an equal skill in weaponry, suggesting that the retention of a somewhat larger canine tooth was associated with its functions of grasping and tearing. This in turn suggests a diet that came to consist more of meat than did that of *A. boisei*. But the human canine is nevertheless relatively small, and the relative size of this tooth which we associate with the Hominidae seems to be the main reason for the characteristic parabolic dental arcade.

7.8 Dentition: the premolars

THE PREMOLAR TEETH have served a variety of functions in mammalian evolution. In general, they have slowly evolved from their peg-like original condition as grasping teeth to flatter teeth for crushing and even for grinding; that is, they have tended to become increasingly molar-like in function, for the molars have evolved mainly as crushing and grinding teeth. In the primates, however, the premolars have retained their peg-like form to some extent, especially in some prosimians (Fig. 7.16). Their evolution toward the molar condition has been termed "molarization," but the teeth become fully "molarized" only in the lemurs. In most species of higher primates the premolars and molars are still clearly distinct.

From an original four premolars in early mammals (Fig. 7.12), we find only two surviving in catarrhine monkeys and hominoids (the third and fourth of the primitive series, known appropriately as P3 and P4) (Fig. 7.13). In Old World monkeys, the premolars are typically *biscuspid* (with two *cusps* or points), which makes possible crushing as well as grasping, for the cusps interlock when the jaw is closed. One premolar has a special function, however, in this group. The anterior lower premolar is somewhat enlarged, has a single cusp, and it shears against the upper canine when the jaw is closed so as to make an effective cutting organ. That is why this modified premolar (lower P3) is termed *sectorial* (Fig. 7.13). The same character is also typical of the Pongidae; the lower P3 is functionally correlated with the large upper canine.

This sectorial lower premolar is not characteristic of the Hominidae, which lack a large upper canine. In the early hominid *Ramapithecus* from Fort Ternan we find a third premolar which is intermediate between the pongid and hominid condition. It is basically sectorial, but bears an incipient second cusp. The characters of the fourth premolar and molar teeth all suggest a hominid. The third premolars of *Australopithecus* and later hominids are all bicuspid, with the cusps more or less equal in size.

In modern man the *lingual* (inner) cusp is slightly reduced. The shape of the premolar crown has been considered an important character in the classification of early Hominidae (Leakey *et al.* 1964). However, in both the known species of *Australopithecus* and in modern man the shape is quite variable, and canine and molar teeth are of more interest in this connection.

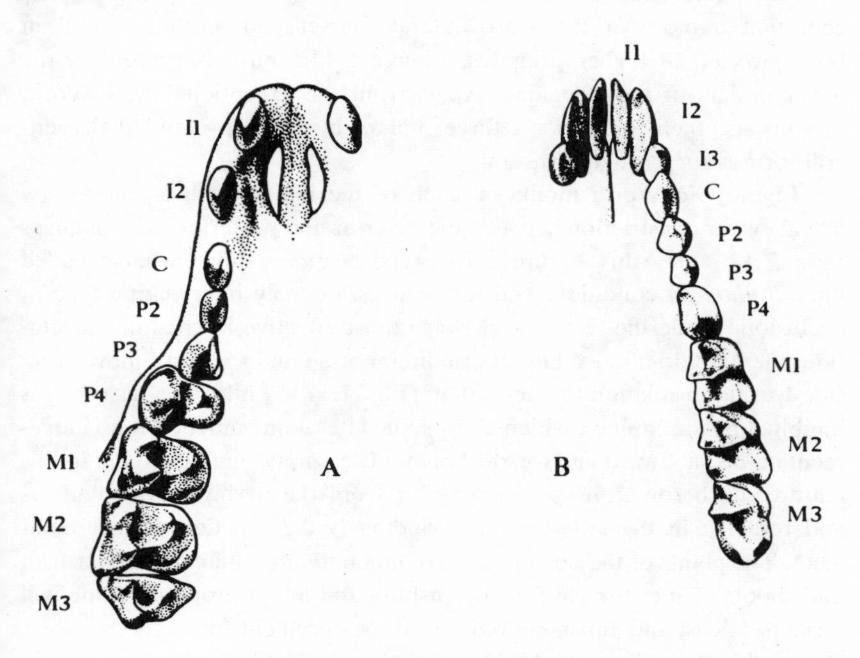

FIG. 7.16. Upper (*A*) and lower dentition (*B*) of the tree shrew *Ptilocercus*. Note the primitive arrangement of three incisors in the lower dentition and their strange shape, due to their use as a comb rather than for cutting food. The three premolars are still rather like canines in form, with the exception of the upper *P4*. (From Le Gros Clark 1971.)

The Pleistocene hominid *Australopithecus boisei* has taken molarization of the premolars to its most advanced condition in the higher primates.* As a result of the grinding action of the jaw, the premolars are large and flattened like the molars. The absence of protruding canines makes possible a rotary movement, which is most effective for the mastication of tough plant material.

* A possible exception is the fossil ape *Gigantopithecus*.

7.9 Dentition: the molars

THE EVOLUTION of the mammalian molar tooth has been subject to a great deal of study and has been recently summarized elsewhere (Le Gros Clark 1971). For our purposes, we need only accept that a more or less quadrilateral four-cusped tooth is found in both jaws of all higher primates, though a fifth cusp is present in the lower molars of the Pongidae. Apart from the exceptional New World marmosets, there are always three molars in each jaw, and their general form is surprisingly constant.

In the Old World monkeys, each of the molar teeth is divided by a "valley" or constriction between the anterior and posterior pairs of cusps (Fig. 7.13 shows this feature). This type of molar shows what is called the *bilophodont* condition, and it produces a deeply interlocking type of occlusion. While the teeth are perhaps most effective in crushing vegetation, the jaws do allow a lateral grinding motion and the teeth move from side to side in relation to each other (Fig. 7.17). Full rotary grinding is inhibited by the canines, which allow some lateral movement but no movement in the backward and forward plane. It is surely significant that in this feature the herbivorous primates differ from the herbivorous ungulates and rodents; in the latter, rotary grinding is the function of the molar teeth. The plants of the open plains are much tougher than the tender fruit and shoots of the tropical forest; crushing and lateral grinding alone will serve to release the nutriment only from the succulent forest foods.

In the Pongidae and Hominidae, the bilophodont condition is not found. Instead, the cusp pattern is more complex and a fifth cusp has been retained in the lower molars called the *hypoconulid*. The pongid molars are perhaps slightly more effective grinders, for in occlusion the teeth do not interlock so deeply, though rotary movement of the jaw is still to some extent restricted by the huge interlocking canines.

Among the Hominidae, there are two functional trends in the evolution of the molar teeth:

1. The evolution in *Australopithecus* of relatively flat and very large grinding molars and premolars, accompanied by rotary action of the jaw, was, as already stated, made possible by a reduction of the canines and was accompanied by molarization of the premolars. This trend appears most clearly in *A. boisei* and is an adaptation to a tough vegetable diet, such as would be available to a plains-living animal that had not

widely exploited animal food (Fig. 7.8). In both species of *Australopithecus* and *Ramapithecus* M2 and M3 are larger than M1 (See Garn *et al.* 1964).

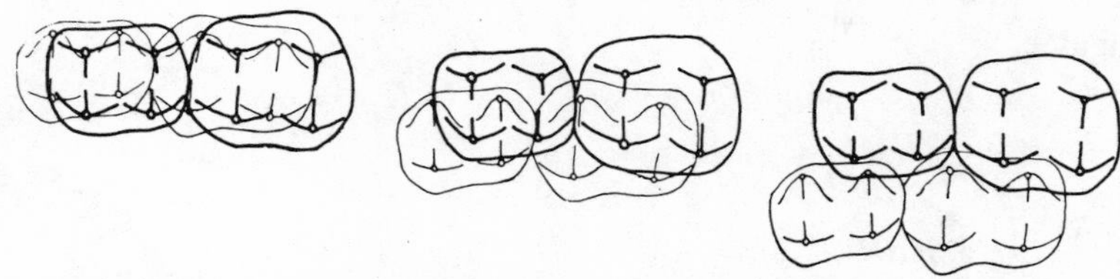

FIG. 7.17. **Occlusion in an Old World monkey, *Macaca*.** The first two maxillary molars are drawn with a heavy line, the mandibular molars with a thin line. The first molar is shown on the left. The central drawing shows the resting position; in the other two drawings the mandible is moved from side to side. The cusps are paired and alternate and side-to-side movement alone is possible. (From Mills 1963.)

2. The reduction of the molar series as a whole and, in particular, reduction of the third molar (the wisdom tooth, M3) in both jaws is a trend seen in the latter stages of human evolution and probably associated with the preparation of food by cooking, when the need for effective mastication is reduced. In modern man, the upper third molar is always considerably smaller than the other two and both third molars are sometimes absent, especially among Mongolians. The lower molars bear either four or five cusps (Figs. 7.11 and 7.14).

A character of some interest in studies of human evolution is the order of tooth eruption. Schultz (1935) has shown that the order in the permanent dentition of the Pongidae and *Homo* is as shown in Table 7.1.

TABLE 7.1 ERUPTION PATTERN OF PERMANENT DENTITION

	Order of Eruption							
	1	2	3	4	5	6	7	8
Pongids	M1	I1	I2	M2	P	P	C	M3
Homo (and rarely among pongids)	M1	I1	I2	P	C	P	M2	M3

The absolute and relative ages of eruption are shown in Figure 7.18. These data together demonstrate the much slower development of the human dentition as a whole (which is correlated with the slower growth rate of the whole body) and the slower development of the molars in particular (which is related to the reduction in importance of the molar series in modern man). Further such data are presented by Garn and Lewis (1963). Unfortunately, information about the tooth eruption sequence of *Australopithecus* is slight, but it does appear to be within the patterns found in modern man, which are distinct from those of the Pongidae.

7.10 Digestion

AFTER MASTICATION, the breakdown of foodstuffs by digestion and their assimilation into the body does not vary greatly among mammals. Studies in comparative anatomy (Straus 1936) suggest that no remarkable changes have occurred in the evolution of the human alimentary canal, but there is of course no fossil record of these soft parts, and such a statement cannot be verified by direct evidence.

There is, however, one variable in mammalian digestion that is of overriding importance: the extent to which specializations have evolved to make possible the digestion of cellulose. (Cellulose is a polysaccharide—though a multiple of a sugar molecule it is normally indigestible because it is insoluble.) Such a specialization as that for cellulose digestion is typical of fully adapted herbivores. It takes the form of an increase in the total volume of the alimentary canal to allow the development of a large population of microorganisms that are able to break down cellulose to synthesize protein and produce sugars. The microorganisms and sugars are in turn digested by the mammal. This development is typical of ruminants (such as cattle and sheep), which have multiple stomachs, and the herbivorous lagomorphs (such as the rabbit), which have a large *cecum,* a special, blindly ending branch of the alimentary canal.

It might be expected that some extension of the alimentary canal would be apparent among the herbivorous primates. This, however, is not very general, though some instances can be found. There is, for example, one unusual specialization in the mouth of certain primates in the evolution of cheek pouches. This character, confined to the *Cercopithecus* group of Old World monkeys, allows food to be retained in the mouth for some time before swallowing, but while it presumably aids the digestion of starch by the salivary enzyme ptyalin, there is no reason to suppose that it is an adaptation for cellulose digestion. It is, however, an adaptive

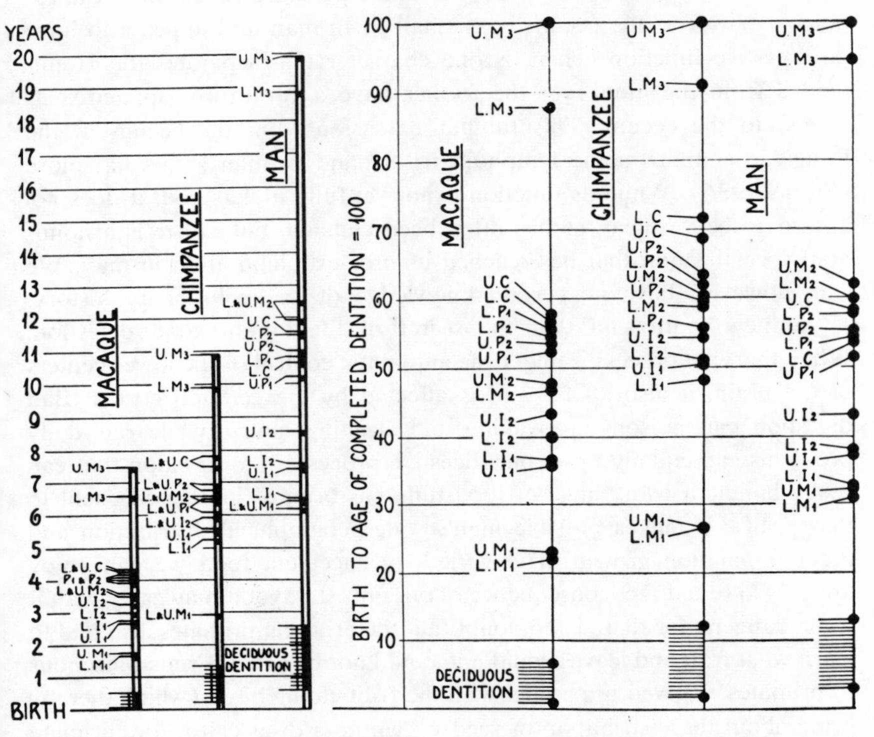

FIG. 7.18. On the left, the absolute ages and, on the right, the relative ages of eruption of the teeth of *Macaca*, chimpanzee, and modern man. (From Schultz 1935.)

character, in view of the high starch content of the diet of these arboreal forms.

The primate stomach is not greatly different from the human stomach, and only in one subfamily of the Old World monkeys is the area of lining of the stomach increased by subdivision into different compartments. This subfamily is the fully arboreal Old World Colobinae, whose diet is confined almost entirely to leaves and shoots with high cellulose content. Schultz (1961) believes them to be the only true vegetarians among primates. The multiple stomach and the presence of cellulose digesting bacteria appears to be an instance of parallel evolution with the ruminants, and it is perhaps surprising that it is not more widespread in the order.

An alternative adaptation is, however, found in the lower primates, where the cecum is often elongated and dilated—perhaps another case of parallel evolution, this time with the lagomorphs. In the Anthropoidea, however, the cecum is as small as in man and appears to have no digestive function. There is one character that separates the Hominoidea from the monkeys: the former have a vermiform appendix attached to the cecum. This tubular extension from the cecum in the Pongidae contains some lymphoid tissue, and in man somewhat more (Straus 1936). While its function is not yet fully understood, it does not appear to be vestigial, as has often been claimed, but rather is a hominoid specialization that has reached its greatest elaboration in man. We can, however, have no certain knowledge of its evolutionary history.

Sidney Smith said that the secret of life lay in good digestion. While this vital process is under the automatic control of the lower centers of the brain, it also affects and is affected by the cerebral cortex. Bad digestion can poison our waking and sleeping hours, while our daily problems can equally upset our digestive processes. A digestion that can easily handle a wide range of foodstuffs has been of immense benefit to the species; it makes possible man's wide geographical distribution and great population growth. The variety of succulent food we can enjoy appears to be a direct consequence of our ancestry, yet it may carry with it some dangers, for there is no doubt that the arboreal primates also had to learn what fruit and leaves could and could not be eaten. Fruits poisonous to primates, evolved in parallel with the fruit eating birds (which thereby bring about the distribution of seeds), can be a disaster for any primate with too much initiative or too little ability to learn.

7.11 Ecology, diet, and behavior

THE OMNIVOROUS NATURE of primates, and of their dentition and digestive processes in particular, made it possible for man's ancestors to evolve from forest-living to plains-living creatures. The flexibility in diet and behavior so typical of primates allowed an adaptation in food-finding behavior and food choice that made it possible for the early hominids to undergo a fundamental change in their environment.

On the basis of present conditions in east and south Africa, we can attempt retrospectively to predict the conditions in the early Pleistocene. During the Pliocene and Miocene, however, the climate is less certainly known. All we can say with some certainty is that in the early Pleistocene at Olduvai, climatic conditions were somewhat similar to those at present (Hay, *in* Evernden and Curtis 1965, p. 383), though the living floors at Olduvai Gorge bordered freshwater streams and a saline lake which are no longer present. (During the Pliocene the climate was probably broadly similar). Extensive excavations by the Leakeys at Olduvai show that *Australopithecus* consumed fish, reptiles, birds, and mammals, most of which were small (Fig. 7.20). Some sites carry remains of much larger game, and these are best considered butchery sites (Fig. 7.21). While *Australopithecus* had the capability of killing and cutting up large mammals, even elephants, the evidence of the living floor suggests scavenging for a very variable diet. We can assume that as in all living hominids, roots, fruits, and other vegetable food formed an important part of the diet, though no evidence of such food is preserved. This sort of mixed diet is nourishing, yet hard to get. Like modern baboons, *Australopithecus* must have spent a large part of each day finding food and scavenging, perhaps with the help of simple tools (Fig. 7.19). *Gathering* is a behavior pattern unique to hominids. It implies the collection of small food objects (such as fruit, nuts, eggs, or small animals) in a container, carrying them back to a larger group at a recognized meeting place, and sharing them. Of the development of food gathering we know nothing, but whenever it occurred, the use of containers must have been an important technological step in man's behavioral evolution. Although the use of stone tools is established by the Pleistocene, it would not be unreasonable to suppose that bones in particular would have already been used as weapons for killing small animals in the Pliocene (see Dart 1957 and Wolberg 1970).

Only in the Middle Pleistocene do we find evidence of a major change in early man's adaptation to plains living, and this change involved *cooperative hunting*—a change in food-getting behavior of central importance to the story of human evolution. It seems clear that man survived the climatic changes of the Middle Pleistocene by a new behavioral adaptation that affected fundamentally and irrevocably his psychosocial character; it was this change in food-finding behavior among other things that finally made man and that, as I shall show (chapter 11), helps to justify our recognition of the change from *Australopithecus* the apeman to *Homo* the man.

For mammals, food-finding and feeding is normally not a social occupation, except among a few carnivores (such as dogs and killer whales). Most primates live in social groups, but such groups are not

FIG. 7.19. Reconstruction of *Australopithecus* living in the grasslands of the Transvaal in Lower Pleistocene times, obtaining meat by scavenging and perhaps by hunting small or young animals. (By permission of the trustees of the British Museum [Natural History].)

primarily based on the need for social hunting and feeding. Social hunting which requires active cooperation is typical only of carnivorous mammals that feed on species larger than themselves. Wolves and jackals hunt in packs, particularly in winter, when small animals are scarce and they have to attack larger species.

An important study of social carnivores of Africa has been made by Schaller & Lowther (1969), which throws light on the possible adaptation of *Australopithecus africanus*. One of their principle conclusions is that all the social carnivores are both hunters and scavengers according to the availability of meat. Meat is obtained in four different ways: (1) scavenging dead, old, or diseased animals, (2) driving predators off

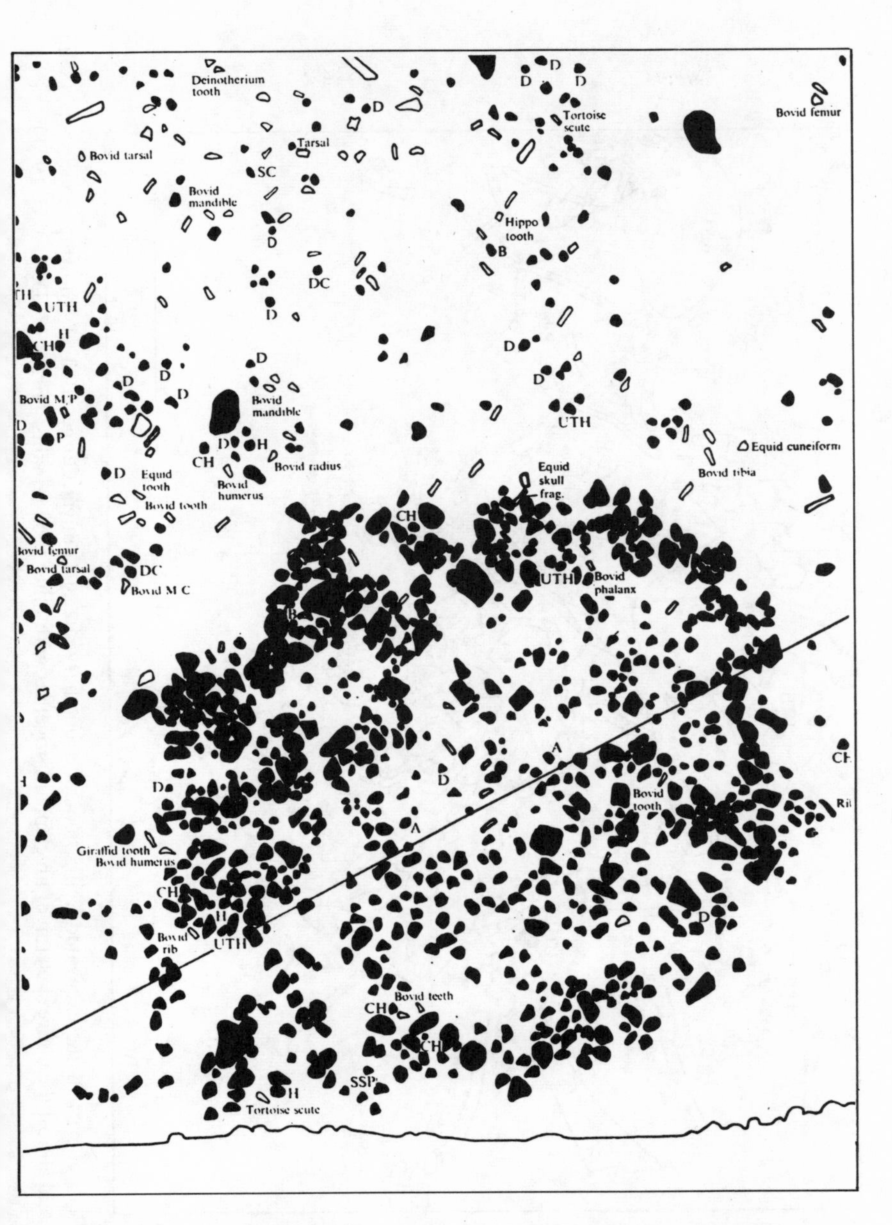

FIG. 7.20. Plan of part of the excavated living floor—an ancient land surface—at Olduvai Gorge (Bed I), showing stone tools, food remains, and a circle of stones which suggests the foundations of some kind of shelter. The age is about 1.8 million years BP. (From M. D. Leakey 1971.)

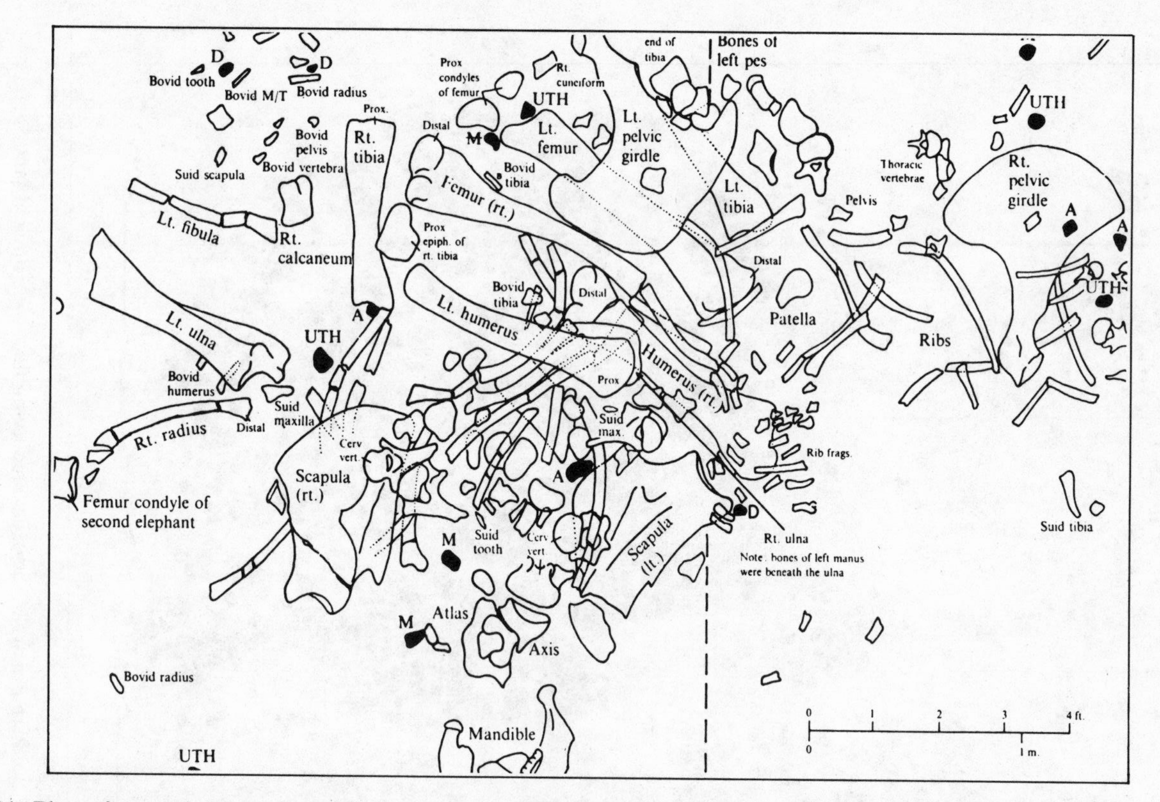

FIG. 7.21. Plan of part of the excavated living floor of a butchery site at Olduvai Gorge (Bed I), showing stone tools (solid black) and almost the entire skeleton of an elephant, together with other food remains. (From M. D. Leakey 1971).

a kill, (3) catching newborn young and other small animals, and (4) hunting healthy adults, as a cooperative team. The archeological evidence suggests that *Australopithecus africanus* probably employed methods 1–3, but that method 4 only came in with *Homo erectus,* something over one million years ago. Under category 4 we find social carnivores hunting on a broad front (dogs and hyenas), and by ambush and stalk (wolves and lions). Early hominids added to this repertoire the persistent chase method which is commonly seen among living hunter-gatherers, in which an animal will eventually be killed by spears after chasing it for many hours or even days. Persistence in the accomplishment of a task is a special character of man.

It was in this way that early man must have evolved cooperative hunting behavior. It seems clear that the prizes available as food make the evolution of cooperative hunting highly adaptive. If a group of males could cooperate to corner and kill an antelope, they would have enough meat for more than one individual and for more than one day. Not only was cooperation essential, but food-sharing followed. The sharing of a small kill serves little purpose; the sharing of a large kill is necessary and highly adaptive. It is clear that cooperative hunting and food-sharing changed the social and physical attributes of early man in a number of very important ways:

1. Success in hunting required cooperation among males.

2. Nursing and pregnant mothers would not have the endurance to carry their young long distances, so the females would have been left together, possibly guarded by a few males, thus involving the establishment of a home base.

3. Traveling long distances and carrying heavy burdens put strong selective pressure on the evolution of really efficient bipedalism.

4. The exertions of the chase may have been responsible for the diminution of subcutaneous fat deposits, loss of body hair, and the considerable development of sweat glands (which are not found in many forest animals), all of which aid the rapid diffusion of metabolic heat.

5. Catching and killing large animals required ingenuity and technological skill. The earliest implements were probably stones and clubs (sometimes used by living pongids). In the Middle Pleistocene, round stones are found that may have been used for throwing, possibly as a *bolas,* an ingenious and deadly device consisting of three stones strung together with long leather thongs. On impact the thongs will wrap round the animal and probably bring it to the ground. At the Torralba-Ambrona

site in Spain we have the earliest preserved evidence of a fire-hardened wooden spear: an implement which was surely essential to the early hunters. Later we have the introduction of the spear-thrower, or atl-atl, and finally the bow and arrow, perhaps 10,000 years ago.

6. The need to cut up large mammals must have stimulated the development of better chopping and cutting tools; hominid teeth were clearly unsuitable for this function.

7. The meat would be in sufficient quantity to be shared by all males and carried back to females. Bipedalism makes hand carrying possible and frees the mouth for speech. In the absence of a suitable alternative diet, bringing meat to the females would be strongly selected as a behavior pattern, and indeed there is evidence for it from the Middle Pleistocene, especially at Choukoutien.

8. Food-sharing within a social group would certainly bring the group together in a very intimate way, a situation that probably encouraged speech, especially for planning the hunt and in accounts of what happened during the hunt. As Roe (1963) has said, speech was perhaps born of the need of intercourse as a result of the division of labor. For the first time, primates had something essential to communicate (see 10.5).

9. Geographical knowledge obviously became of prime importance; the locality of waterholes and herds of game must have become a vital subject for communication. Clearly, improved powers of perception, memory, and prediction were of immense value and probably evolved conjointly with speech.

10. The absence of males from the home base on hunting expeditions must in turn have increased the overall division of labor between the sexes. The females would have taken over all other group activities. Sexual dimorphism might have changed its character in this new way of life (see 9.6).

11. The females may have learned to attract and receive the males sexually at almost any part of the month if the males were away hunting for a considerable portion of the time (see 9.3).

It would be possible to list many other important changes that might have followed the evolution of social hunting, but even from those listed it seems clear that the exploitation of the larger herbivorous mammals as a source of meat required cooperation, endurance, intelligence, foresight, and a precise means of communication. At the same time, it initiated the selection of an evolutionary trend toward further socialization with the development of speech and technology, all of central importance in our survey of human evolution.

We shall return to these socializing factors again in more detail in the following chapters. Here it is necessary only to emphasize, first, that primate feeding habits preadapted bipedal hominids to a terrestrial existence (as they preadapted baboons); second, that the exploitation of plains mammals was made possible by, and also immensely stimulated, the evolution of hominid society. Thus the final determinant of the direction of human evolution was necessarily environmental and ecological; not obscure, but striking and fundamental. Like all great evolutionary radiations of animals, man's appearance on the earth involved the exploitation of a new food source, and a very rich one.

7.12 *Food and fire*

FIRE HAS A NUMBER of functions in human culture, and cooking is by no means the most important. However, the use of fire has affected the digestion of food, and therefore this chapter may be taken as the place to discuss the history and functions of fire in human evolution.

The history of man's use of fire is being pieced together by archeologists and has been reviewed by Oakley (1961). The first traces of manmade hearths come not from Africa but from Europe and Asia. In a cave in southern France (Bouches-du-Rhone), called L'Escale Cave, we have traces of hearths which date from at least the earliest part of the Mindel period and may be older: they are therefore well over half a million years of age (Howell 1967). Somewhat later at a site in Hungary called Vértesszöllös (Oakley 1969) we have traces of hearths together with hominid remains dating from the middle of the Mindel glaciation of Europe, that is, the very beginning of the Middle Pleistocene (Fig. 7.22). Hearths of about the same date have also been found at Choukoutien in China, the site and level of which has also revealed the fossils of Peking man, *Homo erectus pekinensis* (Black *et al.* 1933).

At a much later date, fire is recorded in Africa at the famous Cave of Hearths (Lowe 1954), but the oldest trace there is probably a natural combustion of bat guano. However, above the thick, ashy layer formed by the burning bat guano, there is a succession of manmade hearths dating from the final Acheulian period. Although of relatively recent date (perhaps little more than 50,000 years B.P.), it does show us how fire might first have been used by man. Without an appropriate technology, man must have relied at first on capturing fire from natural conflagrations caused by lightning, or possibly by seepages of mineral oil or gas, or even by deposits of coal revealed by landslide; all these have been known to show spontaneous combustion. The myth of Prometheus describes how man stole fire from the gods; by means of myth, this achievement is recorded in our racial memory and celebrated in some tribes (among Bushmen, for example). It was an achievement of vital importance to mankind.

Only after man had learned to handle fire could he have learned to make it, possibly as a result of striking stones in the manufacture of tools. However it came about, man learned to make, conserve, feed, and handle fire, and it became a precious tool in the advance of his culture. By the time of *Homo erectus,* fire began to assume many important functions, for it was able to provide protection against predators, a center point at the home base, warmth, light, a means of cooking, and an aid to tool-making.

It is interesting that according to the present evidence the use of fire arose first in Europe and Asia at about the time of the Mindel ice age but probably during the warmer phases of this icy period. No traces of fire have been found on the open living floors of East Africa or in the only known rock-shelter used by *Australopithecus,* at Sterkfontein in South Africa (Robinson and Mason 1957). It appears first in the north temperate cave-dwellings. It seems highly probable that fire made cave-dwelling practicable because it gave man warmth and protection at night from wild animals (such as the cave bear), which in cooler climates were dangerous as predators and at the same time competitors for shelter. Fire is still used in parts of Canada for driving bears out of caves.

The achievement of handling and making fire would have resulted in a less nomadic existence, and the cave would have become not merely a temporary base but a home. The fire might have been maintained day and night by the women, and it would have supplied vital warmth and light against the long winter evenings and the cool, wet climate of the glacial period. The effective day would have been lengthened, and work—toolmaking and the preparation of skins and other animal products—could have been continued by firelight in the evening.

The established fire must have led to cooking, perhaps inadvertently at first. Heat will break down the tough structural components of vege-

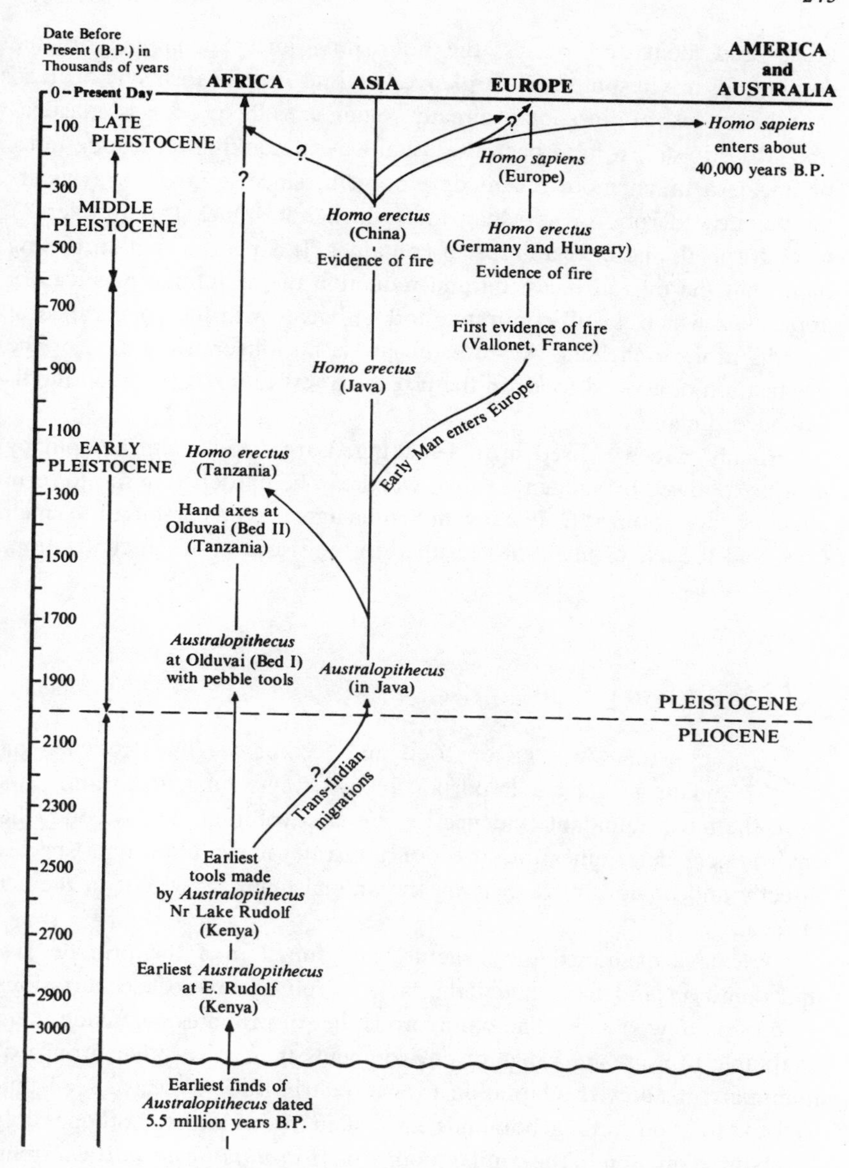

FIG. 7.22. This chart indicates approximately some of the dated archeological discoveries from the Pleistocene and their possible relationships in terms of the migrations of human ancestral groups. From at least one million years ago migrations between Africa, Europe, and Asia were probably continuous and no attempt has been made to show the complexity of these movements.

tables and meat and release the nourishing juices. Complex organic compounds are in some cases broken down into simpler forms by heat, so that the process of digestion is already begun, as is the process of mastication. Roasting must have been the usual way of cooking for a long time; boiling is a much more recent development, since it involves a water-tight hollowed rock or skin bag into which hot stones ("pot-boilers") were dropped, and later a fireproof container. It is perhaps not surprising then, that the overall recession and reduction of the human masticatory apparatus, which is still continuing today, began with the appearance of hearths in cave shelters. A powerful jaw is no longer necessary for the mastication of cooked food. In the process of evolution it has been modified accordingly.

Finally, fire was used in tool-making. Large stones can be split by heating followed by sudden cooling, wood can be hardened in fire to form effective spear-points. Only a few millennia ago, fire was first used to smelt ores, and the use of metal has resulted in the rise of modern civilization.

7.13 Summary

IN THIS CHAPTER on food and feeding we have touched on matters that are important in two ways. First, teeth and jaws form the most abundant evidence for human evolution, and, second, the environment determines the direction and rate of evolution of a species directly and obviously by the way the animal interacts with it in the act of feeding.

We have examined the structure and function of the primate jaw and dentition and have noted the lack of full herbivore specializations except in a few groups. The omnivorous diet of primates has made them adaptable to a greater range of environments than most other groups of mammals; it allowed adaptation to a terrestrial life. We have seen that such adaptation among hominids has taken two courses, both possibly involving reduction in the canine tooth: in *Australopithecus boisei* a trend toward molarization for maximum grinding efficiency and in the *A. africanus-Homo* line a trend toward an omnivorous dentition, with reduction in the molar series and finally in the whole masticatory apparatus.

Such a hominizing trend is accelerated by the evolution of social hunting, by the making of tools for cutting and chopping meat and bones, and by the mastery of fire as well as by food gathering and the use of containers. The social correlates of cooperative hunting have also been briefly considered, and it seems clear that this cultural development was the most important single step in the final evolution of man. Human beings and their society are no exception to the general observation that animal groups evolve as homeostatic adjustments to changing environmental conditions, relating primarily to the interaction of organism and environment that is involved in getting food.

SUGGESTIONS FOR FURTHER READING

The best general account of the evolution of the primate dentition is again W. E. Le Gros Clark, *The Antecedents of Man* (3rd ed.; Chicago: Quadrangle Books, 1971). For a more detailed study, see the classic account by W. K. Gregory, *The Origin and Evolution of the Human Dentition* (Baltimore: Williams and Wilkins, 1922).

The ecological and dietary implications of the hominid adaptation have been discussed in an important article by C. Jolly, "The Seedeaters: A New Model of Hominid Differentiation Based on a Baboon Analogy," *Man* 5 (1970): 5–26. Beyond this, the most interesting paper on these aspects of human evolution is C. F. Hockett and R. Ascher, "The Human Revolution," *Curr. Anthrop.* 5 (1964): 135–68. An important account of some physiological adaptations of modern man are to be found in G. A. Harrison, J. S. Weiner, J. M. Tanner, and N. A. Barnicot, *Human Biology* (Oxford: Oxford University Press, 1964). The adaptations of living hunter-gatherers are discussed in R. B. Lee and I. DeVore (eds.), *Man the Hunter* (Chicago: Aldine Publishing Co., 1968) and reviewed in C. S. Coon, *The Hunting Peoples* (Boston: Little Brown, 1971).

8

The Evolution of the Head

8.1 Factors determining head form

IN STUDYING the evolution of the human head, we must examine not only the masticatory apparatus but also the evolution of the senses and the correlated changes that took place in the brain, all of which have contributed to the present form of the human skull.

In many marine and fresh-water animals the head is streamlined for swimming, and its shape is therefore dictated primarily by hydrodynamic factors. In terrestrial vertebrates the overall shape of the head depends on two factors: first, the form of the masticatory apparatus, sense organs, and brain; second, its position in relation to the body and in relation to gravitation. The shape reflects the function and size of the different parts and by examining the evolution of those parts we can understand the evolution of the skull.

The sense organs situated in the head are listed in Table 8.1. Not all of them affect skull form. The ones that do, together with the other factors mentioned above, add up to the following major determinants of head form: 1) the masticatory apparatus, 2) the eyes, 3) the nose and nasal chamber, which make up the muzzle, 4) the brain, and 5) the position of the skull on the vertebral column.

The evolution of the masticatory apparatus has already been dis-

TABLE 8.1 THE SENSORY RECEPTORS OF MAMMALS

Organ	Sense	Stimulus	See Section
Eyes	Vision (optic)	Radiant energy: light	8.3
Nose	Smell (olfactory)	Certain volatile chemicals	8.4
Ears: cochlea	Hearing (aural)	Vibrations in air (or water)	8.5
Ears: semi-circular canals, etc.	Balance and movement	Direction of gravity and rate of change of direction of movement, i.e., acceleration	8.5
Taste buds	Taste	Certain nonvolatile chemicals	7.3
Skin, lips, and tongue	Touch, temperature, and pain	Local pressure change on skin, temperature change, damage to body tissues	8.9

cussed (see 7.4, 7.5). It remains only to consider the influence of this apparatus on the form of the head as a whole.

8.2 The masticatory apparatus and the head

THE SIZE AND POWER of the masticatory apparatus obviously affects head form. A large mandible needs a large maxilla to carry the upper dentition, and together they will make an animal appear *prognathous* (with its muzzle protruding) if the other parts of the skull are not expanded equally. Not only will the bony parts of the apparatus affect head form, but so will the related muscles and the extent of their attachment areas.

All the muscles of mastication have their origin on the skull, and the extent of their attachment areas has changed in evolution according to the degree of their development, a fact that is particularly relevant to skull architecture when we study the origin of the *temporal muscle* on the skull. Its origin is on the brain case (*neurocranium*) in the area of the *temporal* and *parietal* bones (Figs. 8.1, 8.2; see also Figs. 7.2, 7.3).

The extent of this area of attachment varies in accordance with the power developed in the masticatory apparatus. In the large-jawed primates, such as the gorilla, the area of attachment is so large that the areas on each side meet on top of the cranium, and the bone grows to form a ridge at this meeting point to increase the anchorage still further (Fig. 8.2, *A*). The height of the ridge varies according to the development of the jaws, so in the lighter-jawed female gorilla it is

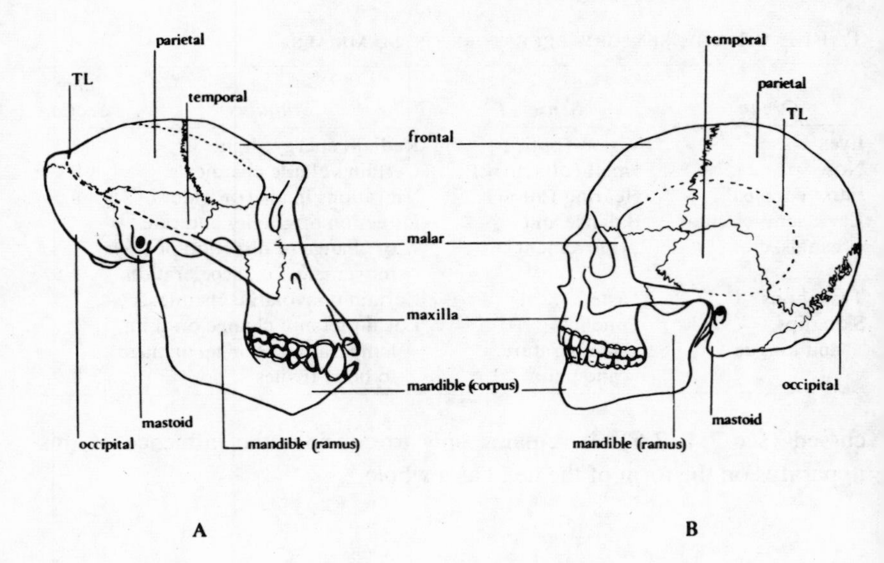

FIG. 8.1. Skulls of a chimpanzee (*A*) and a man (*B*) showing the different bones mentioned in this chapter. *TL* = temporal line.

smaller than in the male or absent altogether. This ridge is the *sagittal crest,* and it is peculiar to primates whose jaws are proportionally larger than their neurocrania—those in which the temporal muscles cover the neurocranium and actually meet in the sagittal plane along the top of the skull. A second crest, the *nuchal crest*, develops in heavy-jawed primates where the extended temporal attachment area meets the proportionally large attachment area of the neck muscles (see 4.4 and Fig. 4.15). We have seen how the size of the masticatory apparatus is not only genetically controlled but also depends on the use it receives (see 7.5), and it therefore follows that these two factors similarly affect the size of the attachment areas of the temporal muscles and, ultimately, the size of the sagittal and nuchal crests.

Many of the fossil skulls of *Australopithecus* carry a marked sagittal crest (Fig. 8.2), but the attachment areas were reduced in evolution along with the masticatory apparatus, so that in *Homo erectus* the areas have retreated down the sides of the skull. The areas of attachment are bordered by the *temporal line,* which shows distinctly on the surface of the skull bones, and in the evolution of man this line moves down the side of the skull as the relative size of the jaws is reduced and the size of the neurocranium increases (Figs. 4.15, and 8.2).

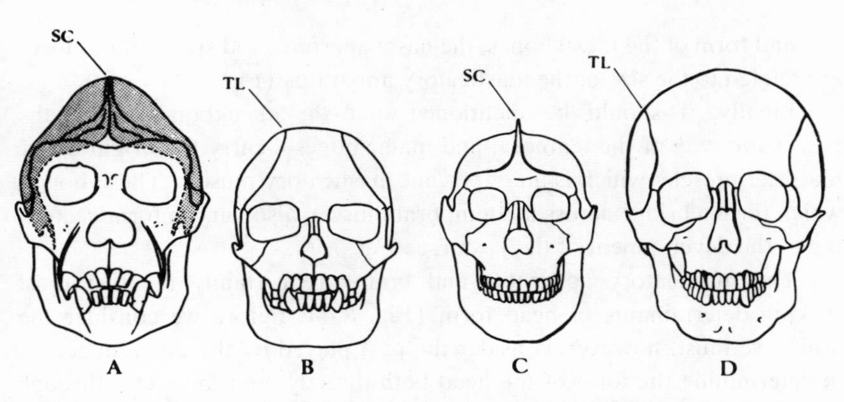

FIG. 8.2. Facial views of gorilla (*A*), *Dryopithecus* (*B*), *Australopithecus* (*C*), and modern man (*D*); skulls drawn approximately the same size. Note the different development of the sagittal crest in each genus. Upon the gorilla skull have been superimposed the temporal muscles that originate on the sagittal crest and surface of the neurocranium and are inserted upon the mandible. *TL* = temporal line; *SC* = sagittal crest.

Since the activity of all the powerful masticatory muscles occurs within the perimeter of the head, the stresses developed by their action must be entirely absorbed by the skull architecture. When the muscles contract and the jaw is closed, the force exerted for chewing pulls the mandible and neurocranium together. The force is very considerable and the bony structure between the roof of the neurocranium and the maxillary teeth—that is, the face and brows (all of which receive the force from the lower dentition)—must be very strong. Thus the thickness and form of the facial bones is related to the power developed by the masticatory apparatus (Endo 1966).

Nearly all this force is transmitted by the molar and premolar teeth and passed up the side of the face to the cranial vault. Owing to the position of the nasal cavity and orbits, the force must be carried round these openings by the maxilla and *malar* bones. Like the mandibular torus and chin, the supraorbital torus (which is the anterior edge of the *frontal bone* of the skull) acts as an essential cross-member at the top of the face to transmit the masticatory forces across the whole vault of the neurocranium rather than at each side only. Where present, the torus also resists other forces developed by the contraction of the temporal muscles. As their points of origin and insertion are closer to the midline and each other than the widest point between them, they develop a considerable horizontal compression at the sides of the frontal bone. Thus the size and form of the facial bones, the nasal aperture, and supraorbital torus are related to the size of the masticatory apparatus (Fig. 8.3).

Finally, it should be mentioned that the cheekbones—each the *zygomatic arch* of the temporal and malar bones—carry the origin of the masseter muscle, which is an important masticatory muscle. These bones, which form a bridge across the temporal muscle, also vary in form according to the development of the jaws (see Fig. 7.2).

The masticatory apparatus and brain are certainly the two most striking determinants of head form (Fig. 8.4). Before we consider the brain, we must, however, consider the part played by the different senses in determining the form of the head both directly, and indirectly through their influence on the evolution of the brain.

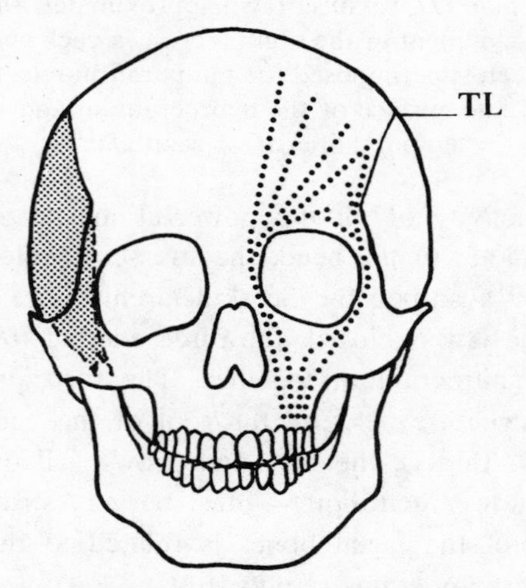

FIG. 8.3. Diagram showing how the forces developed by the masticatory muscles, in particular the *temporalis* (shown on left passing behind the zygomatic arch), are transmitted from the lower jaw to the vault of the skull in man. The lines of compression that pass through the facial bones are shown as dots on the right side of the skull. *TL* = temporal line.

8.3 The eyes

THE IMPORTANCE of eyes in the evolution of the primates has already been mentioned; we have described how vision evolved in response to the demands of the forest environment, where a sense of smell was of less value. Receptive to radiant energy of a certain range of wavelengths (380–760 mμ), the eyes came to provide information at a distance about size, color, texture, movement, pattern, spatial relationships, and distance—information not accurately obtainable by other means. Here we must discuss three aspects of the evolution of vision: (1) the retina, (2) the neural correlates of vision, and (3) the orbital region of the skull.

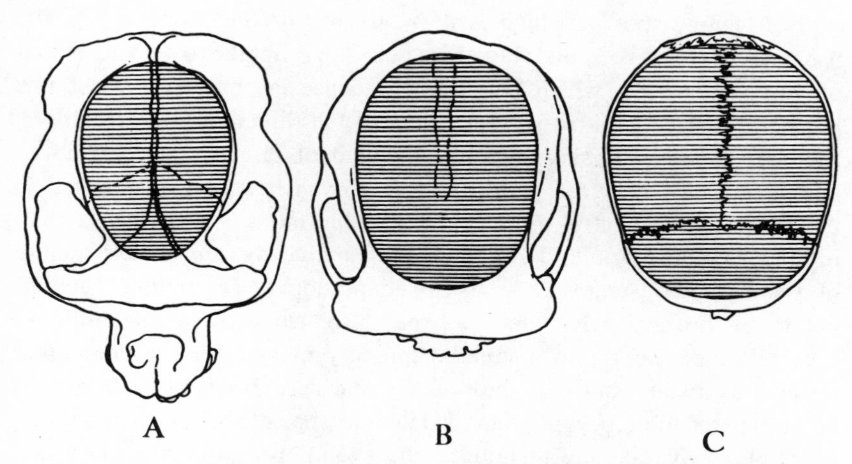

FIG. 8.4. Skulls of gorilla (*A*), *Homo erectus* (*B*), and modern man (*C*) from above showing the relative sizes of the cranial cavity (*shaded*) and masticatory apparatus protruding below the neurocranium. (From Weidenreich 1939–41.)

1. Cats have relatively large eyes, but, apart from a few such carnivores, the eyes of primates are outstandingly large. The amount of light that will enter the eye depends upon the aperture of the lens, and discrimination and sensitivity depend upon the concentration and kinds of photoreceptors on the surface of the retina. A large eye will allow the entry of more light, and thus permit greater sensitivity. The first stimulus to increased eye size may have been the nocturnal habits which were characteristic of some early primates, and today many prosimian eyes are specially adapted for nocturnal vision, with a high density of photoreceptors in the retina of the very sensitive "rod" type. The amount of information received by the brain from the eyes is probably approximately related to the size of the optic nerves. The dog and cat have about 150,000 fibers in each nerve; man has 1,200,000 (von Bonin 1963).

The eyes of the higher primates, however, are adapted for daylight vision, and their structure is fairly consistent (Woollard 1927); the differences between the eye of man and that of monkey are slight. Both have a high proportion of the so-called "cone" type of photoreceptor, which is specialized for discrimination rather than sensitivity. Fine visual discrimination seems to have assumed great importance in the forest, for in the primate we find the evolution of the *fovera centralis* ("yellow spot"), which is a small area of the retina where the usual layer of nerves and blood vessels does not lie over the actual photoreceptor cells, which here form a dense layer of cones. At this point the surface is pitted so that light will fall directly on the receptors. In man, the *fovea* makes possible the kind of discrimination required in reading and other fine work. But the fovea had already evolved in the forest and in fact is more marked in the monkey *Cercocebus* than in modern man (Woollard 1927). It is clear that the remarkable quality of vision that we enjoy arose as an adaptation to forest life. There is no doubt, however, that changes were also necessary in the brain as a correlate of the greatly increased quantity of visual information that became available, and it is these changes which above all have carried human perception beyond that which may be attributed to monkeys.

2. We have already mentioned the evolution of stereoscopic vision among the primates (see 3.4 and 3.7). This important development has involved changes in the position of the eyes, in the structure of the optic nerve paths and brain, and finally in perception.

The evolution of the primate visual sense has involved a large expansion of the part of the brain concerned with vision, the visual cortex of the cerebral hemispheres. In Figure 8.5 the brains of three primates are drawn the same size. The visual cortex is already well developed in the tree shrew, and this development is especially striking when the brain of an insectivore with a minute visual cortex is compared with it. However, in the tree shrew the visual sense has not yet replaced the olfactory sense, as can be seen from the relatively large size of the olfactory bulbs. In the higher primates, such replacement has occurred, and, although the visual cortex gets smaller relative to the rest of the brain, it is still absolutely expanding. This expansion is even more true of the visual association area, the part of the cortex nearest to the visual cortex, which, as we have seen, carries the connections between the visual cortex and other parts of the brain. In primates the expansion of the brain begins with the expansion of the visual cortex, and throughout pri-

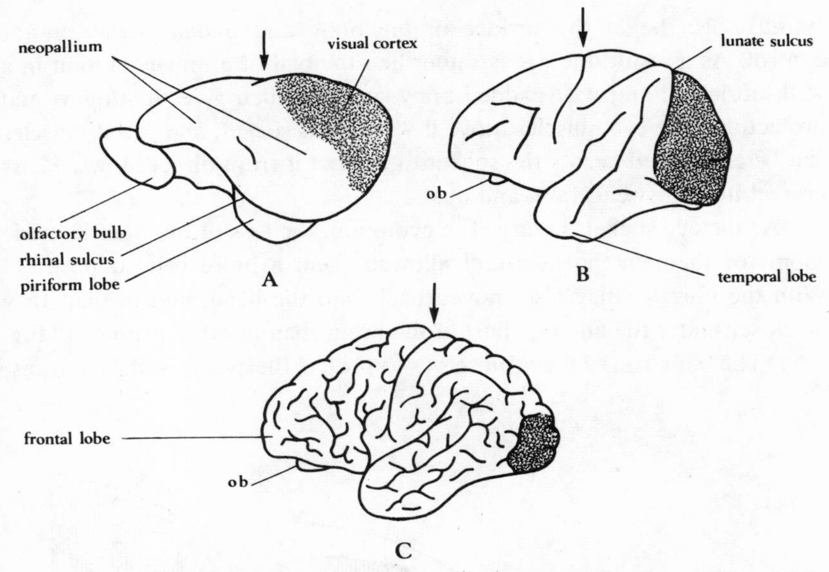

FIG. 8.5. Cerebral hemispheres of the tree shrew *Tupaia* (*A*), *Macaca* (*B*), and man (*C*), drawn the same size. Note the variation in extent of folding of the cortex and the relative size of the olfactory bulbs (*ob*) and visual cortex (*shaded*). The temporal and frontal lobes of the neopallium hide the Piriform lobe of the archipallium in the higher primates, which is visible here only in the brain of *Tupaia.* The arrow points to the central sulcus, which separates the frontal and parietal lobes. The rhinal sulcus separates the archipallium and neopallium, and the lunate or "simian" sulcus demarcates the visual cortex. (For a diagram showing the different lobes of the brain, see Fig. 8.8.)

mate evolution the visual system is increasingly integrated with total brain function. The fact that the *occipital lobe* (that part of the brain which includes the visual areas of the cortex at the back of the head) of the hemispheres expands does not, however, mean a localized expansion of the neurocranium: it means merely an increase in the total cranial capacity, reflected in the size of the neurocranium as a whole.

3. The eye has played an important part in molding the skull, in addition to its psychological and neurological importance in the evolution of man. We have seen that primate eyes have increased in size and have come to face forward (Fig. 3.4). The eye sockets—the *orbits*—have appropriately enlarged and moved into a more central position in the face. The eyeball is a delicate structure, and its form is of necessity optically precise, so it must at no time be subjected to stress or pressure.

It must also be at the surface of the body and under precise motor control. As a result, the eye is rather like the ball of a universal joint in a well-lubricated and well-padded bony socket, which gives it support and protection. The eye muscles move it within this socket, and facial muscles can be contracted across the socket to protect it from direct blows. Hairs protect it from sweat, rain, and dust.

As the eye sockets enlarged in evolution, the loss of the muzzle made room for them in the face and allowed them a more central position. With the muzzle, they also moved back into the head, and in man they lie closer under the anterior part of the brain than in other primates (Fig. 8.6). The bony rim of the orbit serves to protect the eye as well as to trans-

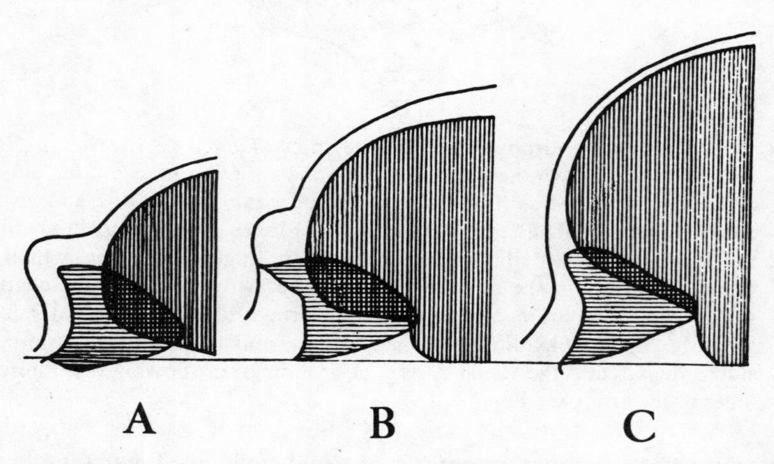

FIG. 8.6. The relation between the orbit and the frontal part of the cranial cavity in chimpanzee (*A*), *Homo erectus* (*B*), and modern man (*C*). (From Weidenreich 1939–41.)

mit the forces developed by the masticatory apparatus. The upper part of this rim assumes the greatest size (and gets the name of supraorbital torus) in big jawed forms, in which the eyes have not fully retreated under the forepart of the brain. Such is the case in the gorilla and chimpanzee, while in man and the gibbon, the torus is less well-developed, since the jaw is smaller and the eye is better protected by the forepart of the neurocranium. In the orang, the relationship between the face and the neurocranium is different and this results in a much more lightly built supraorbital torus (Biegert 1963). It is of interest that the torus is quite heavily built in all the Hominidae except modern man, for it was only in the last stages of human evolution that the jaws finally retreated and the neurocranium finally expanded over the top of the eyes. This expansion of the brain has resulted in the flat vertical forehead of man and his more or less vertical face, which carries the deep eye sockets.

8.4 The nose

THE RHINARIUM, the sensitive skin around the nostrils, was an important sense organ in the lower primates, and its wet surface no doubt enhanced its sensitivity to air currents, which would be interpreted directionally. In the open plains, the direction of the wind, coupled with the scent it carries, is a datum of first importance. In the forest, the sense of smell will be of significance only in relation to an object seen or touched, because the nose is unable to receive directional information. It is no longer a directional distance receptor, so, with the ascendance of the visual sense, the olfactory sense has come to be relatively less important in primates. In the arboreal environment it supplies information limited to the quality and intensity of odor, with no direct spatial implications.

The air required for respiration is drawn into the nostrils and over the olfactory *mucoperiosteum* (the mucous membrane bearing olfactory receptors) within the nasal cavity on its way to the lungs. By means of a complex maze of fine bones, the *turbinal bones,* the actual area of mucoperiosteum may be very considerable, and in the rabbit (for example) it has been estimated to carry 100 million chemoreceptor cells. In the higher primates both the size of the nasal cavity and the complexity of the turbinal bones has been much reduced, together with the total area of mucoperiosteum. That both the muzzle and the jaws are reduced in modern man is coincidental. In some animals they have receded independently: in the gorilla, the muzzle is slight, the jaws large; in the little elephant shrew (*Elephantulus*), the muzzle is far more developed than the small jaws.

It is the reduction of both the muzzle and the jaws in man that has resulted in his flat face. This change has altered the center of gravity of the head and contributed to the balance of the skull upon the vertebral column. The reduction of the muzzle came early in the evolution of the primates; the reduction of the jaws came late in the evolution of man (see Fig. 4.13).

A visit to the zoo will, however, reveal the human nose to be rather more prominent than that of most monkeys and apes—perhaps partly due to the recession of the jaws and the expansion of the brain, which left only a small space for the nasal cavity. However, the olfactory epithelium covers in fact only a small part (2.5 sq. cm.) of the lining of this cavity, so it is not unreasonable to suppose that some other function is performed by the human nose besides smelling. The fleshy surroundings of the nostrils contain fatty tissue, and, within, they are lined with a thick, moist mucous membrane with many blood vessels, suggesting that the function of the nose is to warm and moisten the air on inspiration. The fatty tissue insulates the nasal cavity, and the blood maintains the temperature within. The mucous membrane itself acts like flypaper to catch dust and small insects, and to that end there is a clump of hair near the opening of each nostril. Clearly, the warming of air could be important to the north-temperate peoples not only to protect the delicate membranes lining their lungs but perhaps also to increase the effectiveness of their olfactory organ.

The moist mucous membrane also acts to humidify the inspired air. It has been shown that it secretes up to one liter of water per day and maintains the relative humidity of the inspired air at 95 per cent (at body temperature). Its function is clearly to protect from desiccation the delicate inner surfaces of the nasal and oral passages and the lungs. The nasal index of modern man (which indicates the shape of the nasal openings) is highly correlated with the absolute humidity of the air of the region he inhabits (Weiner 1954). Thus many of the tropical races of man have flatter, more open noses than have those occupying dry areas, such as Mongolia. The prominent human nose is an organ for filtering, warming, and moistening the air and has perhaps evolved in size for that purpose since the movement of man into the dry plains and later into the temperate zones at the beginning of the Middle Pleistocene.

The sense of smell is much reduced in the higher primates, but it retains some value in modern man (though less than in the apes and monkeys). It has already been pointed out that its functions are closely related to the basic drives of feeding and sex. In modern man, smell is an important stimulant to feeding, but individuals in which this sense has been destroyed can survive with little difficulty if dependent upon the civilized preparation of foods. In its second function, sex, smell has again been replaced mainly by sight, but scent is still used by men and women as a sexual stimulant.

The relative size of the olfactory "bulbs" of the brain, to which the olfactory nerves run, can be seen in Figure 8.5. Analysis of the olfactory impulses begins in the olfactory bulbs, and from them, fibers pass to the lower lateral parts of the cerebral hemisphere (the *piriform lobe*), where further analysis presumably occurs. In mammals, electrical stimulation at this point results in such reflexes as licking, salivation, chewing, etc. From the piriform lobe, connections run to the areas of the cerebral cortex, where patterns are memorized and recognized. Though olfactory bulbs have been reduced in primate evolution, in man the olfactory sense is still capable of considerable discrimination and the nose is capable of remarkable sensitivity in spite of the fact that the importance in behavior of the olfactory stimulus is much reduced. This final loss of importance has almost certainly been a recent event.

8.5 The ears

STUDY OF THE EVOLUTION of the head must include consideration of the ear and the other soft parts of the face. The ear and the organs associated with it in the head form a complex mechanoreceptor that is sensitive to sound waves in air, to gravity, and to movement. Like light but unlike scent, sound waves travel in straight lines and therefore can be used in direction-finding. For that reason, the ears are separated (as are the eyes, but not the olfactory organ) and can function as a stereophonic organ. They can supply data about the direction, frequency, and amplitude of sound waves.

The different parts of the ear serve the following functions among primates:

1. The *outer ear* is primarily a directional receiver of sound waves. As in many mammals, the primate external ear can be moved by both its own and associated muscles to receive the maximum amplitude of sound waves and to aid discrimination among different sources of sound. The outer ear also plays some part in thermoregulation, as it does in

mammals generally. There are a large number of blood vessels in the outer ear, and like the rest of the skin they are subject to dilation and contraction to regulate heat loss. This is important to animals with heavy fur (like lemurs) but also naked ears. At the same time, ears held close to or away from the body can adjust the effective surface area for heat loss. The different sizes of the ears in different primates may be due as much to their function as thermoregulators as to their function as organs of hearing.

2. *The middle ear* receives the sound waves at a diaphragm (the ear drum) and transmits them mechanically (with a reduction of amplitude) to the *cochlea* through a second diaphragm (the *fenestra ovalis*).

3. Sound waves of reduced amplitude are transmitted to the liquid-filled cochlea in the *inner ear,* where the vibrations in this liquid are detected by the mechanically stimulated cells of the *organ of Corti*. The small hair-like processes attached to these cells respond selectively to movements in the surrounding liquid and send nerve impulses to the brain by the auditory nerve.

4. In connection with the cochlea are other liquid-filled vessels; the *utricle* and *sacculus,* organs that are lined with similar mechanoreceptors and transmit information to the brain about the posture of the head in relation to gravity.

5. Associated with this system of mechanoreceptors are the *semi-circular canals,* which lie in the three planes of space. Sensory cells here detect the direction and acceleration of movement of the head (Fig. 8.7).

It is certain that an arboreal animal like a primate needs a very efficient sense of movement and posture, but in fact the organs described are really not very different from those found in fishes, reptiles, and birds. Such organs are necessary to any mobile vertebrate. In its function as an organ of hearing, the ear has also changed very little in the evolution of the primates.

In the evolution of man the functions of the ear have not changed greatly. The ear still occasionally functions as an organ of thermo-regulation when, as a result of the dilation of blood vessels, the ear turns bright pink (sometimes caused by blushing). The musculature of the ear has been greatly reduced, however, and few people can move their ears more than a small amount. This reduction of motility begins among the higher primates, and the reason for it can only be guessed, though it may be associated with increased motility of the head. Presumably, an acute sense of hearing is of less vital importance to noisy diurnal animals with few predators than to silent creatures that hunt by night or walk in fear of death. The tendency in primate evolution has been to lose acuity of hearing but to gain discrimination in detecting sound frequency patterns. This change is probably correlated with the development of communication within the group, using sounds

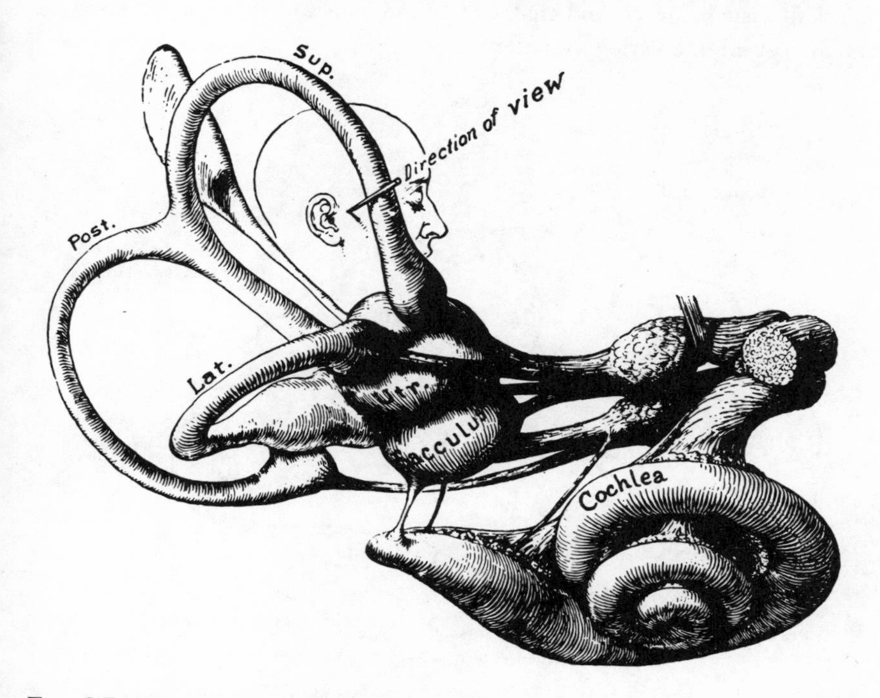

FIG. 8.7. The mechanoreceptors of the inner ear are a system of interconnected liquid-filled tubes. They are shown here with their nerve supply. *Lat.*, the horizontal semicircular canal; *Post* and *Sup.*, the two vertical canals at right angles. The utricle (*Utr.*), sacculus, and cochlea are labeled. (After Hardy 1934.)

of widely different kinds. It is certain that, in spite of the great auditory discrimination of man (and in particular of musical man), he may not be able to detect low level noise as well as his dog, who guards him by listening for unusual sounds.

The high degree of auditory discrimination that man possesses enables him to understand the complex sounds of language and music. This discrimination is probably due to some improvement in the cochlea itself, where it concerns frequency, but is certainly and more emphatically due to the evolution of the auditory cortex, where the incoming signals are analyzed. The auditory cortex lies in the temporal lobe (Fig. 8.8), and it may be no coincidence that memory, which also appears to be to some extent localized in this area, is closely concerned with language as well as with visual images and that words and music may be memorized in an almost infinite variety of patterns.

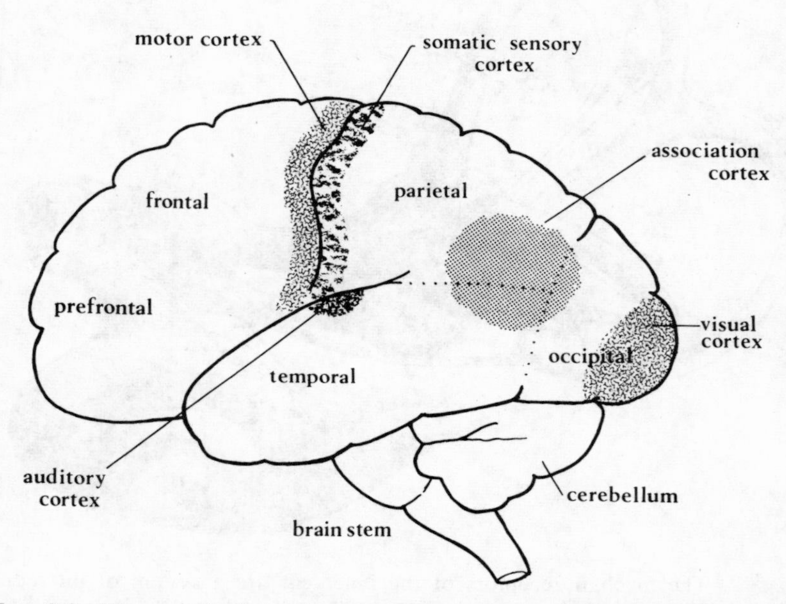

FIG. 8.8. View of the left side of the human brain, showing the main subdivisions of the cerebral cortex, together with the somatic sensory and motor cortices. The dotted area in the inferior posterior parietal region shows very approximately the secondary association cortex discussed in the text. (Compare Fig. 10.8.)

In summary, it may be said that the ear has become concerned with investigation, not so much of the general environment, as of the part of the environment that consists of other members of the species. That is, it is concerned with the kind of auditory discrimination required in human communication. Appreciation of music is perhaps a by-product of the ability to analyze speech sounds.

8.6 *Changes in brain structure and function*

BY MICROSCOPIC EXAMINATION and electrical stimulation and recording within the living brain it has proved possible to identify to some extent the functions of its different parts. Figure 8.5 shows some of the main functional areas of the tree shrew, macaque, and human cerebral cortex. The evolution of these different areas in the primates may be related to the evolution of the sense organs we have discussed, and may be summarized as follows:

1. The *olfactory bulbs* and *piriform lobes* are concerned, as we have seen, with the analysis of the olfactory stimuli from the nose. The piriform lobes form part of the *archipallium,* which is the most archaic part of the cerebral cortex. The archipallium is separated from the *neopallium* (which lies above it) by the *rhinal sulcus.* The neopallium effects, among other things, analysis of the sensory input other than that from the olfactory organ. In the evolution of the primates, the value of the sense of smell diminishes and with it the relative size of the archipallium. With the increasing importance of the other senses, the neopallium expands, the rhinal sulcus moves down the side of the brain, and the archipallium, including the piriform lobe, comes to lie underneath the brain rather than to one side. These changes are indicated in Figure 8.5.

2. The *occipital* (posterior) *lobe* of the brain has been shown to be associated with the analysis of visual images, and the cortical layers here are, as we have seen, known as the *visual cortex.* As might be expected from what has been said of the importance of sight in primate evolution, the visual cortex has greatly expanded, even in the lower primates. This area of the cortex is clearly defined by a curved *sulcus* (fissure), the *lunate sulcus,* or, as it is often (appropriately) called, the *simian sulcus.* The development of the large visual cortex is one of the most striking features of the primate brain.

3. The *temporal lobe* has also expanded at an early stage in primate evolution. A small area on its surface constitutes the *auditory cortex* (Fig. 8.8), but the rest appears to be concerned with the elaboration of visual analysis and with visual and auditory memory. These functions are clearly important in a group of animals that has increasingly come to rely on the visual memory more than on other imprinted sensory patterns.

Clearly, the temporal lobes also play an important part in the human brain. Not only is memory to a major extent visual, but so is the stream of consciousness. Evidence suggests (Penfield and Roberts 1959) that in man the temporal cortex is involved with the whole record of experience and with the reactivation of that experience.

4. The *frontal* and *parietal lobes* have also expanded in primate evolution, an expansion that has exceeded that of the other parts of the cortex in the latter stages of the evolution of man. It is indeed the enlarged frontal lobes that are always considered characteristic of man, that have given him in the last stages of his evolution a neurocranium that has expanded over his eye sockets, protects them, and forms his vertical forehead. The parietal lobes, which lie behind the frontal lobes, have also expanded, possibly more than the frontal lobes (von Bonin and Bailey 1961); the total effect is very striking compared with other primates (see Fig. 8.5). As has been stated, the actual area of the cortex has increased far more than might appear, since its folding is much more complex than in other animals (Fig. 8.9). In man, 64 per cent of the surface of the cortex is hidden in the sulci (or fissures); in apes 25–30 per cent; in primitive monkeys, only 7 per cent.

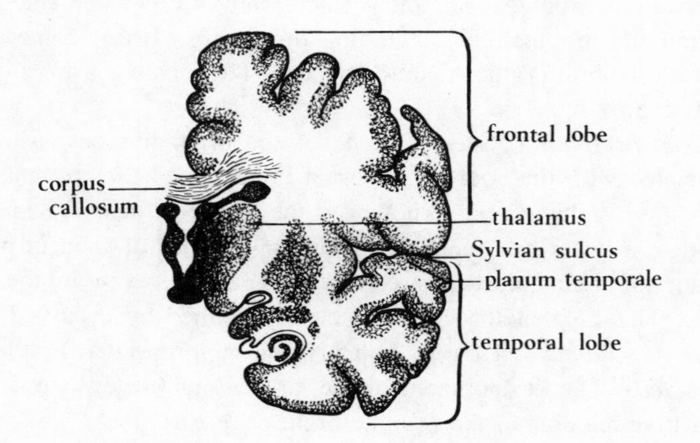

FIG. 8.9. Section through one cerebral hemisphere of man showing the extent of folding of the cerebral cortex.

The function of the frontal and parietal lobes is partly known, partly obscure. Parts of the cortex lying between them are clearly identified as *motor cortex* and *somatic sensory cortex* (see Fig. 8.8). In these areas are located the transmitting and receiving neurons for the motor and *somesthetic* organs (skin sensory receptors) of the body. The "mapping" of these areas has been carried out by electrical stimulation and recording in conscious human beings and other animals (Fig. 8.10).

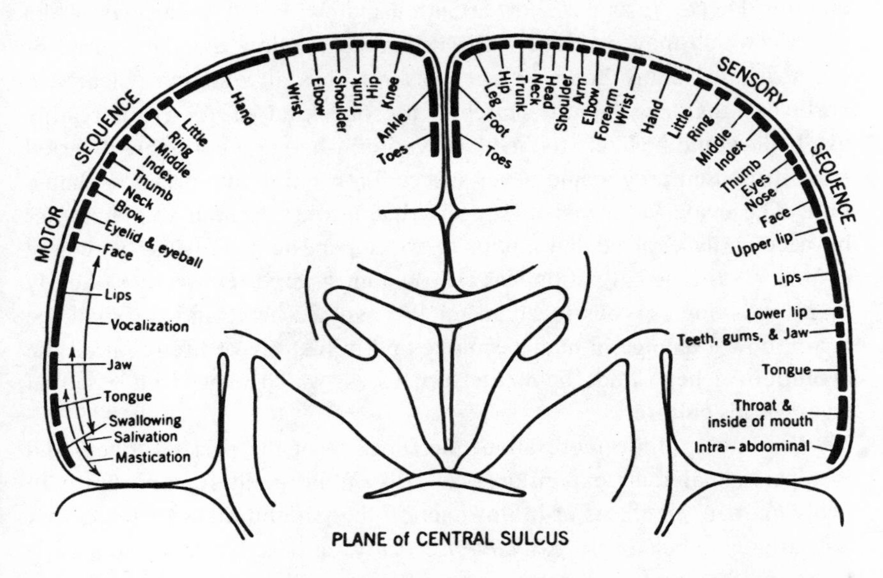

FIG. 8.10. Schematic diagram combining a cross-section of the motor cortex (*left*) and the somatic sensory cortex (*right*) of man, in the plane of the central sulcus. The diagram shows the motor and sensory sequence of representation determined by electrical stimulation of conscious patients. With stronger stimulation, wider zones for a given part might be found, and there is considerable variation from individual to individual. Note that the sensory area devoted to the fingers and hand is greater than that devoted to the much larger skin area lying between forearm and shoulder. (From Rasmussen and Penfield 1947.)

To the front of the motor cortex lie the frontal lobes, also termed the *prefrontal area.* Here we find evidence of a speech center (see 10.6) but in other places no clear reaction can be obtained by electrical stimulation, and the only evidence of function comes from the striking but complex changes in personality that result from its removal in human subjects. Evidence of this kind suggests that these areas are concerned with the maintenance of drive and with its inhibition and restraint. The prefrontal areas appear to make possible both initiative and sustained attention in human behavior.

Man's ability to sustain his attention and concentrate on a particular goal may be an important development in his evolution. Kortlandt (1965) states that as a rule nonhuman primates cannot concentrate on one issue for longer than a quarter of an hour, or, at the most, half an hour. He contrasts this short attention span with that of carnivores like wolves, which may continue to lie in ambush, follow the same prey, or dig a den for many hours or even several days at a stretch. Clearly, a herbivore that persistently searched for one kind of food not readily available would be likely to die of starvation, whereas a carnivore diverted from its chosen prey would never succeed in running that prey to exhaustion. The evidence seems to suggest that insofar as man was a hunter he necessarily evolved the ability to work persistently toward a remote goal, and sustained attention was clearly a most important feature of early man's evolving nervous mechanism. We can see here, and we shall see again, how a change in man's ecology and in his way of life required the evolution of new kinds of mental processes, which proved an essential basis for his culture.

We know a little more about the function of the parietal lobes, and it appears that their expansion was also of the greatest importance to evolving man. In charts of brain function the parietal areas of the cortex are usually labeled the *association cortex,* a descriptive term arising from the fact that the area receives intracortical connections from the three sensory receiving areas: the auditory (in the temporal lobe), the somatic sensory, and the visual (in the occipital lobe), concerned with hearing, touch, and sight, respectively. In monkeys and apes these areas connect directly with the surrounding cortex, so the nearby areas are labeled "auditory association," "somatic sensory association" and "visual association." But in man we find an expansion of the parietal lobe between these three areas to form what has been described as "the association cortex of association cortexes" (Geschwind 1964). This secondary association area is very small in apes but immense in man. All its connections are with the neighboring cortex rather than with the lower centers of the brain.

The association cortex of association cortices is the inferior posterior parietal region approximately delineated in Figure 8.8, and it is of the greatest interest that part of it, at least, seems to coincide with the major speech area (see Fig. 10.8). In this area of the cortex, so particularly human, as von Bonin and Bailey (1961) point out, it seems possible that we may identify at least one place in which integration of the signals from the different sense organs takes place. There is not only some evidence of integration here, but, as in the speech area, there is also evidence of abstraction. As we shall see (10.6), this so-called "speech area" is to be found only on one side of the brain; on the other side there is evidence of neural activity associated with the elaboration of notions of space and body movement. Such elaboration can again result only from integration of sensory input.

The striking result of this expansion of the frontal and parietal lobes in the last stages of human evolution is that they have covered not only the archipallium but even the large and quite newly evolved visual cortex. The simian sulcus, which demarcates the visual cortex in monkeys and apes, is almost obliterated in man, and the visual cortex comes to lie for the most part underneath the brain, rather than on top, and is hardly visible from above. The primary motor, somatic sensory, visual, and auditory areas, which bulk so large on the surface of the cerebral hemispheres of a monkey, are in man crowded down into the sulci.

While the primates have developed advanced sense reception and motor control, it looks as though the specifically human achievement lies in the special and evolved integrative functions of the brain. These functions appear to have developed with the anatomical correlates of cortical expansion in the secondary and association areas and extensive intracortical connections. These increased intracortical connections are particularly apparent in the greatly increased area of cross-section of the *corpus callosum,* the body of fibers connecting the two hemispheres of the *cerebrum* (Figs. 8.9 and 8.11).

One of the most interesting and until recently least understood developments in the evolution of man's brain is the appearance of cerebral bilaterality. In our discussion of the primate visual system (section 3.7) it was mentioned that each side of the brain is associated with the opposite side of the body, and this feature is characteristic of vertebrates generally. Where coordination of input and output is required between the two sides, it is achieved by means of fibers that connect the two hemispheres, the corpus callosum, mentioned above. It has now been established that in human evolution, the redundancy in those brain functions of the two hemispheres that are not directly related to input and output has been reduced. For example, memory storage is duplicated in the cat; learned behavior is stored in both hemispheres. Such brain functions which do not require spatial duplication are present in man on one side only; a "division of labor" between the hemispheres has evolved. In man, speech and word memories involving symbolization and classification are a specialty of the left hemisphere in most people, while concepts of proportion

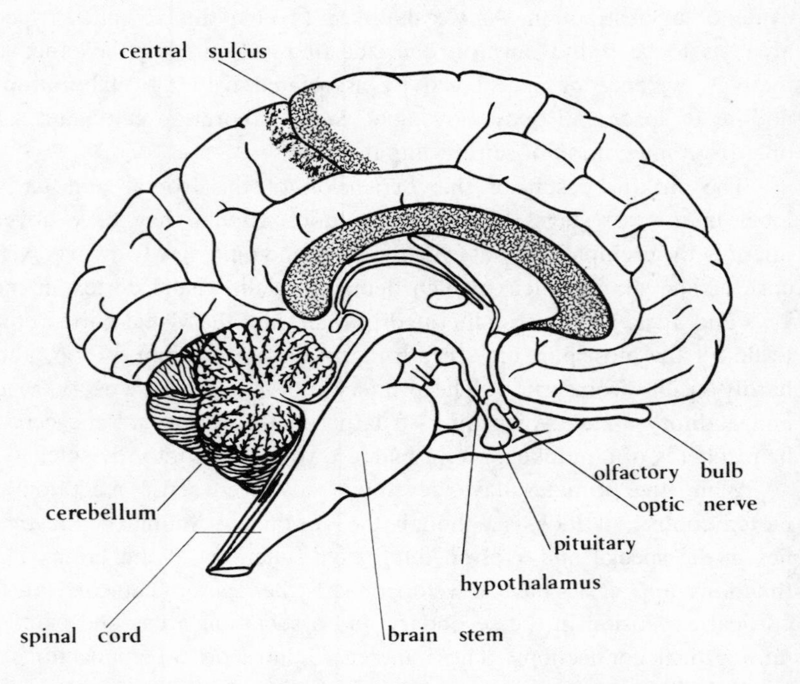

FIG. 8.11. Median sagittal section of the human brain showing (*upper part*) the internal (medial) surfaces of the left cerebral hemisphere, below which lie the transverse fibers of the corpus callosum that connect the two hemispheres (*shaded*). The *central sulcus* separates the somatic sensory and motor cortices (*shaded*, as in Fig. 8.8), which continue on the median side of the hemisphere. The left olfactory bulb and optic nerve are shown at the base of the brain, together with the thalamus, hypothalamus, and pituitary. Like the cerebellum and brainstem, these organs lie medially and are cut in this view.

and relationship in space and sound are stored and analyzed in the other side (Ornstein 1972). This division of function is very extensive in man and possibly unique; by reducing redundancy, it increases the capacity of the brain considerably without further increase in its size. It possibly evolved when the expansion of the brain itself came to an end toward the end of *Homo erectus* times. The differentiation of function is also coincident with differences in anatomy of the two hemispheres, which are no longer mirror images of each other (Geschwind 1972). Thus the human brain has not only increased in size and complexity, but has also undergone a novel kind of bilateral specialization (see also sec. 10.2).

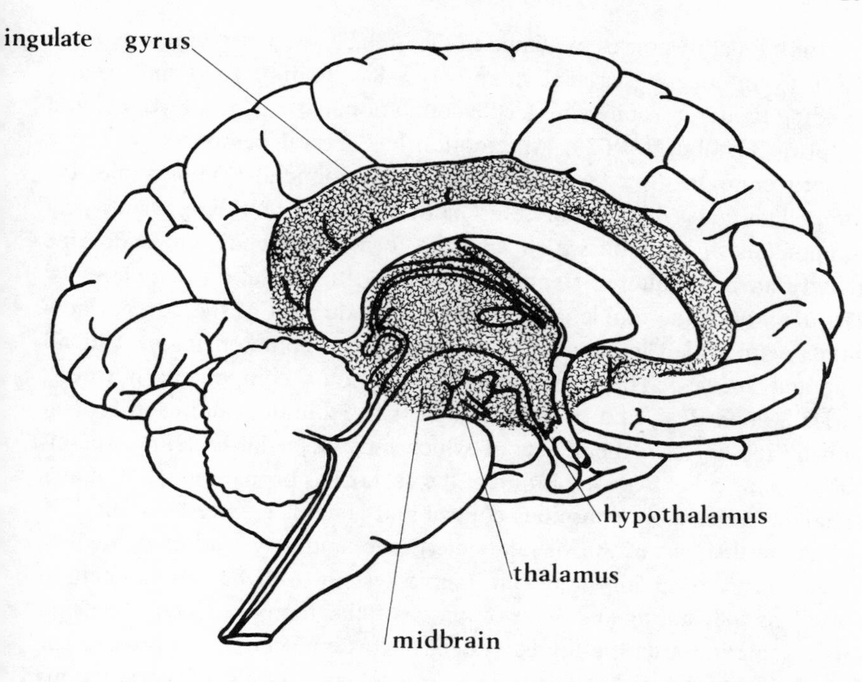

FIG. 8.12. Cross-section of the human brain showing the limbic system (*shaded*). The cingulum is part of the cingulate gyrus. The amygdala and hippocampus are not visible as they lie on each side, on the medial surface of the temporal lobe.

It would be wrong, however, to suppose that it is in the cortex alone that we must look for the neurological correlates of humanity, for the other parts of the brain, the so-called "lower centers," have evolved and expanded in conjunction with the cortex, with which they function in close reciprocal collaboration. Lying along the inner surface of the temporal lobes and toward the lower center of the cerebral hemispheres are a group of structures of extremely complex anatomy. They consist broadly of the *cingulum,* the *hippocampus,* the *thalamus* and *hypothalamus,* the more complex masses of the *basal ganglia,* the *mid-brain* and the *amygdala* (Mark and Ervin 1970), and constitute the *limbic system* (Fig. 8.12). The function of this system is to mediate certain types of sensory input and motor output in a manner recognized subjectively as emotional. This ancient part of the brain constitutes the "emotional brain," and it has been inherited from our early mammal and even our reptile ancestors.

Emotional responses mediated in the limbic system serve motivational purposes in meeting crucial adaptive tasks: finding food and water, avoiding predators, achieving fertile copulation, caring for the young, and so forth (Hamburg 1963). An emotion has several components: (1) a sensory component, (2) a physiological component, (3) a subjective component, and (4) a motor component. It is clear that the physiological components of emotion which take the form of nervous and endocrine activity have not altered significantly during the evolution of mammals. The activity of the limbic system and the production of the adrenal hormones with which its activity is associated, are common features of all mammal species. The expression of the motor component, however, varies a good deal, and in man is subject to inhibition and even repression by the cortex—the effects of which have been far-reaching. In this and in many other ways evolution of the cortex has brought much of man's behavior under more conscious control and given it greater flexibility.

As in the case of the visual, somesthetic, auditory, and other regions of the cortex receiving input, the limbic system itself has an association cortex in the lateral and inferior parts of the temporal lobe. Through this association area the limbic system makes connections with the other association areas, and brings to sight, touch, and sound, the associations of fear, pleasure, rage, and pain, whichever is appropriate (Geschwind 1964). Learning in animals that depends upon trial and error (and is equivalent to the experimental conditions of learning by reward and punishment) results in association being formed between sensory input and limbic response. Experience with a high emotional component will be recorded more deeply in the memory and it is biologically important experience which is accompanied by emotion. The learning and memorizing of biologically useful experience is factilitated at the expense of biologically useless experience and behavior. Thus all animal learning carries a limbic component.

In man, the secondary association area in the inferior parietal region of the cortex allows intracortical connection to be made without a necessary association with the limbic system—which may be only marginally involved. In this way associations between different external events may be recognized and memorized without emotion, and this possibility forms the basis of *detachment*—the specifically human potential for memorizing events which are not of immediate biological importance. Detachment from the limbic system also allows human responses to be relatively free of emotional overtones; it is the basis of cool reasonableness.

Another important structure is the *cerebellum,* which, like the cerebral cortex, has become deeply fissured in evolution (Fig. 8.11). Also like the cerebrum, it can be separated anatomically into a central *medulla* and a cortex (mantle), and its function is the regulation of the action of all skeletal muscles. The pattern of muscular activity is determined elsewhere; the cerebellum, like a servomechanism, controls the strength and timing of contraction of the separate muscles involved in any movement. Its activity is involuntary but of immense importance in any vertebrate with rapid movements requiring accurate control. Clearly, primates of all mammals, depend on the cerebellum to integrate and control their muscular activity. Man has inherited a highly evolved cerebellum from his primate ancestry, which no doubt plays an important part not only in his locomotion but also in the delicate motor control required in toolmaking and in the more sophisticated activities of modern man, such as piano-playing.

From this brief summary we can see that different parts of the brain have expanded at different stages of primate and human evolution. Each expansion has presumably been related to a new functional need of the primate order, and each has played a part in the creation of *Homo sapiens.* But it appears that the expansion of the temporal, frontal, and parietal lobes, which came last, has contributed to make man the creature he is. The temporal and parietal lobes appear to have given man new possibilities of integrating his experience, the frontal lobes have given him new control over his behavior. Bilateral function has increased the effective capacity of the brain and reduced redundancy. The new association areas of the cortex have made possible some freedom from emotion in the learning experience. It is important to stress, however, that in no sense did the brain *cause* the evolution of human behavior. The human brain and human behavior evolved together as necessary correlates in response to environmental change and natural selection.

8.7 *The brain as a determinant of skull form*

THE IMPORTANCE of the brain in the maintenance of life at all homeostatic levels necessitates its protection in a bony case, the neurocranium. In order to give maximum strength for a given amount of material and weight, the neurocranium has tended in evolution to be spherical in form, which is mechanically the most efficient use of protective material. The brain itself is soft and does not appear to dictate greatly the form of the neurocranium, only its volume. But the neurocranium is in fact normally far from spherical, since the actual form is dictated by the development of the various bony appendages to it that we have discussed—in particular, the orbits, muzzle, and jaws, which make up the face. The relative size of the neurocranium and these components, together with the neck musculature, establishes the form of the head; the brain itself establishes only the volume of the cranial cavity.

The neurocranium itself, with which we are here concerned, consists of two layers of bone, the so-called *inner* and *outer tables.* The inner table surrounds the brain itself; the outer table follows the structural demands of the other components of head form (Moss and Young 1960). Between them is a light-weight, webbed bony structure, the *diploë,* and, where the inner and outer tables do not very closely coincide, air spaces are formed called *sinuses.* Any section through the neurocranium shows the diploë, which can be seen in Figure 7.3. According to mechanical principles, the strength of such a structure can be increased, without increasing the weight of the structure significantly, by moving the two outer components apart, and that is what has happened in the evolution of the neurocranium; skulls with a thick diploë weigh little more, yet give more protection to the brain than do thin ones.

The size and shape of the brain is well known in the living primates, and in the fossil forms it can be estimated by making a cast (an *endocast*) of the inside of a complete fossilized skull (Fig. 8.13). It should not be supposed, however, that the cranial capacity is equivalent to the volume of the brain: owing to the presence of associated structures, such as membranes and blood vessels, the human brain occupies only two-thirds of the total cranial volume, and in other primates the proportion of the brain itself may be even less (Mettler 1956). The cranial capacity varies a great deal in modern man and will have varied in more primitive forms as well, so the figures we have, which are based on a few fossils, may be misleading but are unlikely to be very far from correct (Table 8.2).

Among mammals generally, the larger animals have, of course, the larger brains, and a large brain as such is not necessarily an evolutionary novelty, nor does it necessarily indicate special mental power (the brains of elephants and whales may reach 4000 and 6700 cc. respectively). In order to discover whether a brain is unusually large in an

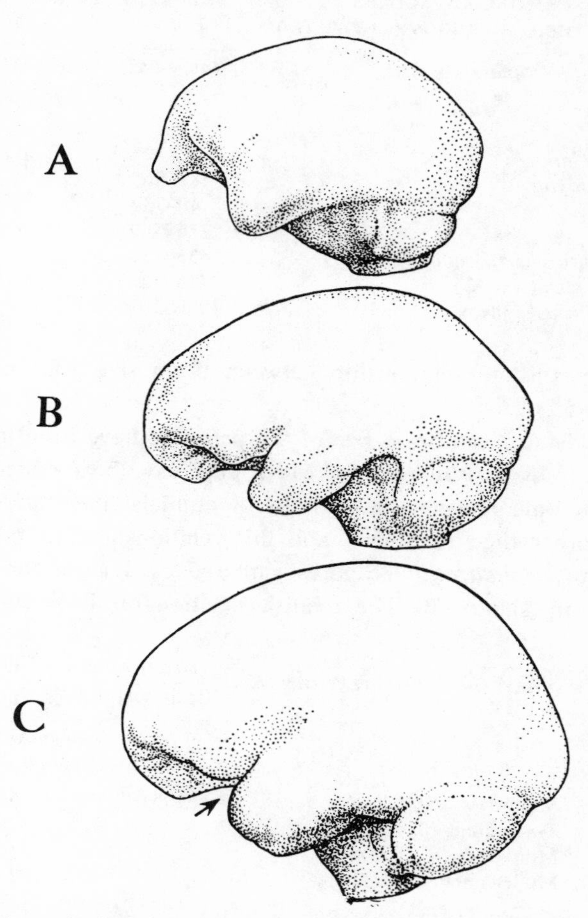

FIG. 8.13. Lateral view of the endocranial casts of the gorilla (*A*), *Homo erectus* (*B*), and modern man (*C*). The arrow points to the notch that indicates the position of the sylvian sulcus between the temporal and frontal lobes. (From Le Gros Clark 1971.)

animal, it is necessary to relate its size to that of the animal's body. Study of evolutionary lineages in different orders of mammals has shown that the brain increases not in simple proportion to the increase in body size but by the power of approximately 0.66 of that increase. Thus, if the body size doubles, the brain will increase to $2^{0.66} = 1.6$ times its original size. It is interesting that this relationship is about the same as that relating the increase of body surface area to body size, so there seems to

TABLE 8.2 CRANIAL CAPACITIES OF SOME PRIMATES, INCLUDING SAMPLE RANGES OF FOSSIL HOMINIDS (AFTER TOBIAS 1971)

Primate	*Range (cc.)*	*Mean (cc.)*
Macaca	—	100
Papio	—	200
Hylobates lar	82–125	103
Pan troglodytes	282–500	383
P. gorilla	340–752	505
Pongo	276–540	405
Australopithecus africanus (n = 13)	435–815	588
Homo erectus (n = 12)	775–1225	950
Homo sapiens (approx.)	1000–2000	1330

be a fairly constant relationship between brain size and surface area (Rensch 1959).

The body size and brain size of the primates have been investigated most recently by Jerison (1955). Much data have been collected showing that in some lineages and groups of animals the brain is proportionately larger than in others, and this relationship has been termed the *index of cephalization*. Figures gathered by Jerison for this index are shown in Table 8.3. The relative position of the great apes and

TABLE 8.3 INDEX OF CEPHALIZATION: $K = \frac{\text{Brain weight}}{\text{Body weight}^{0.66}}$

Faunal Group	*Index of Cephalization*
Mammals (excluding primates) ($n = 108$)	0.06
Three great apes ($n = 35$)	0.29
Monkeys* ($n = 50$)	0.41
Modern man ($n = 50$)	0.92

n = sample size.
Data from Jerison (1955).

* Including *Ateles, Cynopithecus, Macaca, Papio, Presbytis.*

monkeys in this table may at first sight be surprising; it is due to the relatively large bodies of the great apes, and it is highly suggestive: monkeys are nearest to man in their relative development of the brain. We do not know the body weight of fossils and therefore do not know their index of cephalization, but it seems certain that *Australopithecus* was already beginning to show this *Homo*-like increase in brain size, which is not associated with an appropriate increase in body size (such as is found among the great apes). The relationship between brain growth and body growth was no longer the same, for the index of cephalization itself has increased in hominid evolution.

From our knowledge of the dating of fossil remains (see Appendix to chapter 3) and their cranial capacity (Table 8.2), it is clear that the great spurt in brain size occurred in late *Australopithecus* and early *Homo erectus* times. We have, however, noted that there may have been at the same time a slight increase in stature, from a mean stature of about 5 ft. to possibly as much as 5 ft. 6 inches, though the sample size is extremely small. However, this increase, if it occurred, is not enough to lower very greatly the ratio of brain size to body weight, so the increase in the index of cephalization appears to be very close to the absolute increase in brain size. It is significant that the increase in brain size coincided approximately with the beginning of cooperative hunting by hominids and the invasion of northern latitudes (see 7.11).

The great increase in man's brain size, which was occurring during the Middle Pleistocene, has involved increase in all its parts but perhaps most obviously in the cerebral cortex, as we have seen. The surface layers of the cerebral hemispheres are in man deeply folded (Fig. 8.9), so the total surface area of the cortex is far greater than that of the inside of the neurocranium; the surface area of the human brain is four times as great as that of the gorilla's, although the volume has increased by a factor of only 2.5. The expansion in endocranial volume has meant that in man the neurocranium has enlarged at the same time as the facial bones and jaws have been reduced. The effect of the jawbones on the form of the head is therefore less than that of the brain, which as a result in man comes quite near to its "ideal" spherical form. Man is an "egghead" because his brain is the prime determinant of skull form: the face and jaws are a relatively minor appendage. At the same time, the reduced nuchal and masticatory muscles leave their areas of anchorage rather small, smooth, and no longer bordered by crests. With this change in emphasis of head function and structure, we see the rounded skull vault rising above the orbits in human evolution, and this development has been quantified by Le Gros Clark (1971) (Table 8.4 and Fig. 8.14).

When, therefore, we visualize the overall form of the neurocranium in the final stages of human evolution, we see that its shape becomes increasingly spherical: in the horizontal plane the cranium becomes less dolichocephalic (elongated) and more brachycephalic (round-headed). This brachycephalization has been demonstrated in the most recent stages of human evolution as well as in earlier times, and it seems to reflect the

TABLE 8.4 SUPRAORBITAL HEIGHT INDICES* (FOR MEANING SEE FIG. 8.14)

Primate	Mean Supraorbital Height Index	and Source
Pan gorilla	48	Le Gros Clark (1950)
P. troglodytes	47	
Pongo	49	
Dryopithecus africanus	55	Davis & Napier (1963)
Australopithecus boisei	52	Tobias (1967)
A. africanus	61	Robinson (1963)
Homo erectus	63–67	Ashton and Zukerman (1951)
Modern man	64–77	

* A useful discussion of this index is to be found in Rosen & McKern (1971).

continuing trend toward reduction of the jaws and face and improvement in cranial balance. As can be seen from Fig. 8.14, the supra-orbital height index (FB/AB) measures the height of the skull vault above the level of the top of the orbit, and so indicates the height of the neurocranium and cranial cavity relative to the orbital region of the face. From these figures, it is clear that in this character *Australopithecus boisei* forms a group with the great apes (index 47–52) and *A. africanus* forms a group with *Homo* (index 61–77). There is a considerable gap between these two groups which reflects the fundamental difference between the two species of *Australopithecus*. The gap is occupied by the Miocene fossil ape *Dryopithecus*.

8.8 *The face*

THE CHARACTERISTIC features of the human face we owe both to its underlying bone structure and to the form of the soft parts, which now require consideration.

The eyes have been discussed; their sensitive and delicate movements give the face a liveliness that no other feature can impart. The surrounding muscles serve to protect the eye and at the same time add to muscular activity in this region—particularly that associated with frowning and smiling. The nose is a most prominent feature of the face; its form and function have already been described. The muscles that move the nostrils are rather less developed in man than in other primates, as are those of the ears.

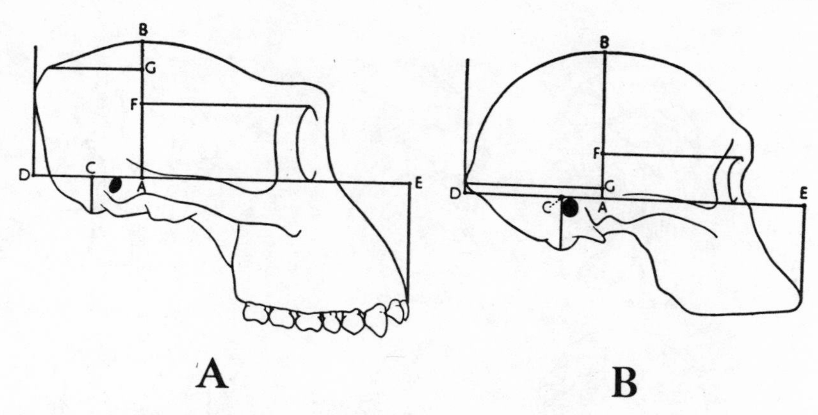

FIG. 8.14. Outlines of skulls of female gorilla (*A*) and *Australopithecus* (*B*), with construction lines to calculate the skull indices introduced by Le Gros Clark: *supraorbital height index FB/AB, condylar position index CD/CE.* (From Le Gros Clark 1971.)

The lips serve both a sensory and a motor function. As a tactile sense organ, not only do they carry a high density of sensory nerve endings, but the surrounding skin also carries hairs, around the follicles of which are wrapped further sensory nerve endings—a condition reminiscent of the facial vibrissae of mammals. This great sensitivity relates to their function in food investigation and is reflected in the area of cortex devoted to them (see Fig. 8.10).

The first motor function of the lips was in eating and swallowing, but even in the lower mammals they play a part in communication, as one of the most mobile parts of the face. In man, the lips have come to have immense importance in speech as well as in facial expression. The area of motor cortex associated with the lips is therefore also very large, as can be seen from Figure 8.10.

The facial musculature as a whole is not only concerned with the masticatory apparatus and the protection of the eyes but also plays a most important part in communication through facial expression. The differentiation of the facial musculature is peculiar to mammals and strikingly advanced in primate evolution. In many mammals, facial muscles are involved in behavior patterns not directly concerned with feeding. Teeth are often displayed to frighten away intruders, as well as in response to pleasure. The faces of the baboon and mandrill carry features that are not merely mechanical but *sematic* (that is, concerned with signaling, with communication within the group); they clearly contribute effects that

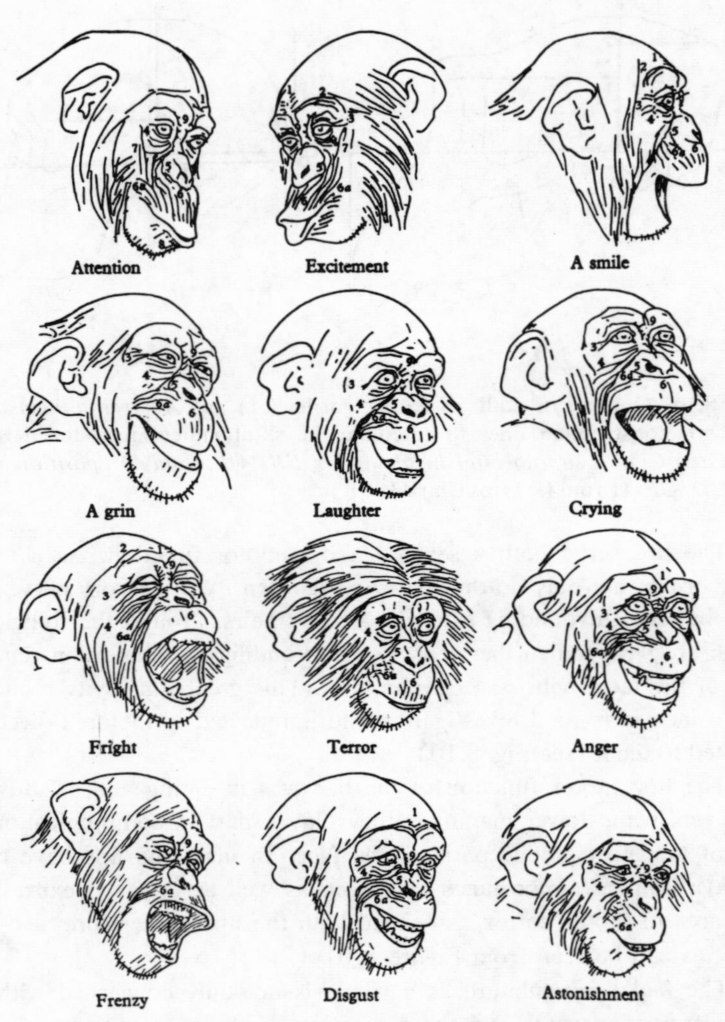

FIG. 8.15. Facial expressions of a young chimpanzee in various moods. Some of the creases are marked with numbers to emphasize that each is by no means confined to one expression. (From Kohts 1935.)

may be imposing and even terrifying in appearance. There is no doubt that natural selection controls not only the mechanical function of the head but also its appearance, through its shape, color, and distribution of hair (8.9), which is important in a social context.

In the higher primates, facial expressions function primarily as signals between members of the same species. Most animals perform a courtship ceremony, which is a signaling device to bring the sexes together with proper timing and stimulation; it is hard to say exactly how large a part facial expression plays in these rituals in primates, but in man it seems certain that it is important. Although dance is often a preliminary to sexual communication, speech and facial expression are paramount.

Studies of primates have revealed that in the monkeys and apes facial expressions are varied and important, and their increasing importance seems to be correlated with a reduction in facial hair. Examples are manifold when photographs are examined (DeVore 1965), and part of the range of expression visible in the face of the chimpanzee is reproduced in Figure 8.15. The origins of facial expression in man are discussed by Andrew (1965); they are an important means of communication in all social primates (see 10.5).

It seems clear then that, in assessing the determinants of head form, we must not omit the soft parts that effect facial appearance and expression. Natural selection operates on the form of the head to evolve a structure with varied functions. The function of the head as an organ of communication in a social primate is not by any means less important than the functions that determine the form of its bones.

8.9 Hair and skin

MAN'S NAKEDNESS is one of his most striking features. The loss of body hair that has occurred in his evolution powerfully suggests his origin as tropical. But his nakedness cannot be the result of his origin alone, since the other tropical primates have retained their hair. The minority of tropical mammals that have lost their body hair, including the elephant, rhinoceros, and hippopotamus, are large animals with thick, heavy skin and subcutaneous fat. The majority of mammals are protected by their hair or fur from the sun's heat as well as from cold in temperate regions. The advantage of nakedness must lie elsewhere; it has been postulated as the need to lose very rapidly the metabolic heat developed in daylight hunting activities, for most carnivores hunt by night or during dawn and dusk. There is no better explanation, and it gains strength from the evidence of the sweat glands given below. It is relevant that we have no evidence of man in north temperate regions before the early Middle Pleistocene, when it is most probable that he had learned to utilize skins as well as fire for protection against cold. The evolution of nakedness has not involved any major evolutionary novelty. Man has about the same number of hair follicles as have the great apes; the difference is that the majority of man's hairs grow very little (Montagna 1965).

The fact that man retains certain areas of hair is perhaps more surprising than his loss of it. The hair on the head gives protection to the brain in tropical climates. Metabolic heat is not generated in the brain and scalp, and there is no advantage in nakedness here; indeed, it is surely a disadvantage. It is possibly relevant that in north temperate peoples baldness is stated to be on the increase (Montagna 1965), which may be because hair serves a less vital protective function in cool climates. Yet it will serve to keep the head warm, and thermoregulation in the brain is very important. There seem to be other factors at work here, especially when we remember that among women baldness is rare, yet they have to survive the same climatic conditions as men. Man, incidentally, is not unique in his baldness; one species of macaque and the orang both go rather bald in middle age, while the uakari monkey has a completely bald head when adult.

In fact, it is certain that the presence and absence of hair on the head, as well as elsewhere, is controlled by other than purely mechanical factors. Hair has been retained around the eyes (brows and lashes), the ears and nose, where its function is primarily protective, but elsewhere it appears to be mainly sematic in function, acting as a signal to other members of the species (Goodhart 1960). A glance at Figure 8.16 suggests that this is indeed the case, for many species of monkeys have striking patches of hair that correspond to our own. In monkeys it is *epigamic* in function—that is, it is used in signaling in relation to sexual dominance. The guenon illustrated here has not only a very smart hairstyle but also a white beard and mustache. The beard is quite common among species of *Cercopithecus,* and a flowing mustache is found in the South American emperor tamarin. Such hair patches, which develop at puberty and are found in males only, are likely to have some social function and, in that case, are related to intermale rivalry; it is not that the female selects the most impressive male but that the most impressive-looking male frightens away his rivals. This hierarchy of dominance in social prestige is based on appearance as much as physical force, hence the evolution of epigamic hair (see 9.8). There is considerable variation among different races of modern man in the development of facial and head hair. Fig. 8.17 suggests that facial hair has an important and recently evolved function in man. Some modern races have much more facial hair than the chimpanzee and gorilla. Shaving is a cultural convention that uses the presence and absence of facial hair as a means of modifying its epigamic function.

Pubic hair and *axillary* (armpit) hair must have a different function, however, since they are found in both sexes and axillary hair is hardly visible unless the arms are raised. These two hair patches coincide with areas of skin containing scent glands (see below), and it seems probable that the hair has some function relating to scent. Goodhart (1960) points out that the scented apocrine secretion (to be described below) is at first odorless and develops a scent only on exposure to air. He suggests that the pubic and axillary hair serves to provide a surface for oxidative reactions and to facilitate the dispersion of scent in the

FIG. 8.16. This *Cercopithecus* monkey (the guenon) has striking patches of distinctive hair, corresponding to man's head hair, mustache, and beard. These patches are important, as signals in sexual competition, just as they are in man (see Fig. 8.17). (Courtesy Belle Vue Zoological Gardens.)

air. However, there is reason to believe that pubic hair also has an epigamic function in men, especially when it grows right up the belly, and it may have been selected as an attractive adornment in women too. There is a useful confirmation of the theory of epigamic hair patches to be derived from warrior dress. The British guardsman's bearskin, the field marshall's plumed hat, and indeed the Highlander's sporran—the ornament of animal fur worn over the pubic region—all serve to accentuate our natural epigamic hair distribution. The most recent to evolve perhaps was the pubic hair, which tends to be sparse in primates and which also would be visible to other individuals only in a bipedal primate.

The glands of the skin have evolved with man's loss of hair. The *sebaceous* (fat-producing) glands are more numerous and active in man than in most other mammals, but their function is still obscure. While they may protect the hair and skin from drying out in the tropical heat, their remarkable distribution on the human body (such as inside the cheeks and on the inner edges of the eyelids) suggests a more complex function that is not presently understood (Montagna 1965).

The sweat glands fall into two groups: the *apocrine* and *eccrine* glands. The aprocrine glands secrete the odorous component of sweat and are primarily scent glands that respond to stress or sexual stimulation. Before the development of artificial scents and deodorants, they no doubt played an important role in human society. In modern man these glands occur only in certain areas of the body, in particular in the armpits, the navel, the anal and genital areas, the nipples, and the ears. Surprisingly enough, glands in the armpits of man are more numerous per unit area than in any other animal. There is no doubt that the function of scent in sexual encounter is of the greatest importance even in the higher primates and man.

The eccrine glands, which are the source of sweat itself, have two functions in primates. Their original function was probably to moisten friction surfaces, such as the volar pads of hand and foot to improve the grip, prevent flaking of the horny layer of the skin, and assist tactile sensitivity. Glands serving that function are also found on the hairless surface of the prehensile tail of New World monkeys and on the knuckles of gorilla and chimpanzee hands, which they use in quadrupedal walking. Glands in these positions are under the control of the brain and adrenal bodies, and in modern man an experience of stress may produce sweaty palms.

The second and more recently evolved function of the eccrine glands

FIG. 8.17. Although clothes have augmented the epigamic function of hair in both men and women, hair styles still retain importance in the establishment of sexual dominance and in effecting sexual attraction in men and women, respectively. Both head and facial epigamic hair patches vary in form and appearance among the living races of man. The paintings were made of individuals from Eastern Europe (*top left*), West Africa (*top right*), New Guinea (*bottom left*) and North America (*bottom right*). (Courtesy Hans Friedenthal and Gustav Fischer.)

is the lowering of body temperature through the evaporation of sweat on the surface of the body. The hairy skin of monkeys and apes carries eccrine glands, but they are neither so active nor so numerous as in man. Modern man is equipped with between two and five million active sweat glands, and they play a vital part in cooling the body. The heat loss that results from the evaporation of water from a surface is enormously greater than that which could be expected to occur as a result of simple radiation. The fact that sweat contains salt necessitates a constant supply of this mineral if man is to survive in a tropical climate.

It has been observed that like almost all mammals, primates sweat very little. Even hunting carnivores, such as dogs, lose heat by other means, such as panting. Sweating has evolved as a most important means of heat loss in man, a fact that is surely correlated with the loss of his body hair. The apparent importance in human evolution of achieving an effective means of heat loss indicates without doubt that early man was subject to intense muscular activity, with the production of much metabolic heat; he could not afford even the smallest variation in body temperature. With such a highly evolved brain, the maintenance of a really constant internal environment was a need of prime importance in human evolution.

Skin color is a rather striking human character, which attracts much attention. The basic racial differences in this character are slight; they are due not to a difference in the number of pigment-forming cells (*melanocytes*) but to the amount of pigment manufactured by these cells. The melanocytes inject pigment into the surrounding cells of the *epidermis* (the outer layer of skin), where it forms a shield to protect the cell nucleus from the harmful ultraviolet rays of the sun. Hair is pigmented by similar injection into the cells of the hair follicle. Active melanocytes may be selected for camouflage (as in the lower primates) or, in hairless man, for protection from the sun's rays. The differences of skin color in modern man are probably due to the advantage that fair skin offers in the temperate regions in allowing the sun's rays to activate the synthesis of vitamin D in the skin. The evidence has been recently reviewed by Williams (1973). The activity of the melanocytes is an example of a character subject to both developmental and genetic homeostasis.

The actual quality of the skin varies probably as much within a single population (if not in a single individual) as it does in the whole human species. Different kinds of skin occur in different parts of the body, and the form of the skin appears to be genetically determined. Modern man certainly appears to possess fine, smooth skin, but this effect is probably produced by its hairlessness and the protection that clothing and modern comforts offer. Again, a smooth skin may be subject to sexual selection, particularly among women (Crook 1972).

The skin is a very important and complex organ of the body. Besides enveloping the body and maintaining an effective boundary and container for the body tissues, it is a strong waterproof membrane and a highly complex sense organ, responsive to pressure, temperature change, and damage. It is also an organ of thermoregulation. It is strong, hard-wearing, and develops claws, nails, and hair, as well as protective pigments. Man's skin has evolved with the rest of his body as an adaptation to his environment.

8.10 The human head

THE HEAD OF MAN can be seen to have been formed by the evolution of a number of separate functional complexes. How these complexes have influenced its visible and internal structure will now be summarized.

1. The evolution of the visual sense, accompanied by the expansion of the visual cortex, was the first factor in the evolution of the human head from that of an early primate. In particular, the large eyes, facing forward for stereoscopic vision, changed the appearance of the face and its bony architecture, which was modified to protect the delicate eyeballs.

2. Recession of the muzzle and reduction of the turbinal bones followed, with a resultant flattening and even hollowing of the face between orbits and jaws. The olfactory sense took second or even third place in the primate's perception of the environment.

3. Erect posture and later bipedalism brought with it a forward movement of the occipital condyles, which, together with the recession of the face as a whole, changed and improved the balance of the head on the spine, with the result that the nuchal crest and nuchal area were reduced.

4. At a late stage in human evolution came the recession of the masticatory apparatus and a reduction of the power and extent of the grinding surfaces involved in the mastication of food.

5. The brain had been increasing in size throughout primate evolution; with the reduction of the masticatory apparatus came a final great increase in brain size, which was doubled in about one million years. This final enlargement filled out the forehead and contributed to the vertical face of modern man.

6. Finally, hairlessness revealed a face that contained not only the speech apparatus but also a musculature that was increasingly used in visual communication by facial expression. With the reduction of facial hair the face became an important means of communication and the seat of beauty.

In this chapter we have tried to subdivide the evolution of the human head into the evolution of a series of separate functional parts and then, as it were, to fit them all together again. It is important to emphasize that, since the head is so highly integrated a structure, the evolution of each functional part must have affected the evolution of every other part. A change in size of any part will have effected the balance of the head on the vertebral column and so the development of the neck musculature. At the same time, the amount of development of any muscle (and in particular the masticatory muscles) will produce tensions and stresses in the bony architecture and in turn will affect its form.

In practice, the interactions of the different functional units of the head are very complex and can be fully grasped only by considerable familiarity with the skulls themselves. The basis of such interaction has, however, been presented here and should form a means of approaching a functional study of the cranial morphology of man's evolutionary lineage.

SUGGESTIONS FOR FURTHER READING

A useful account of the structure and function of the mammalian sense organs and the brain is to be found in J. Z. Young, *The Life of Mammals* (Oxford: Oxford University Press, 1957). A more recent text, which develops some of the ideas referred to in this chapter and is strongly recommended, is H. J. Jerison, *Evolution of the Brain and Intelligence* (New York and London: Academic Press, 1973).

9

Reproduction, the Family, and Social Structure

9.1 Reproduction and the placenta

TO WRITE A CHAPTER on the evolution of the human reproductive processes seems a project fraught with difficulty, for fossil remains tell us nothing of this all-important aspect of our subject. However, by comparing the reproductive processes of human and nonhuman primates, we can obtain valuable insight into the important changes that must have occurred during the processes of human evolution and into their probable function. Because sexual reproduction is the mechanism by which a species both survives and varies in evolution, the form of the mechanism is of special interest to us. In this chapter we shall consider some large changes in the behavioral aspects of the reproductive processes, changes that have influenced the form of our society and the structure of our individual lives. Their study is a fascinating aspect of human evolutionary biology.

Although the development of the human *embryo* (the individual before birth) in the uterus, the process of *gestation,* is part of ontogenetic growth, it is convenient to treat it as part of the reproductive processes. As we shall see, while the whole life-span of man has changed and lengthened during his evolution (9.7), this first phase of his growth is an exception. Table 9.1, which relates the period of gestation to the total

TABLE 9.1 PRENATAL AND POSTNATAL GROWTH PERIODS (FROM SCHULTZ 1956 AND OTHERS)

Primate	*Gestation (in Weeks)*	*Menarche (Age)*	*Completion of Growth (Age)*	*Life-span (in Years)*
Macaca	24	2	7	24
Hylobates	30	8.5	9	30
Pongo	39	?	11	30
Pan troglodytes	33	8.8	11	35
P. gorilla	36	9	11	40
Man	38	13.7	20	75

life-span of different primates, shows that man's gestation is relatively short compared with the period up to the completion of his growth. While his total growth period is twice as long as that of the great apes, his gestation is not significantly longer. However, the human infant is far more helpless than the infant ape or monkey, since he is born at an earlier stage in development; yet, at the same time, his growth rate during gestation is much faster. When we consider (in Table 9.1) the figures for man and, for example, *Pongo,* it is relevant that the average increase in weight of the human fetus is 12.5 grams per day, while in the orang it is only 5.7 grams per day. The evolution of the human placenta helps make possible this fast growth rate and needs consideration at this point.

Together with the monkeys, apes, and some few other mammals (insectivores and some rodents), man shares what would appear to be one of the most efficient types of placenta from the point of view of chemical interchange with the mother. In this *hemochorial* type of placenta, which may be compared with the lemur's more primitive *epitheliochorial* type (Fig. 9.1), there are no maternal tissues—only fetal tissues—separating the fetal and maternal bloodstreams. The blood vessels of the endometrium—the lining of the uterus—are actually penetrated by the fetal vessels, and the former break down to form a spongy, blood-filled tissue, with the result that the fetal vessels are bathed in blood and chemical interchange between the bloodstreams is maximal (Amoroso 1952).

As a result of this intimate association between maternal and fetal tissues, the placenta, at birth, carries with it a considerable part of the endometrial lining of the uterus. The placenta is for that reason termed *deciduate*.

The differentiation and elaboration of the primate placenta promotes optimum conditions for the developing embryo. Not only is the supply of food and oxygen enhanced, but waste products are rapidly removed. Another resultant advantage is the improved transmission of antibodies in the blood, which pass from the mother to the fetus and will immunize the new-born against disease. There is, however, the concomitant disadvantage that, when the blood proteins of the fetus are different from those of the mother, antibodies may develop in the mother's blood, "protecting" her from the alien blood of the fetus, but in turn passing into the fetal blood stream to cause clotting of the fetal blood. This dangerous effect, known as *isoimmunization,* probably tends to reduce genetic variability in blood proteins by removing variants from the breeding population.

A review of primate placentation from the primitive New World

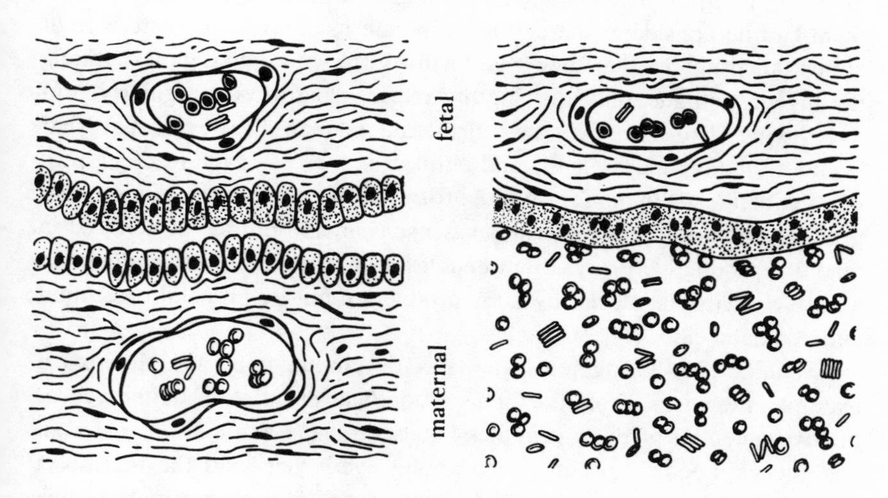

FIG. 9.1. Diagrams of two types of placentae found among the primates. *Left,* the lemur; *right,* the human type, showing the relationship between fetal blood (*above,* corpuscles with dark centers) and maternal blood (*below,* corpuscles with light centers). In the human placenta the fetal tissues are bathed in maternal blood with no intervening maternal tissues. The cell walls of the fetal epithelium have dissolved. (Redrawn from Amoroso 1952.)

monkeys to the Hominoidea shows that the placenta develops at an earlier stage in the course of gestation in the Hominoidea than in the monkeys. The human embryo does not begin its development in the *lumen* (open central space) of the uterus, but at a very early stage it enters the uterus wall and becomes implanted in the vascular tissue (densely filled with blood vessels) of the endometrium. This tissue is already being prepared for implantation from the time of ovulation (see 9.3), even though conception has not yet occurred. The human fetus is able to draw the maximum amount of nutriment from the mother as soon as the fetal circulation develops. In its early development the human fetus has a remarkably fast start and can grow at an exceedingly rapid rate.

9.2 Birth and infancy

THE OVERALL REPRODUCTIVE rate in any species or population is controlled by a large number of factors, but only one of them can be considered here: the birth rate itself. It is characteristic of almost all primates that they give birth to one young at a time. Among the higher primates only the South American marmosets regularly produce twins. The extremely low birth rate in man (one child per year is about the maximum) is the end point of a process of reduced egg production that leads to man from those forms of life, such as the salmon, that produce many millions of fertile eggs per year. Evolution has selected in man a reproductive process that enables him to maintain his numbers in a hostile environment, not by mass production but by prenatal protection and postnatal care.

Schultz (1948) demonstrates this development by the theoretical example shown in Table 9.2. This theoretical calculation of the reproductive potential of three primates is based on the assumption that man and the chimpanzee produce one young every year and the marmoset three sets of twins every two years, with a sex ratio of one male to one female. It shows the very striking discrepancy between potential and actual in the marmoset compared with that in man. Of course, neither animal is able to utilize its full reproductive capacity, but it is clear that a human infant at conception starts out on a relatively safe journey. Its calculated chance of survival (in this example, with the population

TABLE 9.2 THE REPRODUCTIVE POTENTIAL OF HIGHER PRIMATES

Primate	*Period Fertility of Female (in Years)*	*Fertility Commences (Years from Birth)*	*Theoretical Maximum Number of Offspring after 45 Years*
Marmoset	7	3	20 million
Chimpanzee	16	9	408
Man	28	17	64

remaining constant) may be only 1 in 32, but in practice it is greater than that, since the birthrate is never maximal. Observed infant mortality among chimpanzees is about 50 percent and the same figure may well apply to early man.

We have stressed (see 1.3) that competition is a correlate of the evolutionary process and that it is far more marked within species than between species because members of one species require the same food and habitat, while members of different species generally do not. This intra-specific competition is found not only between adults but also between infants and embryos. In animals that produce several young at birth, competition exists among the different fetuses for a limited supply of nourishment and space. Rapid development of the growing organisms may be favored for that reason, as well as for others that we shall discuss. The production of single offspring, typical of the higher primates, is a development of importance because it removes intrauterine competition and allows a slowing-down of maturation in the fetal stages. In man the growth rate of the fetus is striking, as we have seen, but maturation is slow: an important evolutionary novelty is helped by the absence of competition during the fetal stage of growth.

During human evolution, therefore, there has been a slowing-down of maturation and a change in the stage at which birth occurs in *ontogeny* (the development of the individual from egg to maturity). There are a number of factors that affect this stage and require discussion. As a general rule, it is advantageous that the new-born animal be able to run about independently as soon as possible to fend for itself and escape predators, so a certain physical maturity at birth is essential. Operating against this important factor, however, there is another.

In the first place, the fact that the birth canal passes through a bony ring in the mother's pelvis (see Fig. 5.3 and 5.10) means that its maximum size is fixed in any mature female, and this maximum limits the size of the fetus' head (which is the largest part of the fetus' body in its cross-sectional area). Since the brain is the highly evolved controller of all homeostatic mechanisms, including behavior, considerable development of this organ is necessary for any degree of infant maturity. The size of the pelvic birth canal therefore limits the maturity of the newborn. The size of the birth canal itself is positively correlated with the overall size of the mother, so if its size is to increase in evolution (as it has), the overall size of the mature female will also be likely to increase, in spite of the competing advantages of small size (see 9.6). Whatever other factors are involved in the determination of the stage of birth in monkey or man, the close relationship between the size of the fetus' head and the birth canal is a strictly limiting factor (see also 5.2).

We have mentioned that the baby is born at an earlier stage in ontogeny in modern man than in any other primate. It is helpless, and the bones of its body are very incompletely formed. The latter fact, however, proves to have its advantages, for the bones of the skull are not fused together and as a result can survive compression and distortion without damage to themselves or to the soft brain within them. The head is in fact flexible and can be somewhat elongated during birth; it will subsequently and spontaneously regain its characteristic shape.

Birth at an early stage in ontogeny therefore serves to provide a certain mechanical advantage, helping somewhat to overcome the limiting size of the birth canal. It may also be more efficient to feed the growing infant outside the mother's body rather than inside it. But there are other important by-products of this evolutionary change. Experience can influence brain function more directly and at an earlier stage than in other primate species. Whether a great deal of essential learning occurs is questionable, but the infant is soon adapting to its social and physical environment, and this adaptation may determine to a considerable extent the nature of the child's developing personality. It seems clear that it is disadvantageous to risk the infant's exposure to traumata sooner than absolutely necessary—and this is perhaps one of the drawbacks of humanity. To say that the disadvantageous helplessness of the newborn (including imperfectly developed homeostatic mechanisms like body temperature control) has been overcome by the evolution of reliable maternal responses is to assume a quality of motherhood that is rare, at least in western societies.

It is noteworthy that the helpless human infant could not have evolved without a mother able to carry it. The baby baboon, for instance, may be helped by its mother for the first day or two after birth but otherwise clings independently to its mother's body hair during the long daily trips to food and water. Only a bipedal animal is in a position to carry its infant everywhere for at least a year, with the infant unable to hang on by itself. During its period of helplessness the infant is being exposed to the vagaries of the environment—and in particular to the behavior of its parent—rather than remaining fully insulated in the constant environment of the water-filled fetal membranes within the uterus. In man, the environment will start to mold the developing creature at a relatively early stage in its growth, and so environmental influence will be greatly increased.

The influence of the environment upon the young animal begins with its relationship to its mother. The environmental input is in a sense filtered by the mother's behavior and especially her responses to external simuli. The close proximity of the mother's body and the comfort of the milk supply stabilize the first affectional relationship of the young creature. Suckling establishes communication between mother and child; the tactile sense is first in this respect. Close dependence makes possible the transmission of learned behavior patterns. The longer the period of dependence, the longer the period available for such transmission and the greater the long-term effect of the affectional relationship on the family group (see 9.9). Only at sexual maturity does the bond begin to weaken and the young begin to assume adult functions. The lengthened period to *menarche* (puberty) and the completion of growth is shown in Table 9.1. As we shall see (9.7), this lengthened period of dependence on the mother is a factor of the utmost importance in human evolution.

But dependence on the mother has another and more alarming aspect. Among non-human primates, the infant can, after a few days at most, cling by its own hands and feet to its mother's fur on the ventral surface of her body. If momentarily separated from its mother, the infant can at will move toward her and cling to her without her aid. Later, in times of tension and anxiety, it can run instantly to its mother and climb onto her back to ride there in safety. Thus, after about three days, the infant can secure its own survival and initiate and effect the satisfaction of its needs from its mother.

In man the situation is altered in a fundamental way. While the baby can still initiate the satisfaction of its needs by crying, it is entirely dependent on its mother to make an appropriate response. Thus the human mother carries much more responsibility for her child's welfare than does the nonhuman. Unlike the ape or monkey mother, the human mother can choose how to respond to the cry of her infant, whether to satisfy it or not. It is now well established that human babies can be deeply and permanently disturbed by a continued failure by the mother to satisfy the child's needs, and especially by failure to respond to the infant's cry in an appropriate way. In a young infant such response should take the form of intimate skin-to-skin contact between mother and young, which supplies warmth, nourishment and security with the appropriate tactile experience. The proper maternal response does not appear to be fully realized if the mother herself was deprived of this experience in her own childhood.

Thus the evolution in man of infant dependency and its product, maternal responsibility, have had a profound effect on the behavior of modern man. We find that social traditions dictate childrearing practices and thus determine to a considerable extent the most fundamental attitudes of individuals.

9.3 *Female sexuality*

THE HUMAN *menstrual cycle* is of course the human version of the *estrous cycle* of other mammals. The two terms arise from two differences. In the menstrual cycle, the most notable event is the monthly discharge of blood from the uterus; in the estrous cycle, the most notable event is the period called estrus ("heat"), in which the female both desires to copulate and stimulates males for that purpose. Ovulation occurs during the phase of estrus. This period of great sexual activity is found in all mammals except man. *Menstruation* (loss of blood) occurs at the end of the cycle, long after ovulation. Slight bleeding occurs in many primates (Zuckerman 1930), but only in man does it amount to a heavy monthly loss of blood. The term "menstrual" means "monthly," and the cycle does occupy about thirty days in most higher primates.

The full cycle, which is properly called estrous (rather than menstrual), is shown in Figure 2.6 and occurs among all primates. Those which have a breeding season, rather than breeding the whole year round, may experience a period of *anestrus,* when the cycle subsides and sexual activity ceases. Only a few of the prosimians and the Japanese macaques are reported definitely to have such nonbreeding seasons. The normal estrous cycle is interrupted in the mature female higher primate only by pregnancy and by a period for *lactation* (producing milk) following *parturition* (giving birth). The discharge of blood, that is so typical of *Homo sapiens,* follows the stage in the cycle that has quite reasonably been described as "pseudopregnancy." Immediately after ovulation, the lining or endometrium of the uterus begins to develop its spongy blood-filled texture, which is in preparation for the implantation of the embryo. If implantation and pregnancy do not occur, this wall breaks down just as it does at birth, when the deciduate placenta is shed. The increased blood loss in women would appear to be correlated with the degree to which the endometrium is prematurely modified for implantation and the nourishment of the embryo. It indicates the speed with which the human endometrium responds to the hormone progesterone (see 2.4), and the speed with which the embryo will come to require nourishment from the mother and begin its growth.

In many primates, estrus is accompanied by variation in the appearance of what has been termed the "sexual skin"—a specialized area of skin on the female, contiguous with the external genitalia. In these genera, the skin undergoes cyclic variation in its external form and color. The skin is thick and carries an unusually rich blood supply; at estrus the sexual skin swells and may assume bright-red coloration. The presence of sexual skin has not been reported among New World monkeys (except the marmoset) and varies a great deal among Old World forms, reaching its greatest development in baboons (Fig. 9.2). Among anthropoid apes, only in the chimpanzee is it at all developed. The *tumescence* (swelling and thickening) and coloration of the sexual skin at ovulation is due to the increased level of the hormone estrogen in the blood. Its function apparently is to act as a visual sexual stimulant to males, thus insuring that copulation occurs at ovulation. As it occurs among non-human primates, therefore, estrus has three components: (1) Its physiological basis in the hormone cycle and in ovulation, (2) The physiological signal of the female's scent or *pheromone,* and in some species the sexual skin's tumescence, and (3) The behavioral change in the female that makes her not only sexually receptive but even anxious to solicit the male's attention.

Among higher primates we find some evidence of receptivity on the part of the female extended beyond the usual 2–5 days of estrus. Female pigtail macaques have been shown in an experiment to be willing to receive male advances at any time of the cycle, while a female gorilla has been observed to respond to manual excitation in a zoo (Schaller 1963). The same has been seen in a wild female chimpanzee (Goodall 1971). Receptivity by the female primate outside estrus is, however, the exception rather than the rule among primates and is normal only among women. But this rather minor behavioral change is accompanied by the apparent absence of physiological signals associated with estrus.

In modern women there is no sign of sexual skin, nor is there any clear evidence of estrus unless it be a very small rise in body temperature

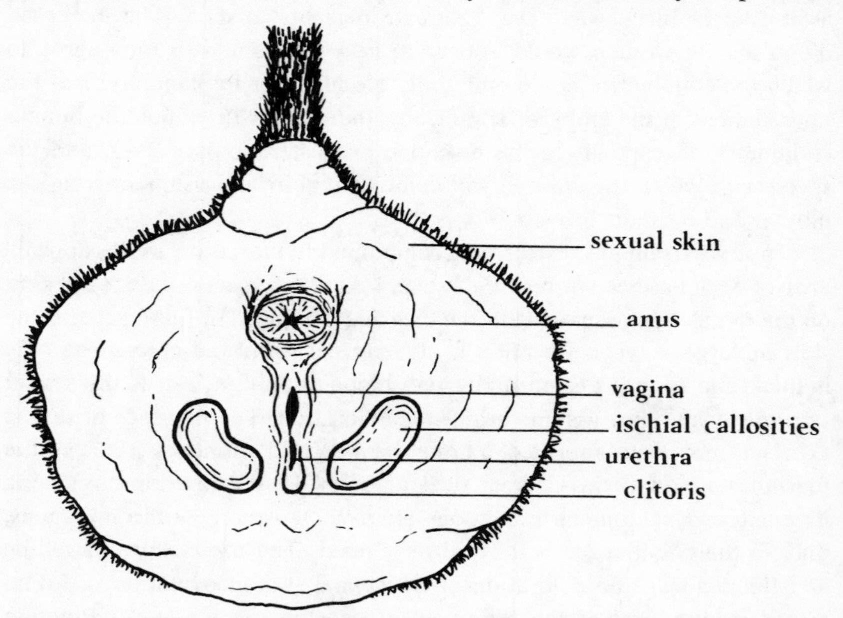

FIG. 9.2. Rear view of a female *Macaca* at estrus, showing the great development of the hairless sexual skin. Note the position of the ischial callosities in relation to the other anatomical landmarks.

that can be detected at ovulation. Though some evidence of a peak of sexual response among women at estrus has been recorded by Benedek (1952) and Udrey and Morris (1968), most of the available evidence suggests a maximum response just before menstruation, not ovulation, so we must conclude that at present we cannot recognize estrus in man. In mammals generally, the phenomenon of estrus interrupts normal behavior, with the result that the female, and in turn the male or males, are almost totally preoccupied with sexual activity. Among primates, estrus has been described as totally incompatible with the nursing of infants, and a period of anestrus necessarily follows the birth of young and accompanies lactation.

The loss of these aspects of estrus in man's evolution has had far reaching results. When a female monkey comes into estrus, the dominant male alone will copulate with her at the peak of her tumescence. When a female chimpanzee comes into estrus, all or most of the mature males will copulate with her. There is almost no element of choice on the part of the female in either group, nor do males show hesitancy in their sexual responses. (Only among chimpanzees do we begin to see the existence of sexual preferences.) Among men, however, any man can theoretically copulate with any woman and any woman with any man, so that what determines who copulates with whom is not a physiological mechanism but personal choice and social sanction. There is a possibility of choice of sexual partners by both sexes that did not exist before, and this opens up the further possibility of male–female friendship—the basis of a more permanent sexual relationship. This may account, at least in part, for the evolutionary selection of females with a long period of sexual receptivity and little or no expression of estrus itself.

The female organs do not otherwise vary a great deal among the higher primates. The protective *labia* (lips of the genitalia, see Fig. 2.4) are rather more fully developed in woman than in most other primates. On the other hand, the sensitive *clitoris* is larger in the nonhuman primates than in women, is exposed and often pendulous (except in the *Colobus* family), and even has a small *baculum* (stiffening bone) in *Cebus* (see 9.4). The small size of the clitoris in *Homo sapiens* is possibly due to one or both of two factors: (1) the absence of any swelling of sexual skin, which would obscure the clitoris and protect it from stimulation, and (2) a slight alteration in function due to a change in position in copulation (see 9.5); in particular, it is so effectively placed for stimulation during ventral copulation that excessive size is unnecessary and could be disadvantageous.

Finally, it is worth noting that the visual epigamic features that the female offers the male have moved in human evolution from the dorsal to the ventral side of her body. In place of the sexual skin, visible from behind, man responds most strongly to features upon the ventral surface of the female, and these epigamic features have spread from the genital region to the face. We find the development of pubic hair, of rounded breasts with their *areolae* (the sensitive skin surrounding the nipples), and of smooth facial skin and lips (Goodhart 1960). From Upper

Palaeolithic times, perhaps 20,000 years ago, we have definite evidence from rock art of man's interest in the epigamic sematic features of the human body (Fig. 9.3). While today many of these features are hidden by clothing, their use, like other aspects of human sexual behavior, comes under conscious control, which, indeed, may be one important reason for the development of clothes. Thus we find voluntary signalling by the female replacing the involuntary physiological signals of estrus. In both sexes sematic features are positioned for face-to-face communication, and the tactile receptors they contain can establish cortical connection to effect erotic response.

9.4 Male sexuality

MALE PRIMATE GENITALIA have been described by Pocock (1925). The evolution of the male sex organs has not involved any remarkable changes, though there are some small differences. The human penis is more like that of the Old World monkeys than like that of the apes; even the gorilla penis is considerably smaller than that of man, which is surprising, considering that a gorilla is about three times the weight of a man.

A small bone or cartilage is found in the penis of all monkeys and apes, called the *os penis* or *baculum*. It extends from the body of the penis into the left side of the *glans* (the enlarged tip of the penis), which it stiffens. This bone has apparently been lost in human evolution, and it appears that the blood pressure in the glans during erection gives sufficient firmness for penetration without its support. The loss of the bone may be related to rather more gentle and careful copulatory behavior among humans.

In general, the length of the penis of Old World monkeys and apes seems to be related to the full thickness of the sexual skin at estrus, which would otherwise serve to keep a short penis out of the vagina. Among monkeys and apes, the females of which do not have a developed sexual skin, the males have shorter penes. Man's long penis is possibly related to the long vagina in women, which with the labia protects the uterus from external infection.

Judging by the presence of a free-hanging scrotum and testes in most of the higher primates, the development of sperm cells is more efficient at a temperature lower than that of the rest of the body. This has been experimentally demonstrated in different mammals, including man. The testes are obviously vulnerable, but apparently this factor is of overriding importance only in the gibbon, which swings through the trees with folded legs (see Fig. 3.8), and whose testes are held close to the body wall and cannot be observed.

9.5 Sexual behavior and copulation

THERE IS LITTLE doubt that the expansion in human evolution of the cerebral cortex and in particular of the frontal and parietal lobes has had as important an effect on human sexual behavior as on other activities. In mammals generally, sexual response depends on reflex behavior stimulated in the first place by harmones in the female and certain olfactory, visual, and tactile sensations in the male. In the evolution of man these subcortical mechanisms have become subject to excitation or inhibition by the cerebral cortex.

On the one hand, cortical excitation increases the range of stimuli that can effect response, so by association they may include many factors of no direct biological sexual significance. The rise of cortical control may also be seen in the behavior of immature individuals. As a rule, immature female animals indulge in no sexual play before the secretion of the appropriate hormones at approaching maturity; immature males occasionally do. The young female chimpanzee does show some interest in sex before maturity, but only among humans will young boys and girls (under the less rigorous repression of Polynesian society, for example) indulge in coital play involving nearly all the elements of adult behavior except ejaculation (Beach 1947).

On the other hand, the cortex may also inhibit sexual activity when a full range of valid biological stimuli is present; response to a real sexual situation may be totally inhibited. This inhibitory action of the cortex is so effective that in women sexual desire related to estrus has almost disappeared. The estrogen still flows as before, but its effect is no longer clearly apparent; instead of three days of sexual mania, sexual receptivity is effected by stimuli that may operate at almost any time.

It has been mentioned that this development has been brought about by a change in the source of the sexual stimulus, which in the female arises not from internal factors (the hereditary hormone cycle) by a lengthening of estrus, but from external factors (such as the advances of men), which may occur at any time but which are associated in the cortex with the arousal and satisfaction of sexual desire. Similarly, males respond not to scent and sexual skin but to features permanently present in mature women and under conscious control. As we have said, sexual response outside the estrous period has been observed in apes, but complete copulation is very rare indeed. However, in this feature, as in so many others, apes exhibit a shadow of the trend which became fully evolved only in man.

A second factor requiring mention in the evolution of human sexual behavior is the adoption by man of his characteristic (but not universal) face-to-face position in coitus—the so-called "ventral" position, with the female supine, the male prone. An approximation to this position is found in all hominoids, whose mating positions are very variable, but, while the female gorilla or chimpanzee may lie on her back, the male will squat between her legs, not lie upon her body. When the ventral position became general among humans is quite uncertain. There are representations of human copulation dating from the late Pleistocene to Etruscan times (Fig. 9.3, *C* is possibly an example), and they seem to show humans behaving like animals in this respect, with entry from behind. However, pictures of animal copulations are equally common, and, in accordance with the magical interpretation of much early art, it seems possible that this animal behavior was specially adopted and represented by man to increase the fertility of the game animals. It would not be unreasonable to assume the adoption of the ventral position as man slowly became adapted to a terrestrial environment. A number of factors must have operated at that time:

1. The development of the buttock (see 5.2), especially the steatopygous buttock (which contains large fat reserves and is found among Bushmen [see Fig. 11.9]), makes entry from the rear increasingly difficult.

2. The adoption of a horizontal resting position in a terrestrial animal would lead naturally to the ventral position in copulation.

3. The ventral position might have been associated with the invention of speech and the development of friendship between men and women and is surely associated with the ventral position of human epigamic features.

Consideration must also be given to the female orgasm, which has been claimed to be a uniquely human experience and which Ford and Beach (1951) suggest may be correlated with the extra stimulation that the clitoris receives in ventral copulation. There is, however, suggestive evidence of strong sexual response in female animals during *intromission* (full entrance of the penis) as well as before. What is perhaps uniquely human is the variability in a woman's sexual response from a regular absence of orgasm to its multiple expression. Again we can account for this by the increased extent of cortical control in sexual behavior generally.

There is no doubt that in some animals the function of the clitoris is to raise the level of sexual excitement to the point necessary for effective intromission. The gorilla is reported (Schaller 1963) to stimulate the female's genital areas both orally and digitally, and in one case even to stimulate the breast. The male chimpanzee has also been reported to stimulate the clitoris orally, resulting in its erection (Ford and Beach 1951), but during copulation it receives no further direct stimulation. However, by this point in copulation a high blood pressure has been developed in the female and the vulva and vagina are tumescent. This blood pressure has been measured in man and dog, and a high pressure subsists throughout coitus (but much longer in the bitch, in which intromission is maintained after ejaculation). Peaks of high pressure are identified by the woman as orgasm, which may be repeated: the experience is accompanied by contractions of the vagina and uterus, as well as many other subsidiary reactions. The female orgasm appears not to be uniquely human, but may involve an intensification of the sexual response seen in other mammals, brought about by increased clitoral stimulation and associated sensations.

It appears that in human evolution the loss of the innate and predetermined sexual drive associated with estrus is, as it were, balanced by the more effective but casual stimulation of the clitoris that occurs not only before but during intromission. In mammals generally, the estrus of the female initiates copulation; the male responds automatically. Among humans, copulation is brought about at any time by mutual interest and stimulation. This gives the human sexes a unique equality not found among other animals. It makes the sexual embrace a balanced experience, which may be an important factor in the structure of human society.

However, perhaps the most interesting result of this complex development was the individualization of sexual relations. Mate selection is of course common among some mammals and birds, but in man the recog-

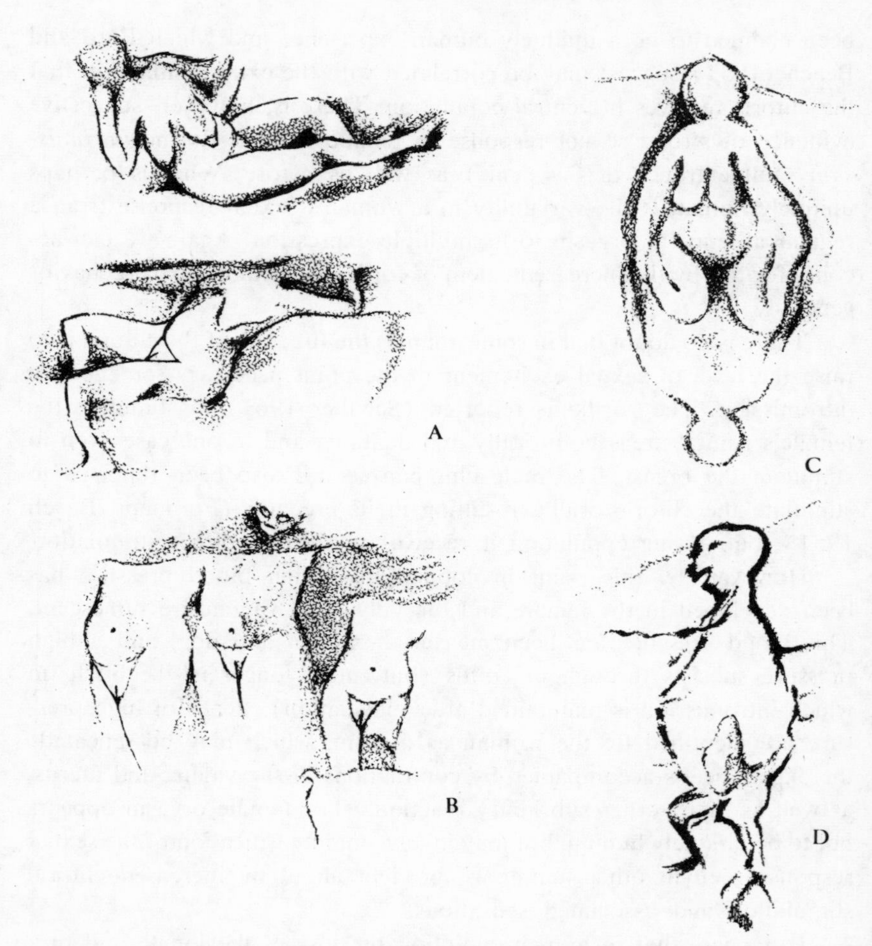

FIG. 9.3. Some of the most ancient known representations of the human figure, of paleolithic age, *A* from La Madeleine, *B* from Angles sur l'Anglin, *C* from Laussel, and *D* from Le Portel—all in France. (Drawn by Rosemary Powers.)

nition of individual qualities goes much deeper. As a result of this and many other factors, more permanent bonds began to develop between man and woman. When female receptivity began to overlap the nursing of children, the man was brought into the mother-child relationship. Thus more permanent sexual ties underlie the structure of the family (but see 9.9).

9.6 Sexual dimorphism and secondary sexual characters

CHARACTERS PECULIAR to the sexes but not directly part of the sex organs are present in the majority of animals. These differences usually develop at menarche and are called "secondary" sexual characters. Among primates, differences in size are common. In baboons, for example, males may have twice the weight of females, while in gibbons the females are the same weight or may be somewhat heavier (Schultz 1963b). When the males are larger than the females, they frequently have larger jaws, larger canines and associated structures, such as sagittal crests.

DeVore (1963) demonstrates that among Old World monkeys these morphological adaptations, which are correlates of large size and appear to serve for fighting and defense, occur primarily in terrestrial forms such as the baboon and gorilla. The defense behavior of the male is less necessary in arboreal genera, since escape is so much easier in the trees and predators are fewer. (A striking exception to this hypothesis is the extensive dimorphism of the orang, in which males are at least twice as large as females. Here we must suppose another factor is at work.) These data mostly suggest that sexual dimorphism has evolved with the male animal's role as defender of the primate troop as well as a result of sexual selection resulting from dominance (see 9.8).

There is, however, in this dimorphism more than survival value in the defense capability of a large male; it is equally advantageous to evolve a small female if food supplies are limited, and the distribution of food supplies in the terrestrial environment could be one factor that limits population density. DeVore and Washburn (1963) have shown that, with female baboons half the size of males, twice as many females can be maintained with a limited food supply than would otherwise be possible. Selection not only favors large males but favors females as small as is compatible with their social and maternal roles.

It might therefore be expected that man, as a terrestrial animal, would show considerable sexual dimorphism. In modern man, however, secondary sexual characters are somewhat limited and the difference in size between sexes is not nearly as great as in gorillas or baboons. Three factors may have operated to reduce sexual dimorphism:

1. In his role as defender, man has been able to turn to weapons. Size in man was not necessarily a condition of survival: a small but intelligent man would be able to outwit his enemy as well as would a more heavily built man. Although men generally are large primates, there is no longer any selective advantage in a heavy masticatory apparatus and large canines. Man's size gives him the strength necessary to use the weapons that he has developed for attack and defense and the ability to run fast when hunting.

2. Fossil evidence suggests that in human evolution individuals may have increased in size during the Middle Pleistocene. What perhaps is most noteworthy in this terrestrial genus is the large size of the females. The large size is perhaps due to two factors: first, the necessity for a large birth canal in an animal with such a large-headed fetus (with cranial capacity averaging 400 cc. at birth) and, second, the very demanding social role of the woman, who must be physically able to raise her large family which involves carrying the baby as well as preparing food and undertaking a certain amount of food-gathering.

3. As we shall see (9.9), there is reason to believe that something approaching a human family may have evolved quite early in human evolution. This development would have meant a considerable reduction in intermale rivalry. Under these circumstances we can posit a reduction in these secondary sexual characters that we believe may have arisen from such rivalry in the establishment of the dominance hierarchy (for example, the epigamic hair patches described in 8.9).

It has proved difficult to determine the amount of sexual dimorphism present in extinct species such as *Australopithecus africanus*. Since we cannot yet determine the sex of fossil bones, we cannot know how much of the known variability is due to variations between the sexes. In theory, a large sample with considerable sexual dimorphism should give us a bimodal curve in the expression of certain characters, but in practice the sample size has not yet enabled us to achieve this degree of precision. At Swartkrans, where the fossils show considerable variability and appear to fall into two groups, they have been attributed to two different hominid species by a number of authors (e.g., Robinson 1963, Tobias 1971, Howell 1967). The differences we see here, which are expressed in degree of robusticity as well as shape, might be attributable to sexual dimorphism, though of this we cannot yet be sure. One thing is clear, however: there is no evidence of a large canine in any skull, nor a sexual difference in the size of the canine teeth as a whole, such as we so commonly find among other primates. However, there is considerable variation in the robusticity of the skull and the development of the masticatory apparatus, and some of this variation almost certainly reflects a degree of sexual dimorphism. Heavier musculature and heavier bone structure suggest a heavier male, which is hardly surprising. What is surprising is that the use of a heavy masticatory apparatus in defense was already a thing of the past. Surely *Australopithecus* must have already been adept at defending himself with sticks, stones and bones.

We do not know, of course, that the ancestor of *Australopithecus* showed sexual dimorphism equal to that of the modern apes—indeed, it is unlikely—and the chimpanzee, for example, is not a very dimorphic species. It seems fairly certain, however, that some reduction in sexual dimorphism was characteristic of evolving hominids (see Table 9.3).

Figures for average heights and weights of English men and women are given in Table 9.3 (see also Fig. 9.4). This size and weight differ-

TABLE 9.3 DIFFERENCES IN WEIGHT ASSOCIATED WITH SEX IN HIGHER PRIMATES

	MEAN WEIGHT IN POUNDS	
Primate	*Male*	*Female*
Baboon	75	30
Gibbon	13	13
Orang	165	80
Chimpanzee	110	88
Man (U.K.)	155 (Height, 67.5 in.)	150 (Height, 62.5 in.)

The figures are approximate, since they are derived from small samples.

ence is the most obvious secondary sexual character that differentiates modern men and women. Besides this feature of sexual dimorphism, we find in modern man a number of minor differences, including such characters as shape of head, shape of pelvis, fat distribution, voice, body hair and skin (8.9), strength, conception rate, infant mortality, growth rate, age at puberty, longevity, energy utilization and many others. Some factors, such as the shape of the pelvis, are directly connected with the reproductive function; others, such as fat distribution and age at puberty, are more remotely related to sexual function. Some may be associated with sexual selection (reduction of body hair and retention of head hair), while others may be merely dependent on one of these factors (such as voice differences). Visible secondary sexual characteristics are certainly selected in modern societies, where a small proportion of individuals who do not satisfy the socially accepted requirements for sexual attraction do not succeed in finding sexual partners. The reproductive rate of such individuals is reduced by sexual selection.

This classic conception of sexual selection, introduced by Darwin (1871), accounts for features selected by the opposite sex. If, among humans, the choice of partner lies with the male, it follows that only the physical characteristics of the female are subject to sexual selection of this kind. In particular, we may mention under this heading the relative reduction of facial and body hair and the overall form of the face and body. Head hair, smooth skin, together with well-rounded buttocks and breasts, are today attractive to men and have no doubt been subject to sexual selection; if they did not increase the chances of a particular female's finding a mate, they did increase the chances of her mating with a genetically better-endowed male, so their joint descendents were perhaps more likely to transmit their genes to posterity. In practice, at least some selection is exercised in most human societies by both sexes, so the situation is more complex than this.

But there are other kinds of sexual selection: there is the selection of the "genetically better-endowed male" to consider. If this selection is not subject to female choice, it is due to the effect of intermale rivalry, which establishes the dominance hierarchy (see 9.8). Referring to animals, Darwin wrote (1871, p. 368): "Many characters proper to the males, such as size, strength, special weapons, courage and pugnacity, have been acquired through the law of battle." As we have pointed out (in 8.9), the establishment of sexual dominance is based on appearance rather than physical force, on threat rather than fight, the effect of which is that males with an impressive appearance (in the form of large teeth, epigamic hair, etc.) pass more genes to the next generation than do those less well-endowed. It is appearances that count. As a result, it is clear that male secondary sexual characters are selected not by females (except to some extent in our own society) but by intermale rivalry, which is reflected in the mating pattern.

Having discussed briefly the physiological and anatomical aspects of human sexuality in comparison with primate sexuality, we must now consider how the reproductive processes fit into the total life-span of the individual (9.7) and the structure of the society in which he lives (9.8–9.10). While sexual relationships are never, among higher primates, the main determinant of social structure, they are always important in that respect, and, at the same time, the socially controlled mating pattern affects the size and nature of the evolving population.

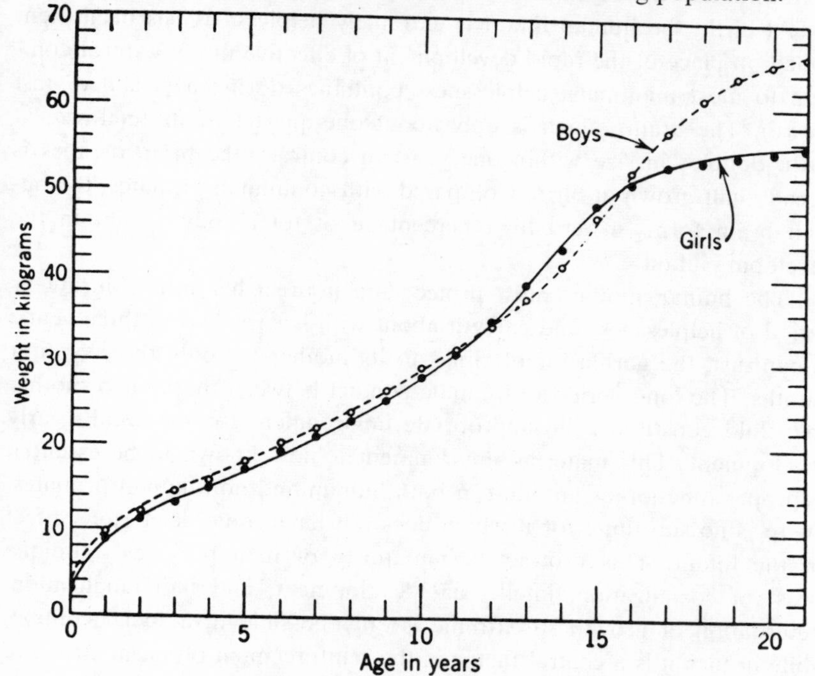

FIG. 9.4. Sexual dimorphism in man: the growth curve and adult size of males and females is clearly distinct. Growth is arrested by the hormones produced at puberty, so the smaller size of women is a function of their earlier physical maturation. (From Shock 1951.)

9.7 The life-span of man

ONE OF THE MOST important and far-reaching differences between man and the nonhuman primates is their difference in growth rate. A slowing-down of the rate of growth and maturity is already apparent in the evolution of the primates as a whole, but the trend is greatly accentuated with the coming of man (Fig. 9.5). The period of infant dependency, which averages just a few months in most mammals, has already been extended to a year in monkeys and two years in apes.

In man it is four to eight years. To these figures we must add a period of cultural dependency, which may continue even beyond sexual maturity.

At birth, the human infant is extremely helpless, as has been mentioned; in place of the rapid development of effective motor control common to most mammals, helplessness continues during a period of fast growth. The brain, which is only about one-quarter of its final size at birth, doubles in size within one year; in contrast, the brain of apes is already half-grown at birth. Compared with nonhuman primates, the human brain forms a very high percentage of total body weight in the developing infant.

The human mother must protect and nourish her infant during its period of helplessness and carry it about with her for two or three years; in contrast, the gorilla infant clings to its mother for only three or four months. The long period of intimate contact between the human mother and child constitutes the appropriate environment for the child's early development. This maternal environment is now known to be essential to proper function as an adult in both human and non-human primates. There is no substitute for it which does not have some deleterious effect on the infant. This protracted immaturity of man provides a unique basis for socialization, intellectual development, and individualization. Retardation of growth so extreme is only possible in a social context, while in turn it is a central factor in the reinforcement of social structure and in the evolution of culture. This very slow growth rate is a unique and quite fundamental character of mankind.

Among girls, puberty arrives in two stages. The first stage is marked by the completed development of the sex organs and the appearance of the secondary sexual characters; it is termed *menarche*. The second stage is marked by the production of mature gametes (ova) and is termed *nubility*. The ages of menarche are shown for some primates in Table 9.1. In girls nubility follows two or three years later (about fifteen years of age), but in monkeys and apes it develops more rapidly.

Most mammals now enter upon their full reproductive life. Among chimpanzees, for example, an estrous female will accept the advances of any number of males in her group, from immature juveniles to dominant males (Goodall 1968). Among many monkeys, however, the female may begin her reproductive life soon after menarche but the male does not effectively do so at puberty. The mating pattern among baboons (which may be typical) is one in which the younger males are permitted by the older dominant males to copulate with the females when they first

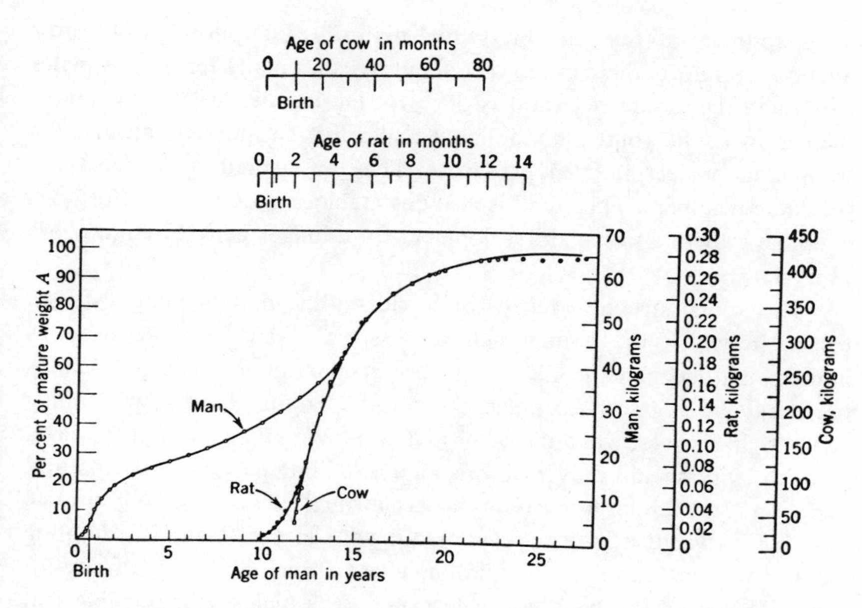

FIG. 9.5. Comparison of growth curves for man, rat, and cow, with the time scales and weight scales adjusted for equivalence. The curve for the cow is simple; for the rat it is an S-curve, and for man a double S. (From Shock 1951.)

come into estrus, but after a few days one or other of the dominant males moves in and acts as consort to the female for the rest of estrus, when tumescence is maximal (Washburn and DeVore 1961). The relationship between estrus and ovulation is not known for certain in different primates, but, in general, ovulation occurs toward the end of estrus, so it appears that the older dominant male, not the younger males, will father the young. The young males appear able to satisfy their sexual drives without serious competition from the older males, but they do not contribute very much to the gene pool of the next generation until they too arrive at a position of seniority. It may be of only marginal importance, but this mating pattern also appears to reduce the incidence of son-mother matings that would result in conception, in the absence of any more direct mechanism for the avoidance of incest.

This delay in the assumption of full reproductive life among baboons is correlated with their sexual dimorphism. The females complete their growth and are sexually and socially mature at about four years, but the much larger males complete their growth and reach social maturity only about four years after reaching sexual maturity. For purposes of reproduction, therefore, there are in fact about twice as many females as males (although the actual sex ratio is 50:50). During the last four years of their growth, the young subdominant males live peripherally around the troop and protect it from predators. They are, suitably, its most expendable members. This trend is also detectable in modern man, for boys continue growth and social development for a longer period than do girls (Fig. 9.4).

This social structure recalls in broad outline that of many African tribes, in which the young men must spend their early maturity as hunters and warriors, while the elders remain in the village. Among present-day polygamous peoples, the young men may have to wait many years to take wives, while the older men are taking their third and fourth. Though by no means universal, this kind of mating pattern allows natural selection a maximum opportunity to act through the males. Only a male who has survived a long apprenticeship both in society and in hunting will be in a position to father children. There is maximum opportunity for a weeding-out of less healthy and less intelligent males. It is possible that natural selection will in this way tend to maintain the evolution of a mating pattern most readily subject to its action, since groups with such a mating pattern will be most readily adapted to environmental change and so gain an increased probability of survival.

We can see, then, how natural selection, acting through society, can delay the full reproductive functions of adult male nonhuman primates as well as of man. Delay in the full reproductive function of adult females is, however, a uniquely human characteristic. It is primarily related to the evolution of culture, for, since one function of human parenthood is the transmission of learned behavior and culture, it is desirable that individuals themselves should have assimilated cultural traditions fully before they are called upon to pass them on to their heirs. This means that in advanced human society education (at home and in institutions) may delay marriage beyond the advent of nubility.

Traditions of learning and of culture are protected by specific (and particularly religious) institutions, and the possibility of cultural transmission is also assured in most societies by various other means. Such protection is assured widely by *rites de passage,* that is, social rites marking stages in the physical and social status of individuals. Initiation ceremonies are a case in point. Such ceremonies mark the attainment of sexual maturity and include a period of instruction in traditional tribal lore.

In Western society, child education is enforced by law, while marriage is prevented by a rule of minimum age. In England, the minimum age has risen from twelve years in medieval times to sixteen years today. Those in our own society whose cultural heritage imposes the need for a long period of education experience a long delay between the coming of nubility and the onset of a full reproductive life.

It is noteworthy that from the standpoint of evolution this delay seems to be nothing new among men, for a period of activity directed to the immediate benefit of the social group has usually preceded a full reproductive role both in nonhuman primates and in humans. On the other hand, the longer period of education now offered girls makes the assumption of parenthood at nubility inappropriate. The delay between nubility and marriage is furthermore greatly extended by the recent startling increase in growth rate, which has brought the age of menarche from fifteen or sixteen years down to thirteen (Fig. 9.6). This trend, which is a complete reversal of the overall direction of human evolution toward a slower growth rate, is presumably due to changed ecological conditions associated with Western society and is perhaps an instance of genetic versatility rather than genetic change.

One clear trend in the evolution of man is the expansion in the actual amount of knowledge and skill that has to be transmitted from one generation to the next, an increase entailing an ever longer period of training and instruction. It follows then that, while the specialist aspect of this burden is taken from the parents' shoulders by the development of institutions of education, an immense responsibility still remains with the parents. The present trend toward earlier marriage in Western countries appears to be a solution neither to this problem nor to that arising from overpopulation. The age of marriage is a cultural trait of great biological importance.

Man's "three score years and ten" takes him through a slow period of development and growth and a long spell of child-rearing into an old age beyond his reproductive years. The end of woman's reproductive term of life comes with the menopause between the fortieth and fiftieth year; her survival beyond that age is an interesting and unique phenomenon of the animal world. Postreproductive primates have been recorded in zoos and very occasionally in nature but are certainly uncommon phenomena. It is not immediately easy to see how postreproductive life could be selected in evolution, since the fate of postreproductive individuals will not affect the composition of the next or future generations

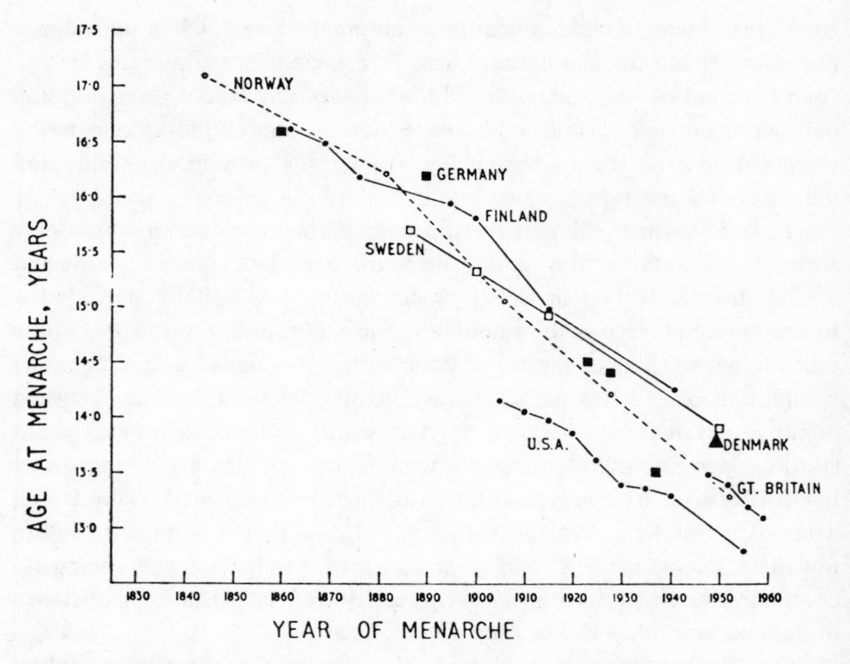

FIG. 9.6. Age of girls at menarche, 1850–1950. This diagram shows clearly the remarkable change in the age at menarche that has occurred in Europe and North America during the last 100 years. (From Tanner 1962.)

in terms of differences in the gene pool. However, there is one way in which a postreproductive individual could affect the birth rate, and that is by any cultural contribution (see 10.1) made to the descendant population in a social context. This possibility requires, however, a well integrated society, one in which old individuals can play a social role other than reproductive. While such conditions could conceivably apply to nonhumans, it is hard to see how a female nonhuman primate could fulfill a socially useful role in old age after her children have matured, since her adult social function lies almost entirely in her reproductive capacity.

There are two good reasons, however, why a postreproductive period of life can be of value to evolving man. In the first place, the slow growth of children means that they will require the presence of their parents until about fifteen years after their birth. Therefore the mother who bears her last child at forty-five will still be occupied in her function as a mother until she is at least sixty. In view of the slow maturing of the children, a postreproductive period will be essential for the survival of the last-born.

There may be an advantage here in widening the gap between the average age of death and the menopause.

A second factor comes with the evolution of culture, and it applies to both men and women. The longer the period of life that must be spent learning, the more valuable to the population will the adult individual be. Among lower animals, in which learning is not an important determinant of behavior, a quick turnover in population will make possible a faster rate of evolution, and so greater flexibility in the face of environmental change. When behavior is innate, it is known at birth, but when it is learned, it has to be laboriously acquired by each generation and cannot be lightly thrown away. Thus man's dependence on learned behavior is a reason for the selection of old age as well as for an extension in the overall life-span. The wise old man will be greatly valued in a social group for his knowledge of hunting, tool-making, and other masculine activities, and in particular for his experience of rare and occasional events, such as flood, drought, locust infestation, and so on. Similarly, the old woman will be valued for her knowledge and experience in childbirth, child-rearing, food-gathering and preparation, and other household arts. In a literate society with good libraries their special value may not be so great in this respect, but generally the old make a unique contribution to the survival of the cultural animal, for they are the storhouse of knowledge and wisdom.

9.8 *Primate social life*

THE HIGHER PRIMATES are social animals, and the adaptive advantages of a social way of life are manifold. The prime advantage possibly lies in the fact that their social life facilitates learning. Clearly, infants can learn from their mother without the existence of social groups, or troops, as primate groups are usually called. But in social groups the amount of learning received through observation by the young can increase and broaden, and young individuals can benefit not only from the experience of their own mothers but from that of their peers as well as the older members of the troop (Hall 1963c). The troop retains a pool of experience and knowledge, which exceeds that of any individual female, and it is available to the growing infant. For example, the identification of edible and inedible foods, the recognition of competitors and predators and what to fear in general, are vital knowledge. Such knowledge can be gained secondhand from other members of the troop and does not require individual experience. Since disaster to one individual is felt as shock throughout the troop, all members may learn from such a disaster. Young troop members may even imitate behavior patterns that have been taken up by the troop in the past and that result from such vicarious experience (Washburn and Hamburg 1965, p. 619). This socially learned behavior is the beginning of culture, and, since it is of great importance in human evolution it will be considered in chapter 10; for the present, we need only note its existence as a fundamental adaptive determinant of the social life of primates.

Beyond its facilitation of learning, social grouping must have been selected in evolution as an adaptation for mutual defense against predators (Chance 1961). The baboon troop tends to be centered on the females and juveniles; the males live peripherally and, as we have seen, are equipped by their secondary sexual characters to defend the troop, and this fact especially applies to terrestrial species. At the approach of predators, such as lions or leopards, the males will always take up a position between the females and the danger and will act in concert to drive away enemies (Fig. 9.7).

Another important adaptive characteristic of the social group is its

FIG. 9.7. A baboon troop moving across country. Dominant adult males accompany the females with small infants and slightly older infants in the group's center. Some young juveniles are shown below the center, and older juveniles above. Other adult males and females precede and follow the group's center. Two estrous females (dark hindquarters) are in consort with adult males. (From DeVore 1965.)

association with a recognizable area of land or forest that contains sufficient space, food, water, and safe sleeping sites for all its members. Behavioral spacing mechanisms operate to bring this about, and it reduces the possibility of over-exploitation of food resources and of conflict over those resources. The area in question—which may be partly shared with other groups, and is certainly shared with other species—is called the *home range.* Within it, one can usually define a smaller area containing resources absolutely essential for survival, which is called the *core area.* Where this area is actively defended against intruders from neighboring groups, it is called a *defended territory.*

Among primates we find an overlapping home range pattern to be most common (Fig. 9.8), characteristic of most monkeys and apes. In areas where the food supply is very rich, and therefore the required area of resources is small, the range may be reduced to a defended territory (as in the gibbons and a few monkeys). In this case, which is the exception rather than the rule for primates, threats are used to maintain control of the territory and its essential resources. In all cases, the relationship between the troop and its environment is ultimately determined by the distribution and density of the natural resources essential to the troop's survival.

Human hunter-gatherers commonly have a home range and core area pattern of territorial distribution. The ranges involved are today very large (Table 9.4), because these people are now generally confined to very poor environments where the population is extremely sparse. There is no evidence of defended territories among them, and it seems unlikely that man was territorial in this sense before the development of agriculture.

The primate troop is also characterized by an internal structure based on dominance, sex attraction, and mother-child relationships. To understand how dominance functions in effecting a stable social structure, it is necessary to consider briefly the part played by conflict in the creation of the dominance hierarchy. Conflict can take three forms: play conflict, threat, and open conflict. The controlled and harmless conflict of young animals at play gives them the opportunity to develop precise motor control and coordination and at the same time to establish their social status. In a similar way, a social structure will develop in a boy's school, both according to seniority and within a group of peers. A dominance hierarchy is established on the basis of strength, agility, and personality, and similar factors appear to operate among

TABLE 9.4 SUMMARY OF DATA ON SOCIAL GROUPS OF PRIMATES (AFTER DE VORE 1963 AND ITANI 1965)

Primate	TROOP SIZE Mean	TROOP SIZE Range	Density per Sq. Mile	Territory in Sq. Miles
Langur	25	5–120	12	3
Papio (Africa)	61	12–185	10	15
Macaca (Japan)	90	4–700	—	6
Orang-utan	3	2–5	—	—
Gibbon	4	2–6	11	0.1
Gorilla (mountain)	17	5–30	3	10–15
Homo sapiens				
Bushmen	20	—	0.03	440–1250
Australian	35	—	0.08	100–750

nonhuman primates. After completion of the period of play conflict, the structure will be maintained by threat, which alone is usually sufficient and is obviously adaptive, since it avoids the risks of wounds and diseases arising from open conflict within the troop. Every species has its own forms of threat signaling—an important form of communication in any social group. An example is shown in Figure 9.9.

Once established, the dominance hierarchy may remain stable for long periods but will of course be disturbed by the introduction to the

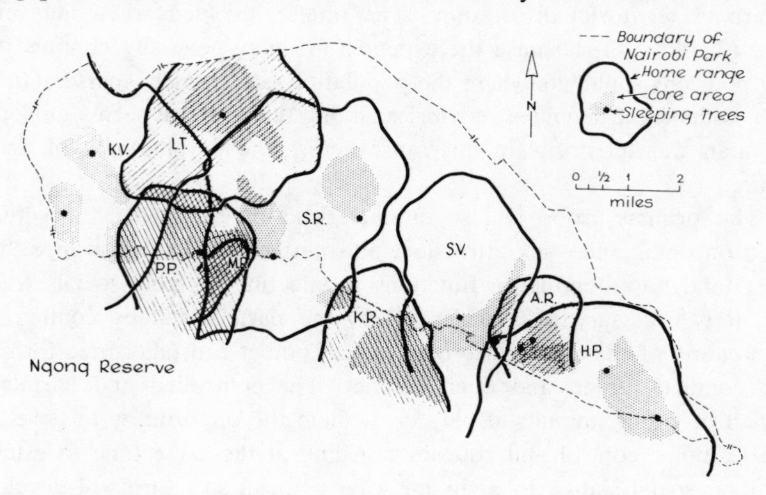

FIG. 9.8. Plan of the home ranges and core areas of baboon troops in the Nairobi Game Park. Note the considerable overlap of home ranges, but the separation of core areas which contain the sleeping trees. (From DeVore 1965.)

troop of a strange animal, in which case open conflict may occur between adults to reestablish a modified social structure. Conflict, usually in the harmless form of threat, is very important in maintaining proper relations with other species and other troops. Where present, defended territory is maintained by threat, and, for defense against a variety of enemies, sufficiently threatening gestures on the part of the males will usually scare away the intruder.

Founded on threat of conflict, the dominance hierarchy maintains

FIG. 9.9. Threat is an important means of establishing and maintaining social structure in primate groups. The actual form taken by the gesture varies among different species, but the "yawn" and lowered eyelids of the baboon are typical. (Courtesy Irven DeVore.)

the structure of the troop and orders its behavior and movement. The dominant males take precedence in the use of limited food supply, so that in a famine not all animals will starve equally but the dominant will more probably survive. Among baboons, but not among chimpanzees, the dominant males will also act as consort to the females at the height of estrus. While the young males determine their status through juvenile play, the females in some species inherit their status from their mothers, though it usually increases with age. However, the female hierarchy is not always identifiable and is never as clearcut as the male hierarchy. It is always rather labile, since each female in estrus comes in turn to occupy the dominant position.

Any observer of primates will be struck by the amount of mutual grooming that takes place, and there is no doubt that it serves an important function in the removal of parasites and in the conservation of salt—both parasites and salt crystals are removed from the fur and eaten. But in some species the amount of time spent grooming is unusually great, and here it carries a second important function. It seems to be a gesture of friendship between individuals, for it is accompanied by a definite reduction in tension. We find grooming of this sort most common between mothers and their young and between siblings, but from time to time females may groom dominant males, so that grooming and sexual contacts intermingle. The action is pacifying, and it acts as a signal indicating friendly submission. Grooming has a bonding function within the troop and acts in a positive way to hold members in affectionate interdependence.

As a result of the early observation of zoo monkeys whose natural behavior was disturbed, it was for a time believed that sexual attraction between individuals was the primary bonding mechanism of the troop. We now know that many primate species have breeding seasons, and that in the Japanese macaque, for example—the most northerly monkey—there is no copulation from four to nine months each year, while stable group organization is maintained without interruption or change; nor is there variation in the strength of the social bond (Lancaster and Lee 1965). Sexual behavior brings about temporary bonds between consort pairs among baboons and some other species, but does not in itself play a significant part in troop bonding as a whole.

Primate troops vary a great deal in size, from two to about five hundred, but the most common range is from ten to fifty (Table 9.4). Variation occurs even within a species in the same type of environment

(Imanischi 1960). Two primates only, the gibbon and orang, adapted as they are to a fully aboreal life, exist in really small family-sized troops. Carpenter (1940) reports that the gibbon troop usually consists of a mature male and female, with one, two, or three growing young of both sexes. This troop has all the appearance of a human monogamous family, but, as we shall see, it bears no true relationship to the human family because its adaptive determinants are different.

9.9 *The nature and evolution of the human family*

IT IS CLEAR that any extrapolations that we may make from studies of primates about the origins of human society are bound to be extremely tenuous and even perhaps misleading. Nevertheless, a comparative approach to this subject is the only one available to throw light on this fascinating problem. Because we believe that different human social structures are adaptations to particular environments, we cannot conclude that one is necessarily more primitive than another unless we have reason to suppose that one environment is more similar to the environment of early man than is another. For that reason we cannot recognize social survivals (that is, social vestigial characters) any more than we can recognize biological vestigial characters, so we cannot draw conclusions about the origin of human society merely on the basis of our knowledge of existing human societies.

We shall therefore use the comparative method with caution, and we shall lay special emphasis upon the equivalence of environmental conditions between the societies that we shall compare and from which we shall extrapolate.

Among living higher primates, the commonest mating pattern is that of promiscuity within a heterosexual group containing numerous males and females. This pattern is found among our closest relatives, the chimpanzee and gorilla. However, groups with only one adult male and a number of adult females do occur among some species, especially the terrestrial monkeys. The gelada, the hamadryas baboon, the patas monkey and the blue monkey are probably the most well-known (Crook 1972). Beyond this we find even smaller groups among gibbons, where one adult male and one adult female form a permanent pair. Finally, and unique among the higher primates, the orang appears to be solitary, since the only recognizable association is that between mother and young. Thus we find a wide variety of mating patterns among the higher primates, some of which are roughly paralleled in human societies.

There are today considerable variations in the human mating patterns, even without considering contemporary experiments in community living. In almost all instances a nuclear family can be distinguished as a one-male unit with one or more adult females (*monogamy* or *polygyny*). In a world survey of human societies the majority are classified as polygynous, but the majority of marriages are monogamous, since polygyny is usually limited to the rich members of a society. The essential characteristic is the one-male group. The one exception to this general rule is *polyandry*—a rare arrangement in which one woman is united with more than one man. It appears to be an adaptation to limited food resources, and is associated with a reduction in the proportion of nubile girls by female infanticide—a most effective if brutal means of population control (Keesing 1958).

How did the present one-male group evolve from an ancestral primate mating pattern? We will begin with a consideration of the chimpanzee pattern because they are our closest living relatives and occupy (in part of their range) a forest fringe/savanna environment which is probably very similar to that occupied by the earliest hominids. It seems highly probable that human sexual behavior and mating patterns originally evolved from a promiscuous pattern such as we find among chimpanzees—a pattern in which most adult males regularly have the opportunity to copulate with most adult females (Goodall 1968).

We know that early hominids moved out from the forest fringe and came to exploit the more open country of the savanna, and we find that most living terrestrial primates have undergone some modification of the chimpanzee pattern. Among the savanna baboons, the dominance hierarchy has come to determine an order of sexual privilege among males and consort pairs are formed by alpha males for a few days when a female is in estrus. Among other terrestrial monkeys, one-male groups are a permanent feature of the social structure. Both these adaptations can be interpreted as appropriate to the resource distribution and predator pressure of the environment in which the species lives (Kummer 1971), and similar factors may have operated on early hominids. It is clear, however, that in the case of modern humans, the evolution toward a one-male group is less advanced than in geladas and hamadryas baboons, let alone gibbons. Where social enforcement of the human marriage bond is weak and divorce is easy, a pattern of serial monogamy or successive consort relationships is more common. In this case dominance may play a part and bonds may be strongly developed, but the evidence is unequivocal that there is no overriding genetic component operating to bring about the degree of permanency in a relationship that we find in some other primates.

The main difference between the unconstrained sexual behavior of humans and chimpanzees is in the regular (but not invariable) appearance of consort pairs, which are only rarely seen among the latter. The absence of estrus and the permanent sexual attractiveness of women make regular pair bonding a possibility, strengthened as it usually is by friendship. Society, in the institution of marriage, comes to reinforce such bonds in order to establish a relatively permanent unit, the nuclear family. The prime function of marriage appears to be to create an intimate and secure social setting which is appropriate for the development of children. It also tends to stabilize and bring order to social behavior in the area of sexual activity and consort bonding.

In fact, we find there is not a great deal of difference between the sexual behavior of man and chimpanzee. These differences can now be summarized as follows:

1) Loss of estrus and the permanent sexual attractiveness of women makes the consort pair more common in humans than in *Pan*.

2) Loss of estrus means the greater possibility of choice in forming consort pairs, so that friendship (social compatability) is a component of pair bonding.

3) Dominance comes to play some part in the selection of sexual partners.

4) The evolving tendency toward relatively permanent one-male units is reinforced by the institution of marriage.

The institution of marriage has been described by Miller (1931) as a device "to check and regulate promiscuous behavior in the interest of human economic schemes." The human family, bonded by marriage, has evolved as an economic and political institution, and we may therefore suppose that it has arisen as an adaptation to economic and political demands rather than through the evolution of permanent pair bonding. With the development of cooperative hunting in the Middle Pleistocene we can suppose that females became finally and fully dependent on males for part of their food supply; this dependence would have become critical during winter months in the north temperate regions.

In turn, males would have become more dependent on the food-gathering activities of women. But the children would have been dependent on *both* for their survival. This interdependence we may suppose also coincided more or less with the evolution of a longer life-span and a more helpless human infant, unable to cling to the hair of its mother's body. In the course of time, any mature female would have had to support two or three dependent young as well as another whose birth was imminent. With the development of social hunting, therefore, we can see the one-male group bonded more closely, with division of labor and economic dependence.

The human family is the simplest social unit with a complete division of labor between adult individuals. It is to the fact that the roles of man and woman are fully complementary that the family owes its continuance and stability. Any interchange of roles, such as we see today in Western society, could threaten that stability. Economic independence for women is bound to have a fundamental effect on the economic basis of the family; its stability will then perhaps come to rest on sexual rather than economic ties.

With the necessity for increasing cooperation between adult males, the unwritten contract of marriage between each man and woman became part of a political bond between descent groups, necessary for social stability when man could survive only through the existence of the social hunt and other concomitant social institutions (see 10.7). The social group was thus enlarged again, not this time as a promiscuous troop, but as a far-reaching political structure uniting descent groups broadly through intermarriage—all brought about by the phenomenon of *exogamy* (marriage outside the kin group).

This final stage in the evolution of the family was certainly dependent on the evolution of language and social institutions, which are discussed in the next chapter.

9.10 *The evolution of exogamy*

IT APPEARS THAT as a general rule different primate troops do not mix and there is limited movement of individuals between them. They are therefore said to be "closed," and since mating takes place between members of a troop they are described as *endogamous*. This type of mating pattern implies that gene flow across a species' range would be slow, and the variability within populations rather limited. It goes some way toward accounting for the multiplicity of species in such a highly mobile kind of animal.

It is perhaps relevant that evolution of the monkeys appears to have been quite slow, at least since the beginning of the Miocene, for *Victoriapithecus,* a Miocene monkey, and *Mesopithecus,* from the Pliocene, are little different from living monkeys. One thing is certain: what the monkeys lost in morphological adaptability, they gained in versatility. Today the behavior of macaques is quite remarkably versatile (Kawai 1965). They can easily follow the movement of the climatic and vegetative belts to which they are adapted across the world's surface; both their versatility and their mobility screen them from climatic and other environmental changes.

In contrast, the fast evolution of the Hominidae may prove to be correlated with the appearance of *exogamy* (outbreeding) and the resultant increase in genetic variability. The rather special conditions of primate social structure must have been broken down, the mating pattern changed, and an immense evolutionary potential released with the usual advantageous characters of *heterosis* (the selective superiority of genetic mixture in individuals).

In modern human society, exogamy is maintained by the evolution of the incest prohibition, which in turn is based on the recognition of kinship. This, in turn, has probably arisen from the deepening and expansion of affectional relationships within the family. These relationships (father-son, mother-son, brother-sister, etc.), though derived from the sexual function of the parents, are reinforced by economic needs and services. The children are economically dependent on their parents for food, shelter, etc., and brothers and sisters have a subsidiary and complementary role to play in family affairs (food-gathering, baby-minding, etc.). Incestuous relations between members of the family would clearly upset the structure of the unit, since sexual and economic roles would be confused. They would also be genetically dangerous in such a closely related group as a family.

One of the simplest types of exogamy is found today among the food-gathering tribes of Africa and Australia. In these peoples, families are usually united by ties of kinship into small local groups of 20–50 individuals. A local group may occupy a well-defined area of land (comparable to the home range of primate groups) and live upon the vegetables and game found in it (Table 9.4). Members of the group hunt together and also have obligations toward one another in sickness and old age. Being kinsmen, members marry outside the group yet within the tribe, so the incest taboo usually applies throughout the group.

While this group of related families has superficial resemblances to a primate troop, the basis of its structure and breeding pattern is totally different, and, as we have seen in the previous section, it has probably arisen as a secondary development from the one-male group. Broadly speaking, it seems that the primate troop expanded into the tribe (an endogamous group), and smaller local exogamous groups developed within this wider structure. If so, we may see the appearance of exogamy as a direct development from the expansion of affectional relationships within a single descent group.

The existence of exogamy requires a social structure that stretches beyond the exogamous group. This factor results in another important difference between nonhuman and human society. Primate intergroup behavior is generally unfriendly and antagonistic; there is, so far as we know at present, no supergroup social structure. Human intergroup behavior can be considerably more friendly, and a social structure generally exists that relates social groups. This broad structure exists as a result of exogamy and the political relationships developed between descent groups that exchange marriage partners. It seems clear that human social life is based on kinship and intermarriage (though this is no longer very striking in modern Western society), for marriage is historically a contract not only between individuals but between different descent groups.

Thus a local descent group may stand in a certain relationship to other local groups of the same tribe for purposes of marriage (as well as *rites de passage,* where such ceremonies exist). Such a relationship involves periodic social gatherings, especially for potential marriage partners to meet. The gatherings commonly involve feasting and dancing, in which nubile girls and marriageable men are enabled to meet in an intimate manner under socially controlled conditions. In many peoples supergroup structures take the form of clans. In such circumstances, however, descent-group exogamy may in turn expand and enclose the whole clan, in which case the social structure will relate different clans within the tribe for purposes of marriage.

On the other hand, a reduction of the extent of the incest prohibition is characteristic of modern Western society as well as of Eskimo society. Because of the special conditions of modern life (a very dense mobile population) and of Eskimo life (a very sparse static population), a more extensive exogamy is either unnecessary or impractical.

9.11 The rise of human society

IN STUDYING the evolution of human reproduction, we have encountered a number of factors that are important in understanding the development of human society. These are now summarized.

1. Improved conditions for individual growth, both prenatal and postnatal, result in a speed-up in growth rate from conception to two years, followed by a slowing-down of growth to maturity. A long lifespan evolves with a postreproductive period, associated with the development of culture.

2. A reduction of sexual dimorphism results most probably from the use of weapons in defense and a reduction in intermale rivalry. Remaining sexual differences are of direct sexual function or arise through sexual selection.

3. Increase of cortical control of sexual behavior is accompanied by an increase in the range of effective sexual stimuli, as a result of response by association. The phenomenon of estrus is lost, and sexual behavior comes under more conscious control. Options in sexual relationships remain open, but social sanctions reinforce the one-male pattern.

4. Highly versatile creatures such as the early hominids can be expected to exhibit striking individual differences in appearance and behavior. Since loss of estrus allows choice in sexual partners, social compatability (friendship) becomes important in the formation of consort pairs. Together with the frontal position in coitus, this emphasizes individualization in sexual relationships, and strengthens pair bonding, which comes to have affectional overtones.

5. Mutual defense and the search for mates are replaced as a prime determinant of social grouping by the search for food. Social grouping is therefore in man primarily economic and only secondarily defensive and sexual in origin.

6. The human family evolves as a one-male group in response to economic needs, reinforced later by kinship structure and the incest taboo. Division of labor arises by age and sex.

7. Economic competition is replaced by economic cooperation, and

marriage is a factor in such cooperation. Food competition is replaced by food-sharing, and status is not so much maintained by greed as by generosity. These changes arise from the move into a new terrestrial environment with a concomitant dependence on mammal meat. Hunting necessitates cooperative group activity, which stabilizes a new, broader social structure.

8. Conflict for dominance is replaced by kinship as a mechanism of social organization (though the dominance hierarchy may persist in all-male institutions). Recognition of kinship is accompanied by prohibition of incest, a universal characteristic of human society; local descent groups are exogamous.

9. Broader social groups develop with an economic and territorial basis, and, with the further expansion of exogamy, social structures spread beyond the immediate descent group to clan and tribe.

The biological needs of living organisms are, broadly, food, sex, and safety. Among primates, these needs are satisfied within the organization of the troop. In the evolution of man we see a new social grouping arise according to this hierarchy, in response to the difficult environment of the open savanna plains, where the need for food was satisfied first by dispersion and later by teamwork. Therefore we find that economic needs determine a new kind of social structure, which is reinforced by political and sexual relations. Modern human society has evolved out of this economic-political-sexual structure, but it has been strengthened still further by other developments. The basic framework that has been outlined in this chapter was enriched and controlled by the rapid development of culture, which has brought man to his present state.

SUGGESTIONS FOR FURTHER READING

Useful reports on the behavior of wild non-human primates include I. DeVore (ed.), *Primate Behavior* (New York: Holt, Rinehart and Winston, 1965); S. A. Altmann (ed.), *Social Communication among Primates* (Chicago: University of Chicago Press, 1967); and P. C. Jay (ed.), *Primates: Studies in Adaptation and Variability* (New York: Holt, Rinehart and Winston, 1968). An invaluable textbook on primate adaptation is H. Kummer, *Primate Societies: Group Techniques of Ecological Adaptation* (Chicago: Aldine Publishing Co., 1971), and a useful review of social behavior can be found in T. Rowell, *Social Behaviour of Monkeys* (Harmondsworth: Penguin Books, 1972). The question of mating pattern and sexual selection in human evolution is discussed in B. G. Campbell (ed.), *Sexual Selection and the Descent of Man 1871–1971* (Chicago: Aldine Publishing Co., 1972). For a well-illustrated article comparing human and nonhuman societies see M. D. Sahlins, "The Origin of Society," *Scient. Amer.* 203 (1960): 76–87. On the social structure of tribal society see (for example) R. Linton, *The Study of Man* (New York: Appleton-Century Co., Inc.; London: Peter Owen, 1964). Examples of tribal kinship structure can be found in A. R. Radcliffe-Brown and D. Forde (eds.), *African Systems of Kinship and Marriage* (London: Oxford University Press, 1950). A very valuable review of sexual behavior among animals and especially primates has been translated into English: W. Wickler, *The Sexual Code: The Social Behavior of Animals and Men* (Garden City: Doubleday, 1972). For a useful discussion of female sexuality, see M.J. Sherfey, *The Nature and Evolution of Female Sexuality* (New York: Random House, 1972).

10

Culture and Society

10.1 Culture

CULTURE HAS BEEN described, discussed, and defined at great length by social and cultural anthropologists. On the other hand, it has received very scant consideration by biologists, for it has been generally assumed to be a uniquely human phenomenon and thus outside the field of biology. But human biologists cannot avoid at least some consideration of the origin and function of this most all-pervading human attribute.

When we try to discover what culture is, we find it almost always defined as a purely human possession; Dobzhansky calls it a species character of *Homo sapiens* (1963). Most cultural anthropologists define culture as specifically human and at the same time dependent on symbolic thought or language. Kroeber and Kluckhohn (1952, p. 181) define culture as consisting of "patterns explicit and implicit of and for behavior acquired and transmitted by symbols, constituting the distinctive achievement of human groups, including their embodiments as artifacts." L. A. White, too, in a fascinating discussion, states culture to be "dependent upon symboling . . . trafficking in meaning . . . which cannot be comprehended by the senses alone" (1959).

Evolutionary biologists must approach the definition of culture differently. Our definition must not be designed *a priori* to distinguish modern human society from nonhuman primate society; it must try to define in its very simplest form that phenomenon which today is recognized to be what we call culture so that we can then see whether it can be found elsewhere in the animal kingdom and what function it performs. Let us turn first to E. B. Tylor, the father of cultural anthropologists, who, as one of the first investigators, defined man's culture as "capabilities and habits acquired by man as a member of society" (1871). More recently, Keesing (1958) has defined culture as "the totality of learned socially transmitted behavior" and Oakley (1963) has defined it as "the sum total of what a particular society practises, produces, and thinks." These definitions may seem very broad, yet at the same time they are more precise. We are not concerned, like White, with defining a discipline or, like Kroeber and Kluckhohn, with trying to grasp an abstraction. We are concerned first with what men and animals do and how they come to do it. Secondly, we must examine how human behavior and human society differ from animal behavior and animal society, so as to discover what social novelties, if any, appeared during the evolution of man.

For us, culture must be defined along these lines, as a kind of behavior acquired in a particular way: as the totality of behavior patterns of a social group of men or animals that are passed from generation to generation by learning. Thus culture is socially determined, learned behavior, and it clearly is not a specifically human phenomenon. Cultural behavior patterns are the outcome of versatility; they arise from behavioral flexibility and a capacity for observational learning. It seems clear that behavior patterns that become part of the culture of a group must be adaptive, though in what way is often not clearly apparent. But there is no doubt that cultural behavior is an evolutionary development of the greatest importance, since a new adaptive behavior pattern developed by one member of a group may be quickly taken up by others through observation, and such a behavioral response to the environment on the part of the population is much faster than that made possible by genetic mutation.

From the reports of Japanese monkey behavior already referred to (Imanischi 1960, Kawai 1965), there is no doubt that cultural behavior patterns can be easily developed by social primates. During the period of careful observation of these monkeys since 1952, they have been observed to adopt a number of quite novel kinds of behavior, which have been invented by one individual, imitated by other members of the group, and then passed from generation to generation (with, interestingly enough, varying degrees of success). Food-washing and swimming in the sea are both patterns of behavior that were invented and became part of the troop's culture during the period under study. Similarly, it seems very probable that the use by chimpanzees of sticks to extract termites from their nests, reported by Goodall (1968), is a cultural behavior pattern. Haldane (1956) has reviewed the considerable evidence for cultural behavior among other mammals and among birds.

Culture has also been described as a "non-genetic adaptation," and as "an organ of human behavior" (Huxley 1958). But the first description will apply to all learned behavior, the second to material culture only. Culture is more accurately to be seen as a set of behavior patterns that are learned by members of society from members of society and that are adaptations of their society to the environment. Culture is a product of society and a property or character of society. Through culture, "behavior has reached a supraorganismal level" (Huxley 1958).

If cultural behavior is to be passed around a social group, its spread must depend on observational learning if not instruction (1.6). It follows, then, that culture will be limited to social animals that can learn in this way; but, given this possibility, any novel behavior pattern with a selective advantage will, under appropriate circumstances, be incorporated into the behavior of the social group in which it arises. Culture will spread by diffusion.

It is characteristic of culture that cultural techniques or traditions are passed down within an evolving group, and during their passage through time they may also change in accordance with the needs of the social group that possesses them and with their changing environment. Cultures, in fact, "evolve," not according to the mechanism of organic evolution, but independently of it. The reason for the speed of cultural change, or "evolution," should be clear. While adaptive organic mutations (that is, new genes and gene combinations) may take generations to become established and spread through a population, new cultural adaptations may spread in one generation; they may be acquired very quickly and passed on by learning. Cultural adaptation, being fast, is also flexible and will allow numerous specializations for different ecological niches.

Huxley (1958) recognizes three aspects of culture (though they are not clearly distinct), as follows: (1) overt acts of behavior, such as rituals (*socifacts*), (2) material results of overt acts of behavior (*artifacts*), and (3) potential behavior, in the form of assumptions, ideas, values, intentions, etc. (*mentifacts*).

"Socifacts" are certainly the simplest and are the kind of culture not uncommonly found among animals, as has been described. We must also include here a considerable amount of primate behavior, including some elements of social structure that we know to be subject to modification. It is clear, too, that there is no hard and fast line between socifacts and artifacts. The manufacture of artifacts is only a particular sort of behavior, involving objects taken from the environment and an appropriate organ of manipulation. Birds use thorns, stones, and twigs as tools, and so do primates. Chimpanzees prepare their sticks for termite extraction. Men fashion objects with great skill to achieve a particular end. Such objects are termed "artifacts" and are indeed a clear-cut and remarkable product of human cultural behavior. Man is often defined as a toolmaker and therefore as the cultural animal, but toolmaking is really only one sophisticated kind of behavior. It is not really different in kind from other overt behavior except in its material product.

Evolution is opportunistic, and any novel behavior pattern with a selective advantage will, under appropriate circumstances, be incorporated into the behavior of the evolving population. Such behavior is necessary to develop and maintain material culture, for the use of external objects as tools increases the effective exploitation of the environment and decreases the danger of predators. The beginnings of material culture arose when objects were fashioned from raw materials—when sticks were intentionally sharpened and stones carefully shaped.

Techniques of toolmaking are not typically learned by trial and error at many different times but are invented by intelligent behavior on rare occasions and then transmitted as learned behavior from one generation to another. Children learn from their parents not merely by observation but also by instruction involving speech. Very simple cultural traits were certainly evolved without speech, but the real cultural advances that we associate with the Hominidae must almost certainly have awaited an adequate language. Speech alone made possible levels of abstraction that were necessary for the development of material culture and human society, for it is an accomplished kind of communication between individuals that makes possible a material culture and a complex economic social structure and maintains its integration. Thus it is that social anthropologists such as L. A. White insist on the symbolic nature of culture, because human culture is indeed dependent upon symbols.

Man's immense capability for culture, which is dependent on his increasing educability (Dobzhansky and Montagu 1947), is clearly a character of great selective value, and it is surprising that its extensive development is unique in the animal world. Capacity for culture is, of course, a character that is as much a part of man and as much a part of his heritage as his hand or his pelvis, and its evolution is interdependent with all his other attributes. It is, however, possible to see how man's capacity for culture may have arisen and to examine the essential prerequisite for its evolution.

Since the transmission of culture depends upon the facilitation of learning in a social context, such learning, which is typical of the social primates and which in turn maintains social integration, is a *sine qua non* of culture. Behavior patterns transmitted in the social context of early primate society were the forerunners of the human behavior patterns that we call culture and, surely, were the origin of man's capacity for culture. Culture itself was a product of man's inventive genius, which was in turn the product of intelligence. We find the clear traces of culture arising in the tightly integrated societies of primates, where what has been described by Etkin (1963) as "social sensitivity" is very well developed, for social cues are cues to behavior, and social stimuli control individual response. It is in these tightly knit social groups, in which there is ample opportunity for imitative learning, that the transmission of acquired cultural behavior grows, and in which more complex behavior only awaits man's inventive genius.

It is a very important characteristic of material culture that in relation to its creator it is not only an extra-bodily organ but also part of the environment, and it is a part of the environment that affects man closely. When man's culture "evolves," therefore, his environment changes, and a new adaptation of the organic body will be selected (see Fig. 10.1). We therefore find man living in a fast-changing environment and adapting to it with a fast-changing culture. No sooner does the culture adapt than the environment has changed, and further adaptation is selected. In face of this ever increasing rate of environmental change, man's physical evolution runs far behind (Dobzhansky 1963), which suggests that man is ill-adapted genetically to his environment. In at least one sense that is true: without the barrier that his material culture forms between man and his environment, he would be unable to survive. But it is also open to question whether man is fully adapted in any sense, for the rate of change in his environment may appear to be outstripping even his cultural adaptations (see 11.10).

Huxley cites "mentifacts" as an aspect of culture; they are the assumptions, ideas, and values that determine man's conscious behavior.

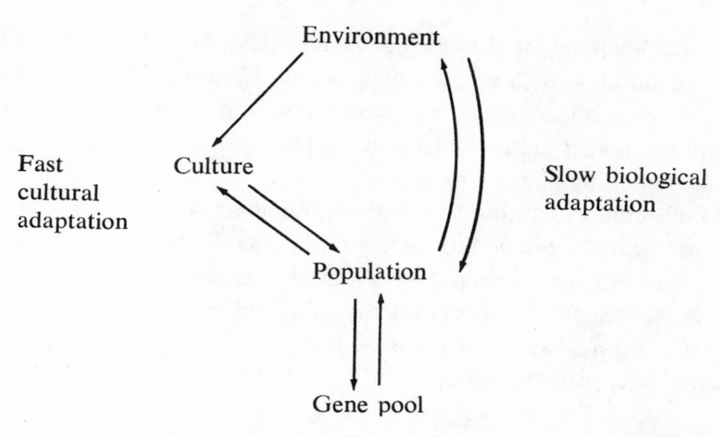

FIG. 10.1. Systemic interaction of population, culture, and environment.

Belief in God, belief in the desirability of moral rectitude, belief in the value of the individual, are the kinds of ideas that form the basis of modern society. Ideas, which form the basis of religion and ethics, are the most sophisticated aspect of culture, for they involve conscious intelligent thought, a uniquely human capacity. This unique characteristic of mankind is of course adaptive, and it too must evolve, like other cultural traits. In general, mentifacts are by nature rather static and stabilizing in their effect; on the other hand, they can be lethal if they do not retain some degree of flexibility.

Surprisingly, one aspect of the subject of this section is well documented in geological deposits. While the remains of early man are rarely preserved, at least one kind of cultural product almost invariably survives: stone tools. The development of stone tools, which have been described as "petrified ideas," while not necessarily correlated with the development of man's culture as a whole, does give us at least some guide to his cultural progress. And this progress is very astonishing, for it is not constant, nor is it irregular, but it is exponential: the rate of change increases regularly.

The ever increasing speed of technological progress is evident. Developments during the three hundred years since the beginning of the industrial revolution have altered the face of the earth. Some six thousand years ago the use of metals, and before that the development of agriculture, had already changed man's environment. Even the rate of progress in agriculture and metals, though slow, was very rapid compared with the evolution of the stone-using cultures that occupied perhaps three million years. The increase of man's cultural resources was at first exceedingly slow, a painful growth of idea and technique. But, unlike evolutionary progress, cultural progress was cumulative, and both the quality and quantity of man's cultural assets have grown and have now surpassed the understanding of any single man.

Man's culture is, however, not a thing apart. It is dependent on the evolution of both speech and society itself. I shall attempt in this chapter to discuss both the internal psychological correlates of culture, such as conceptual thought, memory, intelligence, and speech, and the external correlates, such as social structure. Finally, some of the determinants of human behavior can be summarized.

10.2 Percepts and concepts

IN OUR CONSIDERATION of the evolution of man and his culture, it is necessary to refer to what has been described as man's unique mental* characteristic: conceptual thought. Treatment of this subject is difficult because it involves the discussion of perception, and neither the formation of percept nor that of concept is yet well understood. What follows, therefore, is a personal and immensely simplified account of a very difficult subject, but one that cannot be omitted in any treatment of human evolution.

A *percept* may be considered to be the mental image of the external environment. Just how this mental image comes into existence is not yet known, but it is clear that it is based on two kinds of information; one is the input from the senses, and the other is the memory of previous experience. It can be shown with ease that a given percept is not necessarily a true if limited interpretation of the present environment but that it is strongly influenced by previous experience. It seems clear that in the growth and development of each individual mammal, motor investigation of the environment plays an essential part in building up its appropriate perceptual interpretation. There is no doubt that the content of perception varies beyond our understanding among the different kinds of animals, and we cannot conceive the world of a dog, for example, or of a nocturnal prosimian. Our own perception of the outside

**Mental* is the adjective of *mind*, the functional and subjective aspect of the living brain.

world has evolved with the primates, to which the motor component is of fundamental importance in understanding the environment. Spatial relationships must be first experienced to be later perceived, and we inherit from our primate ancestors a very spatial—a very three-dimensional—perception of our environment.

From these considerations follow two conclusions:

1. If primate perception depends so much on motor investigation, the quality of perception will depend on the extent and refinement of the motor investigation. Primates, more than all other mammals (except bats), live in a three-dimensional world; their eyes are stereoscopic, and their movements are in all three planes of space. They must have very precise ideas of spatial relationships, for aboreal locomotion involves a knowledge of space far greater than that which may be necessary to ground-living forms. The prosimians as well as the higher primates have no doubt come to perceive the environment as a three-dimensional rather than a two-dimensional pattern. The integration of spatial data from the senses to form a composite perception of the environment has clearly gone further among primates than among other groups of animals. Visual, somatic sensory, and auditory inputs will be analyzed and integrated with the motor and proprioceptive patterns. Such integration appears to take place, at least in part, among the association areas of these different sensory inputs on the parietal lobe of the brain (see 8.6).

2. Our second conclusion is that the memory component of perception is of fundamental importance and must almost inevitably come to include not merely an experiential record of events but some generalizations about spatial relationships. These generalizations are immeasurably expanded and deepened by the information obtained about the environment as a result of the manipulation of objects (see 6.6). By manipulation, the higher primate can extract an object from the environment, free it, as it were, from spatial implication, and build up a perception of it as a discrete object, not merely as part of a pattern. In time, the primate will come to perceive the environment not only as a three-dimensional pattern but also as an assemblage of objects.

Man, though a primate, is no longer an arboreal herbivore, a fact that has played an important part in the evolution of his perception. Kortland (1965) records that if food or any other familiar object is given to zoo chimpanzees at a place where they are not accustomed to receive it, the apes are inclined to react to it as though it were something entirely strange, and they may even refuse to consume their meal. For apes, things tend to lose their identity when they are displaced. Among carnivores, on the other hand, a displaced food tray or other object usually causes no problem, as every dog or cat owner knows. As Kortlandt puts it, the perception of apes carries a certain quality of "thereness," that is, its position in relation to other things, for among these primarily visual creatures the patterning of the environment is still the basis of their perception. Carnivores, on the other hand, identify food primarily by scent, and no strong visual pattern contributes to their perception of it. Visual perception in primates functions to identify static objects that comprise their food; in carnivores olfactory perception functions to identify mobile objects that comprise their prey.

When man began to hunt, his perception evolved accordingly. Using his prime sense, vision, man evolved the ability to identify objects on the move without reference to their relationship to the fixed part of the environment; he saw them as totally separate from their environment. Here was a fundamental improvement in perception and something novel among land animals: a carnivore that hunted by sight.

It is clear that, first, manipulation and, later, hunting came to make man's perceptual world different from that of other primates; indeed, different from that of all other animals. Man's analytic perception, more than any other factor, opened the door to the development of conceptual thought and eventually to his remarkable culture.

Let us now turn to concepts. A *concept,* as generally defined, is an abstraction from the particular to the class. The concept "bird," for example, must be abstracted from the perception of, first, "this flying object" and, later, "that flying object." The concept "bird" does not apply to "many birds" or "all the birds I have seen" but to all birds possible in space and time. Similarly, the concept "food" applies not to "this food" but to all possible and potential food—fruit not yet plucked, animals not yet hunted. Consideration of the evolution of perception, such as we have attempted, seems to imply some degree of abstraction from experience, from the particular to the class. The abstraction may not have gone far, and it may not be complete as a concept, for intermediate stages of abstraction can exist. Nor is such abstraction conscious, for perception is not a conscious activity. But it does seem likely that conceptual thought may perhaps have had its origin in the classification of experience that was necessary for the sophisticated interpretation of the environment involved in perception.

It seems likely then, that concepts form part of the mental activity of animals, but there may still be an important distinction in the mental activity of man, for thinking is a conscious process, and so, it follows, is conceptual thought. Apes can deal intelligently with objects here and now—objects they can see and the function of which is clear. They can be trained to carry out activities that appear to show foresight, and they can learn appropriate behavior for future needs. For example, chimpanzees will select straws and prepare vines (by stripping off side shoots), even when they are out of sight of any termite hills, and then set out to visit a known but distant group of hills to obtain termites. Apes in zoos will use sticks to reach bananas too high above them to be plucked by hand and will pile up boxes to get closer to them; yet they have limited creative and imaginative ability. An experiment was performed by Köhler (1925) to demonstrate the limitations of ape mentality. He constructed a box loosely out of sticks, so that they looked like boards. When the chimpanzees were presented with bananas out of reach, they were unable to see, within the structure of the box, the sticks they needed to get the fruit. Thus, an ape is unable to make a tool out of a natural object; he is able only to use a natural object as a tool with, perhaps, slight modification. An ape cannot conceive the tool without seeing it. He cannot see the stick in the plank or a hand-axe within a piece of rock. This man alone can do.

Freedman and Roe (1958) have suggested that frustration may well be one prerequisite in the development of conceptual thought; it is certainly a component of it. If a biological need arises and is not quickly satisfied, the requirement for such satisfaction will appear in the mind, not as a particular object, perhaps, but as a generalized one. Thus frustration may bring with it imagination, and imagination is the consciousness of sets of concepts, which are the classification of experience. Man, having in his possession the concept of a small cutting stone, can look for it in a rock. He can see the possibility of its manufacture in future time, even though he may not have it at present. Abstraction means escape from the present, escape for man's mind from the immediacy of life. It has been said that what distinguishes man from animals is the length of time through which his consciousness extends. In animals, this dimension is small, stretching a little way into past and future; in man, it grows both qualitatively and quantitatively. The evolution of conceptual thought gives man greater power to live in the past and in the future by abstraction from the past.

To summarize, we find in mammals evidence for a classification of experience, which might be called unconscious conceptualization. This sort of classification, at even a very low level, seems to be necessary for the development of perception. As perception improves and incorporates more data from sensory investigation, especially in the higher primates, we find the probability that certain associational patterns of such data are crystallizing as concepts that might be equated with "things" and perhaps with relationships. Human conceptual thought appears to be characterized particularly by its conscious nature, but it is no doubt the result of a steady process of evolution from less conscious and indeed unconscious concepts in primates. Man's achievement was the fully conscious concept of things he does not possess but needs; the recognition of game, weapons, women, or children as classes brought with it the classification of more and more of man's environment and the possibility of foresight of future needs. Man, leaving behind the narrow limits of present time experienced, entered the broad expanse of past memory and future concepts. His foresight depends upon abstraction from the past in a manner that is termed intelligent (see 10.4).

But the conscious concept did not appear unaided. It seems probable that it was finally evoked by the use of symbols, gestures perhaps, but more often words, which make up language. The concept "bird" could be brought into full consciousness only by its identification with the word symbol "BIRD." The word symbol was the twin of the conscious concept, and it seems probable that they were born together and grew together.

As we find language in man today, it is not fully inborn, but the capacity for speech and conceptual thought is certainly innate; only the symbols themselves must be learned. From the kind of non-human primate vocal signaling discussed later in this chapter has evolved a totally new structure of complex sounds to express not only emotions, which animal signaling expresses, but finally concepts; and with them speech was established.

What, finally, are the neurological correlates of conceptual thought? We cannot expect to find a concept in the brain, for the brain is a living system, not a collection of bits. We can, however, point to one fascinating principle, which may in time become more clearly established than it is at present. As we have seen, the brain is a spatially organized and oriented organ in which the two sides are related by contralateral connections to the cpposite sides of the body, and in turn by sensory input to the two halves of the spatial environment. Thus sensory input and motor elaboration within the brain are spatially organized (see 8.6). There are, however, some brain functions that are non-spatial, and they include all physiological homeostatic control mechanisms (such as body temperature, heartbeat, breathing, etc., controlled by the hypothalamus). These are found in the central pre-cerebral parts of the brain. In primates we can add perception of depth derived from stereoscopic integration (see 3.7) and, in man, concept and word memory (see 10.3). Spatial relationships involving depth and distance may appear to be predominantly spatial concepts, but they are not *of* space, they are about space; in themselves they are spaceless and concerned with pattern rather than place.

We do not know accurately the site in the brain of concept and word memory, but we do know that the so-called "primary speech area" is concerned with both these things and that it is to be found on one side of the brain only (see 10.6 and Fig. 10.8). It is shown by Penfield and Roberts (1959) to occur on the left parietal lobe only, in what has been called "the association cortex of association cortexes" (see 8.6). On the other side of the brain, in the same area, notions of space appear to be elaborated, and damage here often leads to disturbances of form and space recognition, body image movement, and spatial interactions.

We have seen that an object that can be manipulated is in some way freed from complete spatial implication and can be "handled" mentally, apart from its environment, as a concept. Such a concept does not need spatial representation in both hemispheres of the brain; representation on one side is sufficient. Equally, concepts of spatial relationships are not in themselves spatial representations, and they too require representation on only one hemisphere. It therefore appears that the neurological correlate of conceptual thought may prove to be unilateral representation in the cerebral cortex. Concepts based on objects, and other abstractions that can be symbolized, are possibly represented on one parietal lobe; spatial relationships (not symbolized) appear on the other. If this correlation proves to be correct, it is a valuable insight into one aspect of brain function (compare sec. 8.6).

10.3 Memory

MEMORY IS A PSYCHOLOGICAL and neurological attribute of fundamental importance in the evolution of animals. By this term I do not refer to the information recorded in the DNA molecule, which is passed from generation to generation, though this might be termed "racial memory" (see 1.3). I refer rather to information recorded in the brain during the life-span of an individual. Penfield and Roberts (1959) distinguish in man three levels of memory, and they will form a useful basis for our discussion. In the brain, however, they may not be neurologically distinct from one another.

1. *Experiential memory.* Experiential memory is a neuronal record of the stream of experience. This remarkable property of the brain appears to preserve *in sequence* all the experiences of which an organism is aware during its life. The extent of development of this phylogenetically old memory cannot be precisely established in animals, but its existence can be postulated as a necessary basis of learned behavior. Learning necessitates recording the sequence of events of any action and its accompanying perception and such a record must be available for comparison with future sequences of experienced events. This memory operates at the subconscious level, and, even in man, learning can be and often is wholly unconscious. Thus, when one meets a person, the degree of familiarity of his face will be instantly known, and an appropriate greeting will follow.

Every experience is automatically compared in an instant of time with all other equivalent experiences, every new face with all other known faces. Normally it will not be possible to recall the previous experience at will; familiarity will not necessarily be accompanied by a full conscious knowledge of the previous circumstances of the meeting. However, it is interesting that ease of recall, both conscious and subconscious, will depend on the amount of emotion that accompanied the original experience. This is a fascinating device, which speeds up the learning process in all animals. Biologically important experience, which is directly concerned with the survival of the individual and the species, is accompanied by emotion—fear, pain, pleasure. In some way, such emotion (typically accompanied by the release of certain hormones) impresses the neuronal record of the experience more deeply in the brain. Thus, the learning of biologically useful behavior is facilitated at the expense of biologically useless behavior. Man, however, has to face the problem of learning facts that are not in a context of emotion; he has evolved further levels of memory record.

2. *Conceptual memory.* Conceptual memory may be considered to be the accumulation of abstract concepts about experience. It is the record of classified experience in no time sequence rather than a full sequence of particular *referents* (that is, individual events and objects). At its

simplest level it is little different from experiential memory, for in the recognition of a face the mind will "present" known faces as a class, ready for comparison. In a dog's experience of rat-catching, the dog will already have all rat-catching experience, as it were, classified for reference, and he will draw on this experience as he tackles his dangerous task (for rats have sharp teeth). But I do not suppose any abstraction is fully made. If a dog is told "rat," he will look excited and ready to spring, but we may guess that he will experience the emotion and sequence of his last, or his most emotional, rat hunt. The abstract concept "rat" is of no biological value to him, but it is interesting that if it was developed it probably resulted from his master's use of the linguistic symbol "rat." For a dog, the meaning of the word is learned by hearing it at different rat-catching experiences, and, if an abstraction is to be made, it is made with the use of that symbol and the assistance of a human brain.

But, in man, conceptual memory goes much further. We can record classes of objects and our experience with them (cars, car-driving), and we can also record emotions without attached objects (love and hate) and classes of experience, such as emotion itself. Beyond this, we can continue to abstract and classify indefinitely to such a point as that of the concept "entropy." Second-order abstractions and others even more remote from direct personal experience are certainly confined to humans, and in view of the importance of symbols in the handling of abstractions (such as $e = mc^2$), it seems unlikely that even the lowest level of abstractions are commonly extant in animal psychology.

3. *Word memory*. The distinction between conceptual memory and word memory is made by Penfield and Roberts on the basis of both clinical and everyday experience. It is not uncommon to recall a concept (for example "stork") but to forget the word symbol, so that one is forced to describe the concept ("you know, those white birds, the kind that nest on rooftops, steeples, etc. What are they called?"). It is even possible to learn the concept without ever knowing its word symbol. ("I have often wondered what those large black birds that nest in groups in treetops are called"). Equally, it is easy to forget the meaning of a word that was once known ("I have forgotten the meaning of 'solipsism' "). Words can be learned that have no conceptual basis ("abracadabra"). The distinction seems clear. In clinical studies, electrical stimulation may result in the recall of words without concepts or, on the other hand, concepts without their word symbols (described with much difficulty), a condition known as aphasia. Evidence strongly suggests that concept and word memory

are closely interconnected and evolved together but are nevertheless distinct.

It appears that concept-word facilitory circuits develop and become fixed as in a conditioned reflex; the nerve cells are quickly habituated to the passage of impulses through a repeated pattern. As Penfield and Roberts (1959, p. 234) point out:

> A man listening to a speaker may follow the man's words, ignoring the concepts: or he may attend only to the concepts of which the words are symbols, ignoring the words. If this listener is bilingual, he may fail to notice what language the speaker had used.

Word symbols may be learned with immense speed by a human child; every normal child learns virtually an entire language by the age of five, and in a bilingual household a child can learn two languages by the age of six. A nine-year-old boy has been known to understand, read, and write nine languages fairly fluently; a mature man many more. It is noteworthy that the units of language are words, not letters; words are heard, and words are read. Chinese writing, which, like ancient Egyptian, is based on ideograms—that is, discrete concept symbols—runs closer to brain function than does Indo-European writing, with its limited alphabet. The basis of reading is word recognition, not word analysis.

The relationship between these different kinds of memory and the sensory input from the environment, which has been discussed in chapter 7, is summarized in Figure 10.2. It will form the basis for our further discussion of brain function.

10.4 Intelligence

A VERY USEFUL DISCUSSION of the distinction between learning and intelligence has been published by Herrick (1956, p. 354). We may define learning as a pattern of constructive activity performed unconsciously or consciously; the act of learning may be a purely empirical process without any end in view, but its outcome will be a kind of knowledge—an adaptive behavior pattern. The capacity to learn has evolved steadily throughout the animal kingdom, and Rensch (1956) has shown that it has evolved with the increase in brain size. The processes of learning and intelligence are closely connected and have something

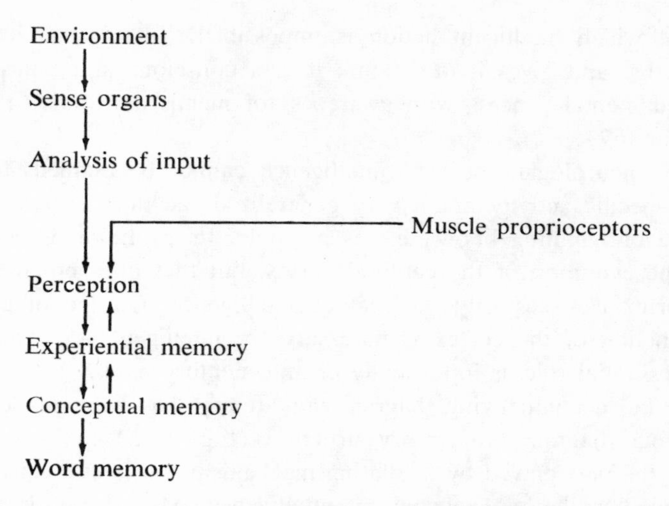

FIG. 10.2. Diagrammatic illustration of the process of perception and its relationship to memory.

in common, but there is certainly some distinct feature of human mental activity that it is convenient to define as "intelligence" and that can be distinguished from a highly developed capacity to learn.

Intelligence has been defined as the relating activity of mind—the ability to realize the connection between discrete objects and events. It clearly involves imagination, which may be considered to be the presentation from the experiential memory of memory traces—knowledge—not obviously directly connected with the immediate experience. It may involve conscious recall of memory, but the significant feature of imaginative thought is that a much broader range of memory traces is brought to the interpretation of day-to-day experience than in the process of simple unconscious learning, for only thus could the imaginative connection between events be realized.

In man, the everyday distinction between intelligence and the capacity to learn is not always immediately clear. For example, expert linguists or stenographers have learned complex skills without necessarily being unusually intelligent, and even professors may be famous for their learning rather than for their native brilliance. In contrast, a comparatively ill-educated man may be highly intelligent. But intelligence cannot exist alone; it is a capacity that interacts with knowledge (which in man may be considered to be the accumulation of experiential and conceptual memory). Acquired knowledge is the fuel with which intelligence burns, without which intelligent action is impossible. "Intelligence integrates knowledge and gives it direction"; it is a conscious and "purposively directed mental process with awareness of means and ends" (Herrick 1956, p. 367).

The neurological seat of intelligence cannot be defined, for it is not a specific activity but a very generalized facilitory system. It is a method of handling knowledge as a whole. Its evolution is correlated with the evolution of the cerebral cortex, but that does not mean that the cortex is necessarily the seat of intelligence. On the other hand, the function of the cortex is necessary for intelligent activity because of its essential role as input analyzer and memory bank.

In the accompanying diagram, an attempt has been made to develop our diagram of memory structure (Fig. 10.2) to indicate very simply the part played by mediating mechanisms in mental activity, and in particular the part played by intelligence. Mental activity means, broadly, the neuronal mechanisms that operate between stimulus and response. It effects response according to the stimulus offered by the environment, and such response may be behavioral or physiological. Physiological responses are automatic and inborn (adjustment of body temperature, salivation, etc.) and do not concern us here (see 1.5). Behavioral responses are often learned and may even follow intelligent thought.

From Figure 10.3 we see that we can consider behavior to be motivated in three ways, by three levels of mediating mechanism. Phylogenetically the most ancient is the innate kind of mediation. This is believed to involve the transmission of the impulse between stimulus and response through an inherited pattern of nerve paths. The response can be modified only by inhibition.

The second type of mechanism is already found in lowly invertebrates (such as the flatworm), and here we have the beginnings of the capacity to learn by the conditioned reflex. In this second kind of mediating mechanism, the relevant cerebral facilitory circuits are activated by the process of learning, the circuits are conditioned by habituation, until the appropriate response follows automatically upon the stimulus (Sperry 1955).

In the most advanced organisms, conscious thought can determine the response according to what amounts (on each occasion) to an original assessment of the environmental situation. Unconscious habit (condi-

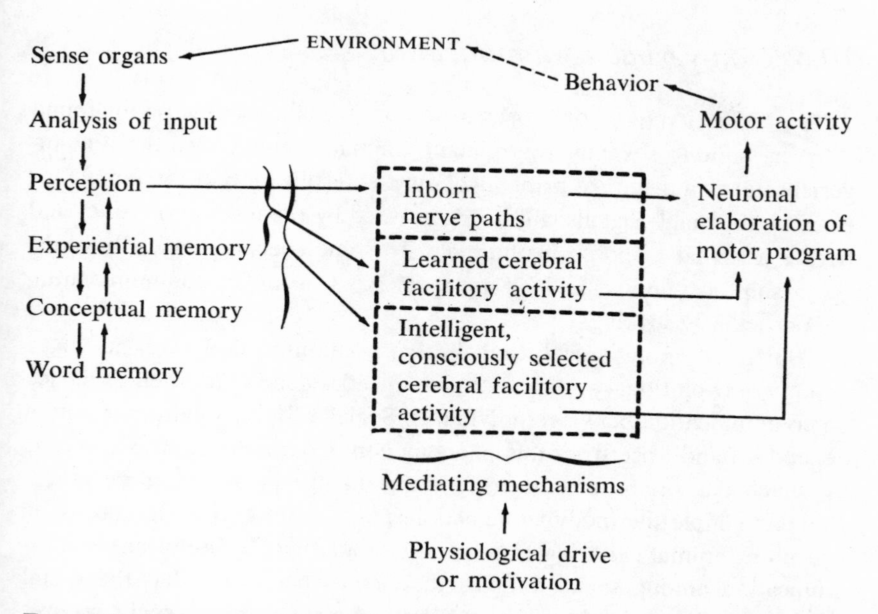

FIG. 10.3. An expansion of Figure 10.2 showing the relationship of perception and memory to mediating mechanisms and motor output.

tioned reflex) can now be superseded by the phylogenetically new and flexible approach that we call intelligent. Of course, men do not always act intelligently; a vast amount of our behavior springs from unconscious conditioned reflexes, but, while activity that was originally intelligent may eventually tend to condition our behavior, it is in its origin at least partially free from previous patterns of activity. Truly intelligent thinking (if such a thing is possible) may be considered to be free from the direct influence of any existing cerebral facilitory activity. When, therefore, intelligence mediates our behavior, it carries our response far beyond that of the conditioned reflex into a totally new realm of flexibility and adaptability. For, on the basis of the refined input analysis that the human brain can provide, a much more precise and delicate assessment of the environment is available, and, in turn, intelligence makes possible a much more exactly and delicately appropriate response. With the considerable development of intelligence that man possesses, the possibilities of novel behavior are practically infinite, and from this remarkable development of the human mind has come man's culture and his genius for invention.

10.5 Non-verbal communication

SOME FORM OF COMMUNICATION is almost invariably found among sexually reproducing animals, both vertebrate and invertebrate, as a means of bringing about the fertilization of the female, or eggs, by the male. Signals will be transmitted by the female to attract and trigger a sexual response in the male. In some species, especially birds, "courtship" ceremonies, which are specialized means of communication, are particularly striking.

But is is among social animals that communication takes its place as a major evolutionary adaptation, for societies depend for their existence on communication between individuals. Social behavior and organization depend on and arise from this phenomenon. Communication is a means by which one organism can bring about change in one or more others (beyond simple thermodynamic action and reaction); it is the means by which one animal can trigger a response in another. So important is communication among social animals that it has been said that the social sciences are the study of the means of communication within social groups. In all animals this means is non-verbal; only in man is language evolved to supplement non-verbal signals. In this section we shall very briefly review the complex subject of non-verbal communication and its evolution among higher primates. In the next section, we shall discuss language.

Non-verbal communication can take a wide variety of different forms; the signals may pass by a number of different channels. The *output channel* is the means whereby the signal is generated and transmitted. The *output effector* may be a specialized organ such as a scent gland or the sexual skin of females; or it may be an organ partly evolved for communication, such as the facial musculature or the vocal tract; or it may be a totally unspecialized organ, such as the skeletal musculature in general.

The *input channels* are the sense receptors and can be classified as shown in Figure 10.4. Signals transmitted via distant non-directional channels are usually received by all members of the group (e.g., calls); signals transmitted by directional or contact channels can be aimed at a particular individual (e.g., tactile or body scent), and these are called *tight-beam* transmissions. Some signals may attract the attention of nonmembers (calls or scent-marking), while others may be limited to reception by members of the group concerned (gesture and touch). Some transmissions

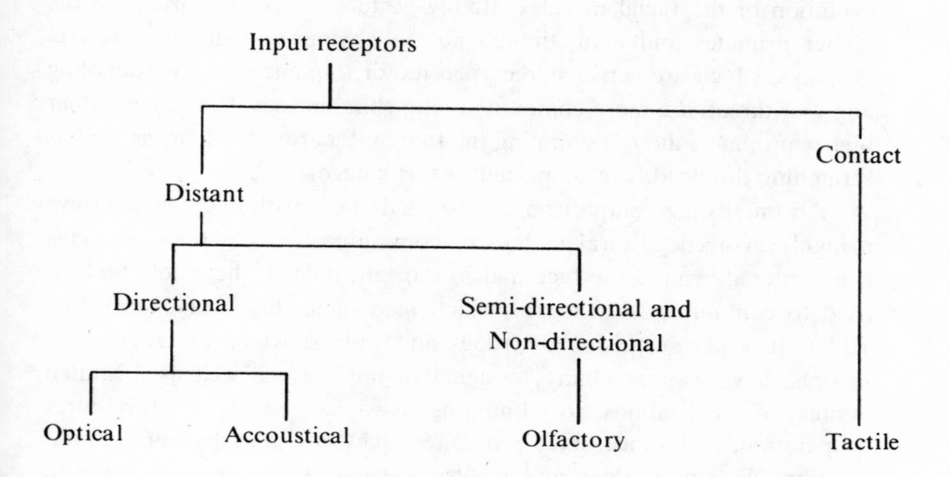

FIG. 10.4. Classification of mammalian sense receptors.

are more suitable for a forest environment where vision is restricted (e.g., calls), while some are appropriate for communication in open country where predators are a threat (gestures and displays). The channel selected and the evolved transmitting and receiving organs are always appropriate to the environment of the species concerned.

Among the higher primates, many channels of communication are exploited, though as we have seen, the emphasis among receptors has changed from olfactory to visual. However, olfactory signals are still important, especially in sexual encounters, where the precise condition of the female is transmitted by *pheromones* (scent) to the interested male. Scent is probably also important in the bonding between mother and newborn infant, but we have no experimental work to confirm this.

Visual receptors receive a vast amount of information in primates, and the optic channel is probably the most important among chimpanzees and terrestrial monkeys. Transmitters vary from the involuntary physiological coloration of the sexual skin of females, which announces estrus, to the facial expressions, gestures and displays that are so well known among baboons and chimpanzees. Facial expression, clearly seen in a relatively hair-free face, is highly evolved among chimpanzees (8.8, Fig. 8.15) and has remained important throughout human evolution. It is interesting that men who allow their beards to grow freely probably use this channel less than chimpanzees. Its importance is related to the evolution of the facial muscles. Bodily gesture is also important to the higher primates and man, though not so obvious in our own culture. Displays, which are active stereotyped motor sequences usually denoting anger or threat, are more common among chimpanzees than among other higher primates, though stamping the foot and tearing the hair, as well as scratching the head, can be placed in this category.

An interesting comparison can be made here with dogs, which have a highly evolved gestural system of communication. In these animals, it is centered around the face and in the tail; indeed, the whole body is used to communicate in a way which men understand very well (Fig. 10.5). It is not surprising that dogs and wolves, which are social carnivores, have a great ability to signal in this way as well as a limited number of vocalizations. Social hunting, as we have seen (7.11), requires cooperation, endurance, intelligence, foresight, and a system of communication. As among dogs and wolves, human communication came to serve an absolutely essential function for the complex organization of the hunt, as well as for the technology associated with it. It is no coincidence that dogs and men understand each other so well today; until recently they both led the same kind of life, that of the social carnivore. As we shall see (10.8), advanced communication is not the only psychological attribute that they have in common.

Touching is also an important means of communication among higher primates. Because of its tight-beam characteristic, it comes into play in interpersonal bonding and is thus more limited in use than the visual signals described above. In the mother-infant bond and in sexual contacts, touch is of primary importance and is vital to the successful outcome of the relationship. Outside of these two basic kinds of behavior, physical contact plays an important part in the establishment of bonds between adults and in lowering tension levels in a social group. Among chimpanzees we find the hand-to-hand touch or caress is as reassuring as it is among humans, and in many other species mutual grooming is an important means of bonding members of the troop and reducing tension. Our own culture seems to have neglected the value of touching as a means of communication essential for full human development (Montagu 1971).

Because our own communication is primarily vocal, research workers have tended to concentrate on the study of vocalizations among higher primates at the expense of other channels. As might be predicted, this channel is exploited most among forest species which have only limited

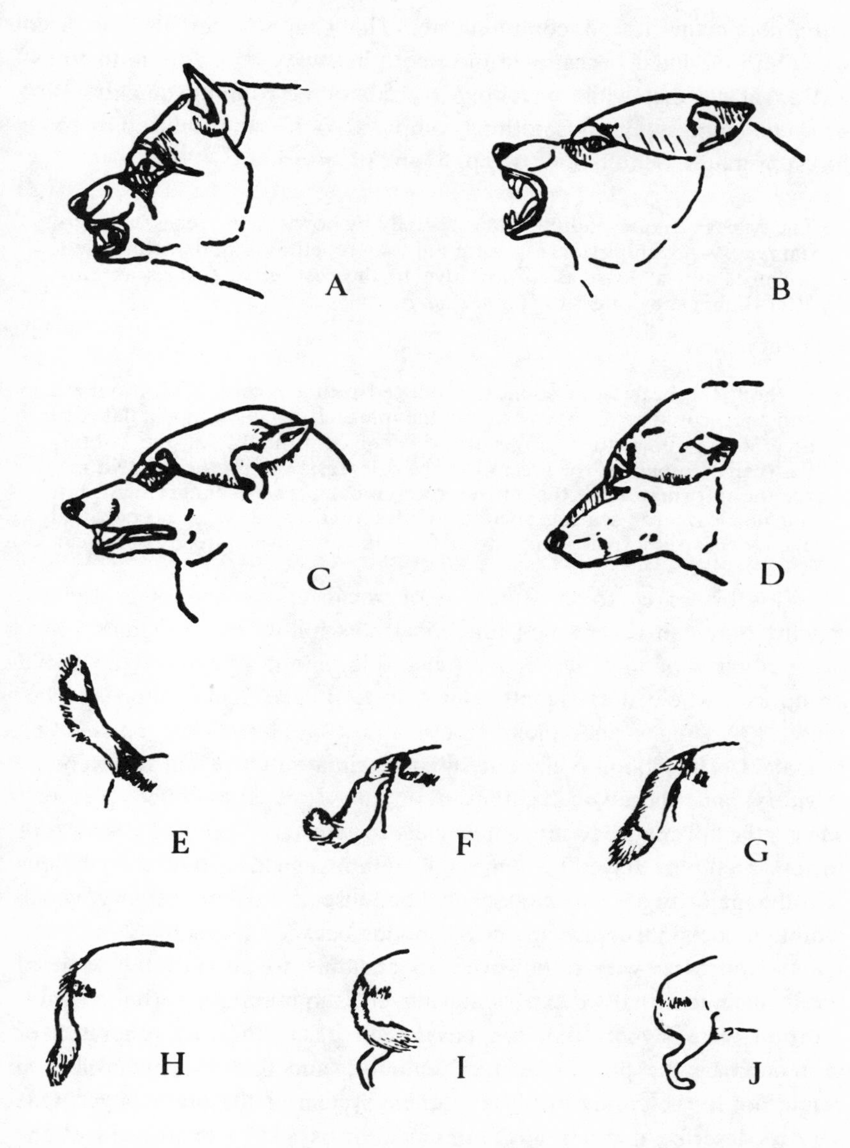

FIG. 10.5. Examples of visual communication in dogs. *Above*, some facial expressions: (*A*) confident threat, (*B*) threat with some uncertainty, (*C*) very weak threat, (*D*) uncertainty and suspicion. *Below*, some caudal expressions: (*E*) self-confidence in social group, (*F*) uncertain threat, (*G*) normal carriage (in a situation without special tension), (*H*) depressed mood, (*I*) and (*J*) complete submission. (From Schenkel 1948.)

visual opportunities to communicate. The channel can also be used when an individual is occupied in some other way, for example to transmit an alarm call while watching a predator or running quickly. Bird song can be noisy, but it is nothing compared to the din produced by some forest primates. Schultz (1961, pp. 61–62) has written:

> The orgies of noise, indulged in especially by howlers, guerezas, gibbons, siamangs and chimpanzees, seemingly so repetitious and meaningless, are probably at least as informative to the respective species as most after-dinner speaking is to *Homo sapiens*.

He comments:

> Without the hearing of sounds, produced by their own kind, monkeys and apes would never have become the intensely social animals that they are. Sounds of a surprising variety serve continually for the contact between members of the group, for the orientation of mother and young, for the information of the entire group about possible danger, and, last but not least, for scaring enemies of different or the same species and even for warning rival groups away from the territories already occupied.

This brings us to the function of vocalizations and other signals. Briefly, they can be grouped functionally as follows: (1) signals which bring about bonding between mother and infant or within the social group as a whole, (2) signals related to territorial demarcation and defense, (3) signals announcing food, and (4) alarm calls on sighting predators. The gibbon is the only higher primate that regularly defends a territory, and whose vocalizations are mainly directed to define its extent. Most other primates confine their calls to items 1, 3 and 4. Because of the essential part played by communication in a social group, it is obvious that the majority of transmissions will be those of the first category, which maintain social structure and social bonds between individuals.

It would be wrong, however, to attribute to animals the kind of conscious intent that we experience ourselves in making a verbal communication. The signals that we have been discussing are generated or motivated by the phenomenon of emotion, and find their neurological origin not in the cortex but in the limbic system of the brain (8.6). One way to describe non-verbal communication is as the expression of the emotions. By expressing emotion in stereotyped and overt ways, animals inform each other of their needs and fears and trigger appropriate responses in each other. Another way of putting it is to say that the expression of the emotions is the basis of social life; it creates and maintains both bond and structure in society. Therefore what we are seeing in the evolution of the higher primates is the adaptive ability to *express* emotion; beyond such physiological changes as in sexual skin or the tumescent penis, this and this alone is their means of communication.

But there is more to communication than transmission, and more than sensory reception. The appropriate response to the reception of a signal is a vital ingredient to communication. This is the social sensitivity we have already described, and which was to become the basis of learning and culture (10.1). Awareness of minute changes in the posture or facial expression of different individuals is an essential component of this adaptation, as is the appropriate response, which may be learned or innate.

In summary, non-verbal communication has evolved in two interlocking fields: in the expression of the emotions, and in the awareness of the behavior of others which alters and directs the behavior of individuals in adaptive ways.

Though a call system and a language are both vocal-auditory means of communication, they are quite distinct. Devoted teaching by the Hayes could not induce the chimpanzee baby they reared with their own baby to learn to speak more than three or four poorly pronounced words (Hayes 1951). Chimpanzees can be taught to use symbols in a somewhat linguistic manner (Gardner and Gardner 1971; Premack 1971). However, since this behavior is dependent on humans to generate it and is basically gestural, this does not imply that the chimpanzee has even limited linguistic ability equivalent to man. Hockett and Ascher (1964) have shown that there are very significant differences between vocalizations and language, so many in fact that it is difficult to trace the possible course of the evolution of one into the other. These differences may be summarized as follows:

1. Calls are mutually exclusive. That is, an animal cannot emit a complex double call with some features of one call and some of another. Only one call can be emitted at a time. This situation is described as "closed"; language, in contrast, is "open," in the sense that we can freely emit utterances that are entirely novel combinations of sounds.

2. A call is emitted only in the presence of the appropriate stimulus; a call system does not show temporal or spatial displacement. Speech, on the other hand, does; we speak freely of things out of sight, in the past, future, or even nonexistent. This capacity of speech is based on thought processes discussed in the preceding sections.

3. The differences between the sounds of any two calls are total. Speech, on the other hand, is the elaboration of a limited number of

components of sound that in themselves have no meaning but, according to their pattern, may have an infinite number of meanings.

4. The call system is innate (generated by the limbic system), so the relationship between stimulus and response is genetically determined and somewhat inflexible. Language, on the other hand, is learned (generated by the cortex), and its transmission must depend on man's capacity to memorize and vocalize. However, this capacity for speech is a genetic endowment, for no amount of effort will teach a nonhuman primate to speak. The actual process by which the primate call system might have evolved into human language has been discussed by Hockett and Ascher (1964) and falls outside the field of this book.

10.6 The evolution of language

THE ACT OF SPEAKING uses a set of body parts of quite diverse primary function. They have been taken over for this new secondary function more or less without modification and without interfering with their previous function. The speech apparatus includes the *larynx,* tongue, and lips, and while these organs must work closely together they are controlled by motor nerve supplies of very different origin in the brain. At the same time, as Spuhler (1959) has pointed out, while such coordinated muscular movement usually requires adjustments from proprioceptors, they are absent from the important laryngeal muscles, and feedback control comes by way of the ear. The only place where the different motor organs and steering apparatus of speech could be connected is in the cerebral cortex. This suggests that the corticocortical connections that make speech possible are phylogenetically new and their appearance is probably correlated with the evolution of the human cerebral cortex.

But before further consideration of the cortical speech areas, it is necessary to understand how the larynx, tongue, and lips came to effect speech. The few slight evolutionary changes that are believed to have effected the speech apparatus of evolving hominids have been described by DuBrul (1958) and Lieberman *et al.* (1972) and can be summarized as follows:

1. As we have already seen (7.5), the chin has been formed in such a way as to avoid any further decrease in the space that forms the floor

of the mouth. This floor is effectively sealed (Fig. 10.6) by a hammock-like diaphragm of the *mylohyoid muscle,* which together with the *genio-hyoid* lies beneath the tongue and can raise and lower it, aided by the tongue musculature.

2. The relative shortening and broadening of the jaw has brought with it a shortening of the tongue. The *genial* (anterior) and *hyoid* insertions of the tongue are brought closer together, and the base of the tongue is thrown back into the pharyngeal area; the whole is quite thick and bulbous.

3. The *hyoid bone* and larynx have also retreated a little down the neck, so the *epiglottis* and *soft palate* are separated by a small space called the *pharynx.*

4. The *digastric muscles* have become slightly rearranged, so that they disturb the stability of the hyoid bone as little as possible, and have moved apart to allow up-and-down movement of the floor of the mouth (Fig. 10.7).

The result of these small changes is that the nasal cavity can be closed off completely so that expelled air can be shunted through the mouth and the tongue and lips can move freely to interrupt this flow of air. The actual sounds of speech are formed by the voice box, or vocal cords, in the larynx. Here vibrations are set up in the expelled air and are modified in the mouth. The shape of the tube formed by the larynx, pharynx, and mouth determines the vowel sounds upon which the consonants are imposed by the tongue and lips.

It seems that the really important physical human character is the possession of the pharynx—the short space between the epiglottis and the soft palate. Here is the tube that makes possible resonant human *phonation* (articulation of sounds). The distinction between man and apes lies not in man's ability to make sounds but in the kinds of sounds he makes and the remarkable way in which he interrupts his resonant notes with the consonants of speech. Movements of the mouth concerned with speech are not greatly different from those concerned with sucking or swallowing. Just as the sucking reflex concerns the lips and tongue, so the consonants are formed: *p, b, m* at the lips; *d, l, n, t* at the tip of the tongue; *g, k* at the tongue base. The proximity of the tongue to the larynx and sides of the mouth may also form gutturals and clicks in certain languages.

As Andrew has demonstrated (1965), the baboons show us how this production of speech may have arisen. In baboons a com-

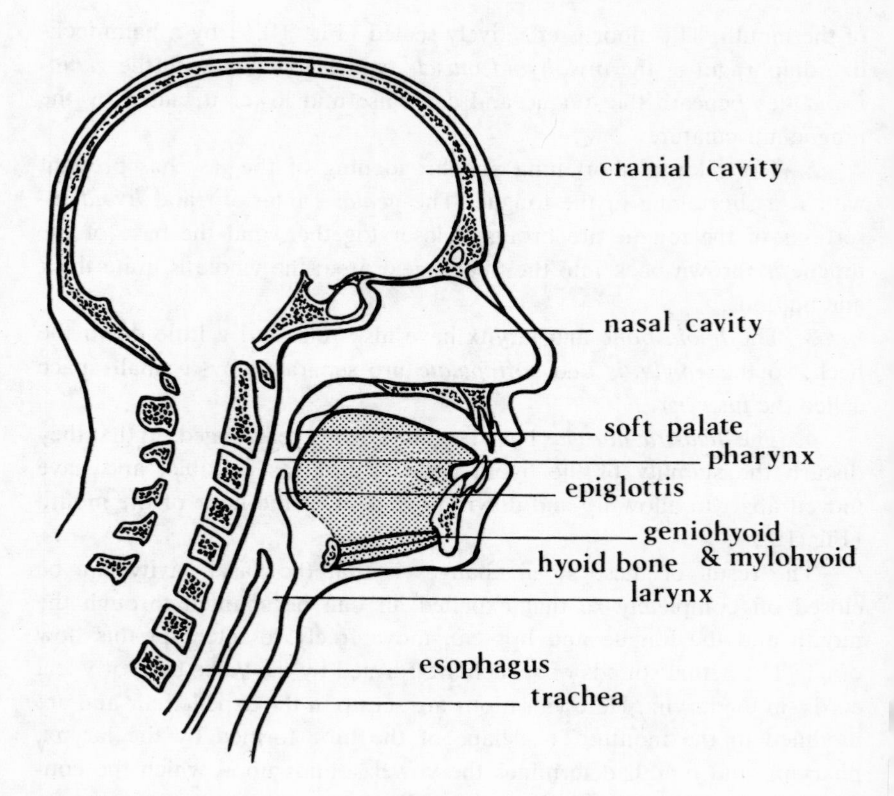

FIG. 10.6. Section of human head showing structures concerned with speech production. The tongue and its musculature are shaded.

mon form of communication is lip-smacking—a display derived from movements of the lips and tongue that they use in grooming. Lip-smacking is not an uncommon greeting among higher primates, but the baboons alone emit a grunt at the same moment. These grunts, modulated by the tongue and lip movements, sound very much like human vowels and are produced in exactly the same way. The plains-dwelling baboons and the plains-dwelling ancestors of man shared a way of life that depended on group cohesion and structure. It is perhaps not a coincidence, therefore, that baboons share with man the ability to make vowel sounds modulated by tongue and lip as an aid to within-group communication.

Speech, however, is more than complex sound. It is the act of codifying thought in sets of controlled and connected sounds, and such

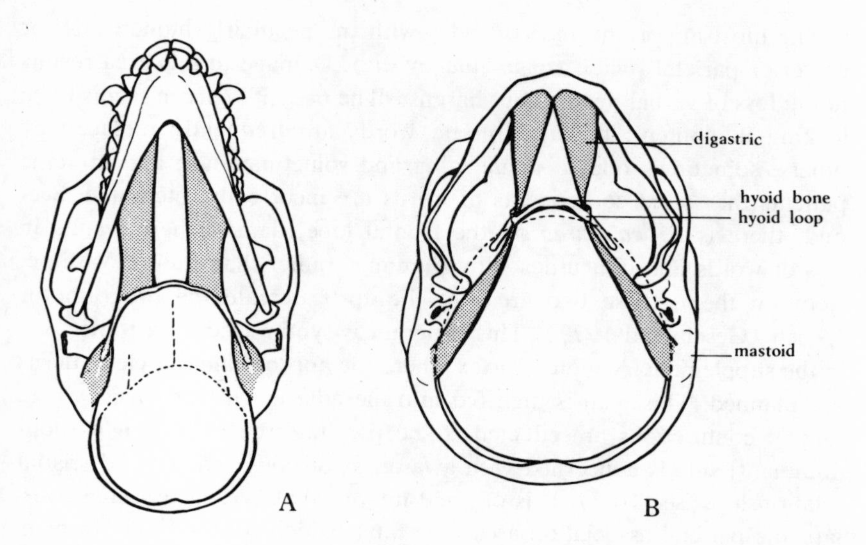

FIG. 10.7. Diagrammatic basal views of the skull of a monkey (*A*) and of man (*B*). In the monkey, the digastric muscles are attached to the hyoid bone; in man, two fibrous loops are formed on the bone through which the muscles pass. In man, the posterior parts of the two muscles are also farther apart to make room for the neck (which has moved forward) and the tongue. The dotted line shows the approximate extent of the ventral margin of the neck in each animal. (Redrawn after DuBrul 1958.)

codification occurs in the cerebral cortex. The primary motor area for speech, which controls movements of the lips, tongue, etc., lies in the *motor gyrus,* as shown in Figure 8.10. In speech and writing, impulses must pass from these areas through the different nerves to the speech apparatus or hand. The resulting movements of the speech apparatus musculature in vocalization and the precise movement of the hand in writing are conditioned reflexes. But the motor gyrus itself is not, of course, concerned with associating words and concepts.

Evidence from cortical excision and electrical stimulation experiments has shown, according to Penfield and Roberts (1959), that the parts of the cortex concerned with the source of speech symbols and concepts lie (as we have seen) for the most part on one side of the brain only, the so-called "dominant" hemisphere, usually the left. (Damage to that hemisphere in early life will result in displacement of the so-called speech areas to the right hemisphere). These speech areas are shown in Figure 10.8. They include *Wernicke's area,* which appears to be the most important and coincides with the peculiarly human inferior posterior parietal region (or *angular gyrus*). Damage to this area results not in loss of verbal fluency, but in sense. The patient's speech seems to be lacking in content; the appropriate words are frequently replaced by others, sometimes related in meaning and sometimes not; and in some patients the sound components of words are incorrectly integrated. Second, there is *Broca's area* on the frontal lobe; damage here results in loss of words and difficulties with grammar, rather than failure of sense. Between them, these two areas of the cortex create the program for speech (Geschwind 1972). This program is synthesized by a third area, in the supplementary motor cortex where the appropriate muscle patterns are planned. This plan is then fed into the adjacent motor gyrus, where muscle contractions are effected. (The parietal area on the right [non-dominant] side is concerned with awareness of body scheme and spatial relationships; see 10.2). It is of great interest that Wernicke's area overlaps the parietal association area (see 8.6), which is found only in man, as well as the interpretive cortex of the temporal lobe.

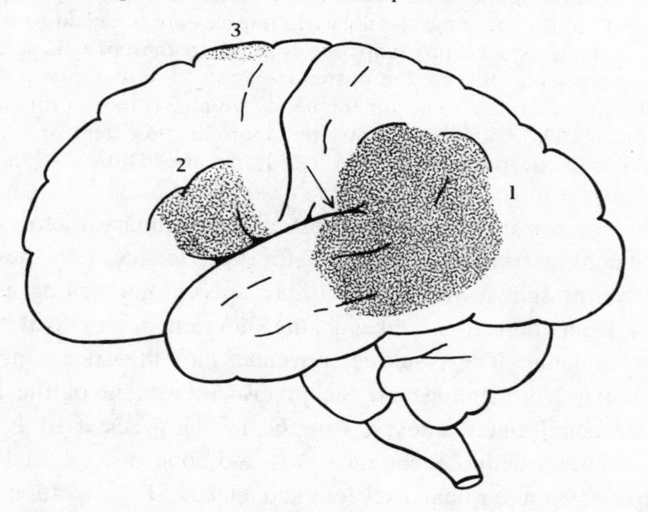

FIG. 10.8. The three speech areas of the cerebral cortex. The angular gyrus and Wernicke's area (*1*) is the most important; the frontal (*2*) Broca's area is the next most important but has been found to be dispensable in some patients; the superior frontal (*3*) is dispensable and lies in the supplementary motor area. The first two are found in the dominant hemisphere only; the third is duplicated on both sides. The arrow indicates the planum temporale of Wernicke's area. (Redrawn from Penfield and Roberts 1959.)

In modern man there appears to be a high negative correlation between the side of the dominant hemisphere and left- and right-handedness. The majority of people are right-handed and have left cerebral dominance, while at least half the left-handed people have right cerebral dominance. Though some degree of handedness is found in monkeys, this hemispheric dominance with preferential use of limbs is a character found only in man. It is also interesting that it is not found at birth but emerges in the course of growth, coinciding with the development of speech from about two to twelve years of age. The brain often becomes asymmetrical in that part of Wernicke's area where it extends into the *sylvian sulcus*. This part of the cortex, the *planum temporale,* is one-third larger on the left side of the brain in 65 percent of individuals (Geschwind 1972) (See Fig. 10.8).

Sperry (1966) has done some remarkable experiments with the help of two patients in which the *corpus callosum*—the nerve tracts connecting the two hemispheres—have been surgically severed. These experiments show that in the right-handed patients only the activities of the right side of the body (with nerve supplies to the dominant left cerebral hemisphere) are reported to be conscious; the left side of the body can be shown to behave quite independently, and the patient is not able to report such activity as conscious experience. We also have confirmation that speech and writing are unique correlates of the dominant hemisphere and that speech is necessary to express consciousness. There is, however, no reason to deduce that consciousness is not also characteristic of the non-verbal hemisphere even though it cannot be verbalized. We are certainly vividly aware of ourselves in relation to our environment without the assistance of symbolic thought.

The fossil evidence is of only limited value in our study of the evolution of language. Attempts have been made to reconstruct the vocal tracts of fossil hominids by a study of the base of the skull and its supposed relation to the hyoid bone. From this work it has been shown that the relatively long bent pharynx peculiar to modern man is not present either in the human newborn, in apes, or in *Australopithecus*. (It is relevant that individuals with Down's syndrome [mongolism] retain the infantile morphology and are unable to speak.) The functional human vocal tract develops during the second year of life. A short ape-like pharynx limits the variety of vowel sounds and the consonants that can be imposed upon them, and alters the possibility of speech in many other ways (Lieberman *et al.* 1972). The human pharynx makes its first fossil appearance in the Steinheim skull, but this does not imply that earlier or less evolved fossil men were not capable of a quite extensive linguistic repertoire that was adequate for their needs.

An equally important correlate of linguistic ability is the size of the brain in hominid evolution. As we have seen, the endocranial capacity expanded from a range of 400–800 cc about two million years ago to a range of 1000–2000 cc today. This size increase was sudden in evolutionary terms and came relatively recently in hominid history. The peak of brain expansion came one-half to one million years ago with the evolution of well-organized social hunting and the technology associated with north temperate adaptations. It is also probable that the most rapid deceleration in post-natal growth occurred at this time. But this cerebral expansion, which is probably a correlate of the expansion of linguistic ability, was the product of a much older transformation—the cerebral reorganization which language implies.

While the large-brained elephant and whale have no language, human "bird-headed" dwarfs, whose brains are no bigger than those of gorillas (400–600 cc), can talk on the level of a five year old child. Evidently they carry the anatomical correlates of language, and they show that a large brain alone is not an essential correlate of linguistic ability. In contrast we have seen in the previous section that apes are not capable of human language though their cranial capacity may reach 700 cc.

It seems reasonable therefore to conclude that language evolved over a long period. The hominid and ape lineages have been independent for at least ten million years if not more, and this seems a reasonable length of time for the evolution of the specialized and novel corticocortical connections that make language possible. Two events may have caused selection for linguistic skills: the primate call-system may have reached a dangerously high level of ambiguity, and simple object naming may have become an essential to further cultural adaptations (Lancaster 1968). No doubt auditory perception and analysis evolved step-by-step with vocalization. The evolution of the modern vocal tract would have followed, responding to the advantages of improved phonation.

One thing is certain, and that is that language does not effectively replace the primate call system, but is a new and supplementary means of communication. While the call system, the expression of the emotions, holds pride of place in the most profound aspects of human relationships, words have evolved to symbolize environmental processes and the abstract concepts arising from the new complexities of social structure. Language has made possible the vast development of man's culture, and brought us the unique human consciousness of ourself and others.

All this evidence points strongly to the fact that language is a genetically determined character of modern man. Our various languages, the actual words we use, are part of our culture; but this remarkable and uniquely human kind of behavior rests on man's peculiar endowment, which gives him his large, slowly maturing brain and his ability to verbalize.

10.7 *Social stability and social institutions*

THE EVOLUTION AND SURVIVAL of a cultural society depends, as does that of a population without culture, on factors of variability and stability. Just as a population of animals evolves by the stabilizing action of natural selection upon a random source of genetic variability to produce a controlled and more or less steady evolution, so society evolves by a process of variation channeled by stabilizing factors. The source of variation lies in the infinite variety of human behavior and invention, and the factors that stabilize this variation are just as important to the survival of society as are those effecting it. The individualistic and imaginative nature of man makes possible an immense variety of behavior patterns, and it is characteristic of social institutions both among higher primates (the dominance hierarchy) and among humans (one-male grouping) that they promote stability in these patterns and in effect protect society from anarchy. Thus the potential plasticity and flexibility of individual behavior, although precious sources of novelty, are controlled in man by social sanctions and other properties of religious and legal institutions. Such institutions protect and transmit both sanctions and knowledge, which are necessary to integrate and maintain social structure and the onward life of the social group.

Some of the most important ways in which social stability may be maintained will be briefly considered, for, although the variety of human customs and beliefs is endless, social behavior and situations are everywhere basically the same. The following stabilizing factors have been selected in cultural evolution because they have helped to insure the survival of human societies:

1. In preliterate societies, traditional knowledge (folklore) is preserved in ritual form by, for example, religious institutions. This function of religious and similar institutions is also carried through into literate society.

2. The integration of the social group, its cohesion, is maintained by the direction of certain sentiments toward a symbolic center.

3. Cohesion may also be maintained by economic bonds of a ritualistic kind as well as by the real economic bonds that are its basis.

4. Individual behavior is controlled and oriented toward social integration by a system of religious and legal sanctions.

Each of these stabilizing factors will be discussed.

1. The preservation of a body of knowledge essential to the social group is an important factor in maintaining continuity in social life and is the basis of tradition, which helps to promote regularity in behavior patterns. Not only knowledge about how to behave but also more abstract knowledge about the supposed bases of human existence are preserved in preliterate societies in the form of ritual (a verbal and behavioral statement) or myth (a verbal statement). Both ritual and myth are means of condensing and encoding knowledge; both are, characteristically, regularly enacted in public as a means of maintaining their power and integrity.

Ritual is a kind of behavior that has been described as typical of both humans and animals, and it is necessary to be clear about the distinction between the two. Although the two forms of ritual, human and animal, have superficial characters in common, their function is really different. Both human and animal rituals are stereotyped behavior patterns used as means of communication; well-known examples among animals are the dance of the bee, which communicates the direction and distance of a source of honey, and the courtship dances of birds. The complexity of these animal rituals has only one aim, to transmit an unambiguous communication signal relating, in these instances, to the whereabouts of food and the prospect of copulation. The signal is related to immediate biological needs. Human rituals may be distinguished insofar as they rarely relate to immediate biological needs, are concerned with social rather than individual communication, and are more than unambiguous signals; in fact, they might better be described as ambiguous statements.

Of course, man himself has predictable courtship patterns that can be described as ritualistic and relate (for example) to the prospect of copulation (drinks, dinner, dancing, kissing, petting, etc.), but there is a flexibility here not found in birds. In a discussion of human rituals we are concerned with stereotyped behavior patterns normally associated with set words, patterns that gain their power by being carried out in an exact form and that are consciously enacted. Ritualistic signaling that precedes copulation in animals and humans is an individual's behavior pattern with an immediate biological end. By human rituals, we refer here to social behavior consciously enacted with a meaning both deep and abstract, relating to no simple biological need. Human ritual serves to perpetuate knowledge necessary for survival and, by combining speech and action, constitutes a much more compact form of social memory than words alone (Leach 1966). Compared with speech, the language of ritual is enormously condensed; a great variety of different meanings are implicit in the same category sets (also an attribute of mathematics). Ritual is a special form of language with a very high information content, but, unlike language, the meaning of ritual depends upon its social context. A ritual act is a social act—it is society's meditation of traditional knowledge and behavior.

The analysis of a human ritual is complex and cannot concern us here. In its remarkable way ritual records knowledge about social origins, about social structure, kinship, and obligations. It records behavior patterns of a fundamental nature, such as hunting, toolmaking, and food preparation. Ritual is, as it were, the DNA of society, the encoded informational basis of culture; it is the memory core of human social achievement.

Myth is the verbal derivative of ritual, and, with the evolution of writing, the informational content of ritual has to some extent taken a verbal form. The written early Sumerian myths are probably only the verbal part of a ritual and as such are only half the language, but, with the development of writing, myth came to incorporate much of the informational content of ritual in its typically condensed form. As literate peoples, we feel more at home with the creation myth of the Book of Genesis than with what seem to us meaningless rituals, but the character of this myth, as that of others, is to have many meanings (one of which is discussed below, in 10.8).

It is impossible to extrapolate to the origins of social behavior from a study of primitive society today, but nevertheless it seems a reasonable postulate that ritual behavior was an early and effective means of recording cultural achievement in the absence of a developed language. With the evolution of language, the trend may have been from ritual to myth, from a more to a less compact means of encoding knowledge necessary for the survival of society. Such knowledge will contribute to the continuity of social norms and the stability of behavior, with its recognition by individuals as tradition worthy of respect.

2. Claim has been made that the word *religion* is derived from the Latin verb *religare* (to bind). If correct, it is not merely a coincidence that an important function of religious institutions is to bind society together, even if it is done by binding men at the same time to God. In primitive societies a personal religion, such as some varieties of Western Christianity, is very weakly developed, and public ritual forms the basis of religious activity. In fact, all rituals may be described as religious, for they not only bind individuals to a core of social knowledge, but, by performing them, individuals are bound to each other in a common activity often requiring much effort and skill.

The binding function of religious ritual is, then, its second important function. Not only are traditions ritualized, but many important social activities appear to take on a sacred and ritual quality. An element in the common life of a society may come to have a special significance and in time become an act of religious observance. Especially important in this respect are the *rites de passage* (birth rites, puberty rites, marriage rites, death rites), which consolidate social roles and social structure, as well as bind members of the society together. But religion does more than that, for it directs social sentiments toward one stable and symbolic center.

In many primitive societies, this further binding property of religion takes the form of ancestor worship, which creates, as it were, a continuum between the living and the dead. The transition from contemporary to ancestor is believed to be a gradual one; the ancestor's identification with the next world is only slowly completed. In time, the sacredness of the deceased increases until in due course great reverence is paid to one common ancestor. Thus worship of one common ancestor binds the members of descent groups together and gives them a common center for their sentiments. Archeologists have given us evidence of burial customs of prehistoric man of perhaps some 70,000 years ago. Skeletal remains of large-jawed men from western Europe and Asia have been discovered with patterned arrangements of rocks, bones, and stone artifacts on and around the bodies (e.g., at La Ferrassie; S. Binford 1968).There are also remains of cave bears (especially skulls) arranged ceremonially in certain sites which were apparently not inhabited (e.g., Regourdou). These people left evidence of religious ritual, buried their dead with great ceremony, and probably believed in their survival after death.

In clans, the totem may fulfill the same function as the ancestor, and in tribes or nations, the ancestor/god may be personified by the living king or queen. Like the ancestor, the king or queen, although alive, is invested with a sacred quality, which enhances the social effectiveness of the institution. Divine kings are known in Africa, and the divinity of the monarch was still recognized in England until the end of Stuart times (1688). The English monarch today, though not explicitly divine, still fulfills a religious function and acts as a symbolic center, especially in times of national danger. Society needs both ritual and symbol to give it form and continuity.

3. Another device that appears to stabilize and maintain social structure is artificial economic interdependence. We have already seen (9.10) that human society is bonded by economics and kinship. What we have to note in this section is that the bonding is sometimes strengthened and channeled by ritualized economic exchange. In societies with a subsistence economy, in which an economic super-structure does not "naturally" exist, we often find a network of economic bonds superimposed in the form of the bride-price or, more generally, in the form of "tribute" or "royalties." The goods thus accumulated by a king or chief priest are in practice redistributed to the members of the society who supplied them; nevertheless the ritual economic transaction serves to bind the society together in a common act and at the same time enhances the role of the king or priest. The modern system of taxes can be seen as a redistribution of goods for the common good and is, in fact, one of very few transactions in which the individual is directly and personally associated with the state.

4. These different means of binding and stabilizing society are clearly factors of overriding importance in the evolution of man. It remains to point out that religious and legal institutions help control human behavior at an individual level for the benefit of the society. What "is done" and "is not done" is a powerful determinant of human behavior; it is dictated by tradition, religion, and law, through each individual's peers and his elders. Religious and legal sanctions control human behavior at its most fundamental level, and the two kinds of sanctions are often so closely allied as to be indistinguishable, as in the Ten Commandments. Among these vital functions of religious and legal institutions falls the

control of the way men express their biological needs, by means of sanctions against theft, aggression, adultery, and so on. From the need for such social control of human behavior arises in turn the mechanism whereby an individual directs his own behavior toward social ends—that is, the evolution of ethics.

10.8 The evolution of ethics

MAN HAS BEEN described as the ethical animal, which is undeniably one striking characteristic that distinguishes him from the rest of the animal kingdom. Its early recognition is recorded in the Book of Genesis, where the story of Adam and Eve describes man's acquisition of the knowledge of good and evil, and how possession of this knowledge resulted in expulsion of man from paradise into the world that we know.

This ancient myth embodies one of the most important events in the evolution of man—the evolution of ethics. Ethics arose as a direct result of the appearance of self-awareness in the growing human consciousness. Cortical control of motor activity slowly came to give evolving man consciousness of his actions and, beyond that, consciousness of himself as an individual being. As has often been said, the animal knows, but only man knows that he knows. Man could see, as a result of his self-consciousness, how many of his activities were directed to satisfy his needs, his basic requirements for life; but he could also see that certain of his actions satisfied only social needs, that they led not to personal satisfaction but to frustration.

It is significant that socially oriented learned behavior is probably most highly developed, not in nonhuman primates, but in dogs and wolves. When a pack of dogs hunts, it performs a cooperative activity that requires considerable discipline. In rounding up their prey, the members of the pack may have to show remarkable self-control in the interest of the group. As social carnivores, dogs have evolved the possibility of self-discipline, which can be learned from and imposed by their peers and elders. As a result, as Kortlandt (1965) has pointed out, dogs are the only creatures besides man that appear to experience guilt—the only creatures to have evolved a recognizable superego, or conscience. Domesticated dogs can be quite easily taught not to touch meat left on the kitchen table. They can generally be relied upon in their master's absence, and they can certainly feel guilt if they disobey, as any dog owner will know. Man and dog understand self-discipline, and that is one reason why they understand each other.

If dogs have a social conscience, as members of a social group, they are certainly not far from being ethical animals, but, as we have seen, ethics is more than this, for it is the resultant of social conscience *and* self-awareness. Without the latter the dog will not consciously make an ethical decision. This is not to deny altogether the possibility of self-awareness in dogs; a certain degree of self-awareness may exist in dogs, as it may in monkeys and apes. It seems likely that it does, for, as we know from our own experience, consciousness is not an all-or-none phenomenon; it varies by degree. However, it seems unlikely that a dog would experience an agony of indecision as to whether or not to succumb to strong temptation. Because guilt follows the act, we need not assume that foresight of guilt necessarily precedes it.

Ethics as we know it may have arisen, then, from the full development of self-consciousness in a demanding social context. The key to man's nature lies in his evolution of social hunting; like the dog, man was a social carnivore. In such circumstances, the actions of every individual affected others in the social group, and what is important is that it was of course the social group (as part of the population) that was evolving, not the individual. In gaining awareness of his own needs and activities, man also gained awareness of the needs of the society in which he lived. It is the conflict between the two needs, those of the individual and those of the society, that makes necessary some device for directing a choice between possible actions. Among men as among dogs, natural selection insured that such choices be made on the basis of their survival value to the social group, so the right choice will more often prove to be altruistic, the wrong choice antisocial.

Ethics therefore arise when man finds that he has to make conscious choices in a social context. That is why the story of Adam and Eve has been described by religious teachers as the "fall of man" and by psychologists as the coming of self-consciousness and self-awareness. Many unselfconscious social animals appear to behave altruistically and to suffer or die for the benefit of their social group or their young. This kind of altruism can be selected in the evolutionary process to become an important and adaptive behavior pattern (Trivers 1971). The agony of the human condition arises from man's self-consciousness and his knowledge of the price he pays as an individual for his social life. The evolution of the human condition is described as "the fall" because the author of Genesis describes the coming of human self-awareness as alienation from a state of holiness and the entry of evil into mankind. Clearly, the evolution of self-awareness brought with it a totally new and terrible situation: that human activity was to be directed, not by the straightforward operations of an unreflecting brain that was a self-sufficient and integrated whole, but by the conscious functioning of the human mind, which can foresee the pleasant and unpleasant, good and evil, results of its actions. Self-consciousness and foresight therefore brought discord to the mind of man.

Freedman and Roe (1958) have accordingly described the emergence of *Homo sapiens* as the beginning of the age of anxiety. The need for values can be seen to arise from his neurological situation—from the fact of man's possessing an "archaic neurological and endocrinological system partially but not completely under cortical control." On the one hand, the innate and unconscious homeostatic system of needs and satisfactions affects the motor system and physiological structures in a broad and relatively imprecise manner, while, on the other hand, the cortical analysis of sensory data responds relatively accurately to fine distinctions and social cues of no obvious physiological significance. In practice, there is a dislocation and delay between biological needs and their satisfaction that is due to social demands. Conscious frustration of biological needs, then, becomes the condition of man and can lead to internal conflict and anxiety.

> Only in man is there simultaneously such a rigidity of social channeling, and such a degree of potential plasticity and flexibility for the individual. Incompatible aims and choices which are desirable but mutually exclusive are inevitable conditions of human development. This discrepancy between possibility and restriction, stimulation and interdiction, range and constriction, underlies that quantitatively unique characteristic of the human being: conflict [Freedman and Roe 1958, p. 461].

Internal conflict and frustration may, however, be among the most important stimuli of cultural progress, of the development of adaptive behavior and technology.

We can see, then, that with the evolution of self-awareness man stands quite suddenly in possession of conscious knowledge, purpose, choice, and values, and, as Simpson (1949) has pointed out, these possessions involve responsibility. Responsibility is man's special burden, and though society supports the individual at every turn, it is the individual who must make the final choice and upon whom the social group always depends. This, however, was not the only way in which the individual assumed a new and magnified importance in human society.

10.9 The rise of the individual

ALTHOUGH IT IS the breeding population that evolves and is the characteristic and permanent feature of organic life, our knowledge of the theory of evolution makes it clear that each individual organism is the functional unit in terms of adaptation and survival. One of the striking features of mankind is the variety seen in individual personality and behavior, which seems to enhance the role and value of the individual in his society. Such individuality is not, however, unique to man, but has evolved slowly like other characters. Most social animals recognize other members of their social group, and the members of a troop of chimpanzees, to take an extreme example, are almost as different anatomically and as variable in personality as are human beings. Nevertheless, variations in personality and enhanced individuality are most striking in human groups. These growing differences between individuals, and the importance of particular individuals to each other, appear to have evolved hand in hand.

The division of labor between the sexes, which we presume must have developed among the ancestors of man, is the beginning of the process of enhanced individualization. The fact that particular males and females are interdependent means that one individual is ultimately important to another, if only at an economic level. With the appearance of one-male groups, the interdependence became sexual and it expanded still further with the evolution of the family. The wife and husband recognized each other as economic and sexual partners. Role names, such as "wife," "husband," "father," "mother," appeared and reflected the division of labor within the family. At this simple level, individuals came to depend on each other more and more, and that interdependence within the family was the basis of love, the expression of need. Individualization appears to have grown through the recognition of an individual's dependence on other particular individuals both within and beyond the family.

At a later stage in human evolution we find a diversity of occupation stretching beyond the family. The important function of the elders, the priest, or the king made these individuals of immense importance to the society that recognized them and their vital role. In Neolithic times we find a new diversity of occupation; individuals made different contributions to social needs, some made tools and weapons and some produced food. Different goods and services were produced by different people in the more tightly knit urban communities that evolved. More people came to depend on more individuals to satisfy their different needs, and individualization developed with the increased complexity of society.

But besides the division of labor there was possibly a more fundamental reason for the individualization of *Homo sapiens*. The more behavior depends on learning and the more complex it becomes, the more it will vary among individuals. Variation in competence in performing tasks is a commonplace in the animal world and is striking among the Japanese monkeys (Itani 1965). It is a very important factor in man's evolution because, the more behavior patterns vary, the more important is individual action—individual achievement—to the population as a whole. By his behavior, an individual can determine the fate of the social group of which he is part, for better or for worse. Society comes to care about the behavior of individuals, especially when they perform their more socially oriented activities, and, as we have seen, society comes to control these activities.

What is especially important in this thesis is that behavior in man is to a great extent intelligently directed, and the application of intelligence is a character of the individual, not of the population. Intelligence itself is genetically determined, and genes for intelligence are selected as are any other genes, but intelligence, as we have seen, is a generalized faculty, and its varied application in problem-solving (and problem-setting) is the property of the individual. The evolution of intelligence will therefore by its nature enhance the individual as an important factor in the survival of the species because the evolution of culture is due not only to natural selection within a varying gene pool but also to selection of a series of imaginative cultural novelties that arise from the intelligent activities of individuals. At the cultural level, therefore, the society looks to the individual, never the reverse. The individual, as the source of invention, is the source of culture.

It is relevant that intelligence and intelligent activity are not directly correlated with ability to learn (see 10.4) or with good citizenship, sociability, or a strict adherence to social norms. As a result, society must attempt to preserve individuals, even if they are not socially well-adapted. Society must value an individual as a person, for upon individual people the evolution of society depends. In a cultural society the contribution of each individual to the population is no longer only via the gene pool but may be direct and immediate, as a result of intelligent behavior. Cultural evolution finds its sources of variability, its "mutations," in individual behavior, and they can spread through the population, if they prove of value, within one generation.

10.10 *Human behavior*

WE HAVE ATTEMPTED briefly to describe and understand the origin and function of culture. Culture appears to comprise behavior and ideas that are the property of a society rather than an individual and are learned by social animals from each other. Human culture has developed from some basic primate characteristics that have been rapidly evolved. In particular, primate visual three-dimensional color perception has evolved into typically human "carnivore" perception, without loss of acuity, and with it have appeared conceptual thought, self-consciousness, and intelligence.

Among primates, culture evolved in tightly knit social groups. Social groups, though of modified structure, remained the necessary basis of human culture and made possible the extensive transmission of this kind of learned behavior. The modifications involved the evolution of the family and of exogamy, which, respectively, improved cultural transmission within the social group and improved cultural diffusion between groups.

Culture must have come at an early stage to contain its own means of transmission. Observation was reinforced by instruction, which was made possible by the evolution of language. Man had something to communicate that could not be transmitted by gesture, facial expressions, or simple sound signals; the development of human culture depended on the development of a symbolic language that was adequate to transmit it in all its complexity.

The newly evolved flexibility of human behavior that arose from the cortical control of behavior patterns might have tended to weaken social bonding, and it seems clear that one of the functions of religious and other social institutions was to stabilize individual behavior for social ends. Social bonding was maintained by these institutions in various ways so that the newly evolved versatility of man's behavior would not disrupt his social group and endanger its survival, yet was available as a source of new cultural invention. Tradition and convention in this way served to control human behavior, especially in the biologically important realms of food, sex, and defense; man became conscious of the different motives of his behavior, and the human condition of internal conflict and mixed motives was present.

Culture now controls every item of human behavior. It is clear that even the simplest basic biological activities (such as suckling or defecation) can be at least partly culturally controlled and that the socially important activities, like feeding and copulating, are under very strict cultural control. In our work we fulfill roles in society determined by society for the benefit of society; at the same time, society aims at arresting the activities of such unethical and antisocial individuals as thieves and gangsters. Only a very stable and secure society can afford much deviation from the social norm, valuable though it may be. Man is a social animal, so his survival as a species depends on the maintenance of his society; as a result, individual behavior, though varied and culturally enriching, must be controlled and directed by social needs. Though man invented culture, it has in turn made man.

Man is a cultural animal, therefore, not because he alone has culture but because culture has come to condition his every act. Man owes his nature to an evolutionary trend in which cultural adaptations have been selected and have maintained on the earth this organism that perceived so clearly the nature of the environment that it learned to change it for its own advantage.

SUGGESTIONS FOR FURTHER READING

A zoologist's view of culture similar to that expressed here is described by H. Kummer, *Primate Societies: Group Techniques of Ecological Adaptation* (Chicago: Aldine Publishing Co., 1971). For a cultural anthropologist's view of culture, see R. Linton, *The Tree of Culture* (New York: A. A. Knopf, 1955). For a recent discussion of perception and consciousness, see R. E. Ornstein, *The Psychology of Consciousness* (San Francisco: Freeman, 1972).

Our understanding of animal communication and its evolution has been discussed in R. Hinde (ed.), *Non-Verbal Communication* (Cambridge: Cambridge University Press, 1972). The biological basis of language is reviewed by E. H. Lenneberg *Biological Foundations of Language* (New York: Wiley, 1967). The evolutionary and adaptive nature of cultural change is clarified by D. T. Campbell, "Variation and Selective Retention in Sociocultural Evolution," in H. R. Barringer, G. I. Blanksten and R. W. Mack (eds.), *Social Change in Developing Areas,* pp. 19–49 (Cambridge: Schenkman, 1965).

11
The Origin of Man

We carry within us the wonders we seek without us;
there is all Africa and her prodigies in us.
SIR THOMAS BROWNE, 1605–82,
Religio Medici, Part I, SEC. 15

11.1 Ape or monkey?

IN THE PRECEDING seven chapters we have reviewed what is known of the evolution of the different great functional complexes of man. We have drawn on the evidence of living and fossil primates in order to gain insight into the evolution of man's functional and behavioral adaptations.

We have seen that both modern man and his closely related and probably ancestral genus *Australopithecus* have clear affinity with the arm-swinging higher primates (see 6.7). This fact seems well established on the basis of the present evidence, but the arm-swinging complex alone does not suggest whether man is more ape-like or more monkey-like, whether he evolved from an ape or a monkey. In certain characters man seems to resemble most closely the living Old World monkeys, and we have noted in particular the high index of cephalization of these primates. Man's hand and penis also show close similarities to those of certain monkeys. In contrast, the apes have hands that have undergone specialization, so they have to a great extent lost their manipulative power.

This situation is typical of many characters of man when they are compared with apes; we find that man is less specialized in many of his locomotor features and shares, instead, more generalized features with the Old World monkeys. The generalized nature of modern man was stressed in a famous paper by Straus (1949) and has been reemphasized by Kurtén (1972); it is of the utmost importance. It used to be accepted without question that man shared a common ancestor with the apes, but, since we recognize that many Old World monkeys use their arms in locomotion quite extensively, we do not need to postulate that man was descended from an ape simply for that reason. As we have seen in the earlier chapters of this book, there are other characters besides those connected with the forelimb skeleton which suggest that man is more ape-like than monkey-like.

Of course, we can always postulate parallel evolution to account for similarities between two groups of living organisms, but, if we do so without evidence, we are going to arrive at a more complex hypothesis than that postulated on the basis of no parallel evolution. A principle of scientific method known as Occam's razor states that "plurality is never to be posited without need" (*essentia non sunt multiplicanda praeter necessitatem*); in other words, the phylogenetic hypothesis should be such that it requires a minimum number of postulates or ancillary hypotheses. For that reason, we cannot postulate parallel evolution without certain evidence for its existence.

Therefore, we can accept man's ape-like characters (other than in the forelimb) as suggesting common ancestory with the African apes; in particular we should recall the absence of a tail, the form of the foot, and the presence of a vermiform appendix. Other characters that man shares with the living apes are less striking but are probably relevant; they include the form of the thorax and abdomen, the absence of sexual skin (though present in the chimpanzee), a retarded growth rate, and great flexibility of behavior. But it is clear that these characters are not so strikingly ape-like as to lead us to postulate a *recent* common ancestry with the African apes (theory *T5*, Fig. 3.14) but merely to deduce that the ancestral hominid was probably slightly more ape-like than monkey-like. If, therefore, such an ancestor bore some characters of monkeys and some of apes, and if we discount extensive parallel evolution, this evidence suggests an early differentiation of the Hominidae from creatures themselves not much differentiated from the common stock of monkeys and apes. Insofar as they were differentiated from such an ancestor, they were beginning to be recognizable apes and so are classified in the subfamily Dryopithecinae. The point of time at which the hominid-pongid "split" occurred is discussed further in the next section.

In the remainder of this chapter, we shall summarize and synthesize the story of man's evolution that has been discussed piecemeal throughout this book.

11.2 The earliest hominids

THE HOMINIDAE ARE a universally accepted and recognizable taxonomic group with a common ancestry shared with the Pongidae. The detailed fossil evidence for human evolution has reduced acceptable theories of the origin of the Hominidae from a large number to just two, and these two have a great deal in common.* Both state that the earliest Hominids split from the apes in tropical Africa, and both would accept that this split occurred about the same time as the apes themselves divided into the two species chimpanzee and gorilla. In one theory, presented in this book, the split is believed to have occurred more than 15 and probably nearer 25 million years ago. In the second theory (Washburn 1968, Wilson & Sarich 1969), the split is believed to have occurred less than 10 million years ago (T3 and T5 in Fig. 3.14). The latter theory is based not on the comparative anatomy of fossils and living species, but on the evidence of comparative biochemistry (of blood albumens in particular), which shows very striking similarities between men and African apes. While an attempt has been made to use this evidence to assess the time of the split, there is no consensus as to how this should be done, nor indeed whether it can be done. The biochemical evidence does strengthen considerably our conclusion that man is more closely related to the African apes than to other apes or monkeys.

The differences involved in these two theories are of degree rather than of kind, and the degree is mainly a matter of timing. The evolutionary process would have been broadly the same in each case, but while in the first theory the split would have occurred among apes which were primarily adapted for arm swinging as a mode of locomotion, in the second theory the split would have occurred among knuckle-walking apes. There is therefore an important difference in the locomotor adaptation that we

* Much of the following material (in this and subsequent sections of this chapter) have been adapted from my essay "Man for All Seasons" which is included in B. G. Campbell (ed.,) *Sexual Selection and the Descent of Man, 1871–1971* (Chicago: Aldine Publishing Co., 1972).

should attribute to the earliest Hominids. The difference of opinion is based on a wide range of indirect evidence, most of which we have reviewed, but in the end we should be able to elucidate which hypothesis is correct by the collection and evaluation of the fossil evidence. In particular, we should ask if there are any fossil Hominidae of 10 or more million years of age. The record of *Australopithecus* goes back in a very clear and satisfactory series to 5.5 million years BP at Lothagam, but before that time we have no *certain* fossil evidence.

From an earlier date we have specimens of the genus *Ramapithecus* (=*Kenyapithecus*) from India and East Africa which have been described as early hominids (see 3.8). The status of these fossils is, however, not a foregone conclusion. As we pass back in time, the fossil record becomes less well documented, and bones of *Ramapithecus* are rare and fragmentary. At the same time, the nearer we come to the split, we can predict greater difficulty in distinguishing the bases of the two diverging families; the fossils will carry more characters of common inheritance and fewer characters of independent acquisition. It may therefore take some time before the matter is resolved. We urgently need evidence of the skull and limb bones of *Ramapithecus*. Extensive fossil evidence may well prove necessary to elucidate the status of these early and closely related groups of hominoid primates.

The faunal context of *Ramapithecus,* in both India and Kenya, implies riverine forest and possibly more open woodland (perhaps bordering on savanna). The climate was warm and moist, and it looks as though ecologically, *Ramapithecus* was not very different from the chimpanzee. The evidence further indicates that there was a moist forest corridor between Eurasia and East Africa, and many of the fauna of the Fort Ternan site are from Eurasia (Simons 1969a). Therefore it is possible that *Ramapithecus* came from Eurasia and evolved from an early Eurasian *Dryopithecus* ape. It is equally possible that the movement of *Ramapithecus* was in the opposite direction, and in this case it is probable that the ancestral form was chimpanzeelike, perhaps close to *Dryopithecus africanus.* Alternatively, the early hominids may have evolved over a wide area of the tropical and subtropical regions of the Old World. But the similarity of man to the chimpanzee, which has been demonstrated through biochemical traits by Wilson & Sarich (1969) as well as by classical comparative anatomy, strongly suggests that we are right to look for man's origin in Africa.

If *Ramapithecus* is a hominid in its total morphological pattern, then

FIG. 11.1. Photograph of forest and grassland savanna on the hills along the east shore of Lake Tanzania, occupied today by the chimpanzees studied by Jane Van Lawick-Goodall. This kind of forest fringe country is probably similar to that occupied by the earliest hominids. (Photo courtesy Harold Bauer.)

we can be fairly sure that the split occurred sometime earlier, *at least* 15 million years BP. If *Ramapithecus* is not an early hominid, the opinion of Sarich and Washburn that the split occurred less than 10 million years ago may hold. It will be necessary to find intermediate fossils from the Pliocene as positive evidence for this theory. Whatever the date of the split, the evidence at present strongly supports Darwin's prediction that our progenitors lived in "some warm, forest-clad land," and it is likely that this land was indeed Africa.

I consider it reasonable to accept *Ramapithecus* as a hominid, as Lewis originally suggested, on the basis of the present evidence. If so, we can suppose that the Hominidae may have become separated from their dryopithecine ancestors in the Middle Miocene (possibly 20 million years ago) at the latest. This is about the time when the various species of *Dryopithecus* were flourishing in East Africa (Bishop 1964), and, as we have seen from the forelimb skeleton, *Dryopithecus africanus* could very well be the arm-swinging ancestor of the Hominidae that we are looking for (theory T4, Fig. 3.14). Yet those who have studied the teeth of *Dryopithecus* (Clark and Leakey 1951, Simons and Pilbeam 1965) proclaim it to be clearly pongid in its dentition, and these fossils have recently been placed squarely in the genus of fossil apes, *Dryopithecus*. It seems clear that while *Dryopithecus* is close to our hypothetical ancestor, especially in the limbs, it is already too ape-like in its teeth to be that ancestor himself.

We must therefore look elsewhere, perhaps to an earlier age. Simons (1965) and Kurtén (1972) have suggested that the Oligocene genus *Propliopithecus,* which has a short face, is more what one would expect to see as a forerunner of the family Hominidae than as an early dryopithecine or gibbon (Fig. 11.2). As we might expect, the further

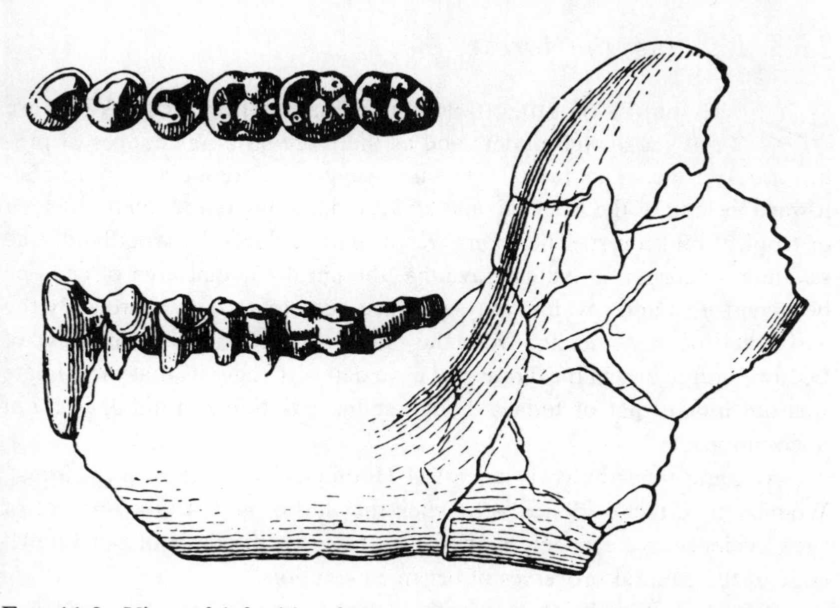

FIG. 11.2. View of left side of the mandible of *Propliopithecus*; above are the crowns of the teeth. Although the tip of the canine is broken, it is smaller than in the *Dryopithecus* species of similar size. Claimed by many to be an ancestor of the gibbons, a recent study has shown the unlikelihood of this claim, and Simons (1965) has suggested that *Propliopithecus* has characters commensurate with hominid ancestry. (From Schlosser 1911.) Approximately twice natural size.

back we pass in time, the more difficult it is to distinguish members of the different families of higher primates, and there is certainly as yet no clear demonstration that *Propliopithecus* should be classed as a hominid. However, this genus, discovered by Markgraf early in this century, may be considered as a possible hominid ancestor from the Oligocene, some 30 million years ago, when the main groups of higher primates were only just becoming clearly differentiated (theory T2).

Turning again to the diagram of hominid ascent (Fig. 3.14), it seems that the most reasonable course would be to accept theory T3, which leads from an early ape-like dryopithecine form in the Miocene, rather than the other theories, which suggest a more recent origin for the Hominidae.

11.3 Leaving the forest

EVIDENCE OF THE CHANGES in climate in prehistoric Africa are not yet so well understood as the more striking changes of prehistoric Europe. The faunal evidence suggests a reduction in rainfall toward the end of the Miocene and an accompanying reduction in the area of tropical rain forest. The forest was in part replaced by woodland, tree savanna, and open grassland savanna, though the actual area of ecotone between forest and savanna may not have been materially altered. By the early Pleistocene, evidence from the fauna and sediments at the base of Olduvai Gorge and in the Transvaal cave deposits suggests that the climate was not unlike that of today, though some variation in rainfall was not uncommon.

At some time the early ancestral Hominidae left the tropical forest. We have no direct evidence as to when they left or why. In the absence of such evidence, we can only make suggestions on the basis of our knowledge of the general processes of organic evolution.

It is clear that the total morphological pattern of the Hominidae is associated primarily with bipedalism. If the interpretation of this pattern that has been placed upon the fossils of *Ramapithecus* is correct, we must suppose that *Ramapithecus* may have moved bipedally more often than living apes do, though clearly the associated adaptations of the pelvis and hindlimbs would be less striking in this animal than those we see in *Australopithecus*. If *Ramapithecus* was bipedal, which hominid status implies, he was most probably living near fairly open country, since bipedalism is not a very useful adaptation in the tropical rain forest. For the present, therefore, it is reasonable to suppose that the Hominidae first began to adapt to more open country in the Miocene (15–20 million years ago), though just why it happened is not known. We do not have to postulate a sudden climatic change to account for a new adaptive radiation, though they may be associated. By Pliocene times many plains-living forms, such as the giraffes, gazelles, and baboons, were evolving under new selection pressures.

As we have seen (1.3), competition is of the nature of organic life, and pressure to radiate is always present in organic evolution. Any character, therefore, that proves to be preadapted to a neighboring environment will allow radiation into that environment if other adaptations do not forbid it. It seems clear that a degree of preadaptation is essential to account for this process, which is the first stage of speciation in evolution. The more striking the contrast in environment, the greater the degree of preadaptation that we must postulate.

We know that various monkeys as well as the hominids have made this move from forest to plain, and, in the case of the monkeys, their arboreal quadrupedal locomotor adaptation was fully preadaptive to terrestrial life. It seems correspondingly clear that the early hominids must have evolved a degree of bipedalism in the forest; this improbable suggestion is acceptable only in the face of the evidence we have of arm-swinging in hominid ancestry. Interestingly enough, the great paleontologist W. K. Gregory postulated in 1928 that some brachiating would have been essential for the hominid ancestors of man before they could have become habitual bipeds. We would much less readily be able to accept the possibility of a change in adaptation from quadruped to biped. The present evidence for arm-swinging gives us a happy solution to the problem of the origin of bipedalism, since it involves an erect posture and the direct vertical transmission of weight from body to feet. It is well known that chimpanzees, gorillas, and especially gibbons can walk for some distance on two legs. Although their posture is stooping with bent-kneed gait, rather than man-like, the possibility of bipedalism is present and is due to adaptation to the changing mechanics of body weight and stress that result from regular brachiation (Napier 1964).

We can suppose, then, that in the early Miocene there were some dryopithecine apes which were already evolved as arm-swingers; their arms were lengthening, their spine was stiffening, and their feet showed adaptation to standing upon branches (see Fig. 5.18). Perhaps some of them were already occupying more open country, such as that occupied by the most western chimpanzees and the (mainly terrestrial) mountain gorilla today. But, unlike the mountain gorilla, they were small, light creatures, perhaps no more than four feet high, so an upright posture presented fewer mechanical problems. At the same time, in the bamboo-covered mountainsides inhabited by the mountain gorilla today, visibility is very limited, and standing up does not improve the view. But in more open country standing erect would have afforded a good view, and bipedalism would also have had other striking advantages. Sometimes a modern baboon (or a patas monkey) will stand erect to get a view over the grasslands, and the protection against predators thus offered to an animal with already excellent sight must be very considerable.

The early hominids were in fact probably preadapted for plains life in a number of ways. Not only could they walk or run short distances, but since they were not very large or heavy they could hold themselves erect without much fatigue and see a long way with great acuity. They had hands, and they possibly could already carry and use objects as tools as well as throw weapons for defense. With an omnivorous taste for food, they could live on anything from foliage and herbage to termites, lizards, and small mammals. Food that was surplus to immediate requirements could be carried until needed or shared. Pilbeam and Simons (1965) claim that the reduced incisors and canines of *Ramapithecus* suggest that tool-using may have become established by the late Miocene, since smaller front teeth require the use of other means to strip bark and prepare both animal and vegetable food, and there is certainly no reason why *Ramapithecus* should not have been equipped with a very dextrous hand even at that early date, and even if tools themselves were not made. At the *Ramapithecus* site at Fort Ternan a battered cobble and depressed fractures on animal bones on an ancient land surface have been claimed as evidence of tool use, and this seems a very real possibility (Leakey 1968).

The pelvis of the early forms may have been far less ape-like than we might suppose from our knowledge of living apes (for example, the chimpanzee pelvis in Fig. 4.7). We have a fossil primate from the early Pliocene, *Oreopithecus,* perhaps 10 or 12 million years old, which is less specialized than the living apes and appears to have a number of hominid-like characters in its pelvis; but so far as we know it was not bipedal (Straus 1963). It seems from such evidence that, as we go back in the evolution of the Hominoidea, we find not only that the hominids get more ape-like but that the apes get more hominid-like, both families are more generalized and more alike.

For the same reason, we need not suppose that the early hominids had either a heavy masticatory apparatus or a large muzzle. Many monkeys have quite short snouts and a few have no sexual dimorphism in that region. Both *Oreopithecus* and *Dryopithecus africanus* were short-faced, and we have pointed out the mechanical relationship between cranial balance and erect posture, another important preadaptation to the hominid way of life.

Ramapithecus had a wide distribution which suggests success, for it is a common evolutionary pattern that the successful exploitation of a new ecological niche will result (where geographically possible) in extensive spatial radiation. Such a successful radiation clearly occurred in the case of *Ramapithecus,* for the specimens from India and Kenya are rather similar, though widely dispersed. It is interesting that their dispersion is paralleled by that of the early ape genus *Dryopithecus* (Simons and Pilbeam 1965).

We need not think, therefore, of a descent from the trees as involving a sudden and revolutionary adaptation resulting, perhaps, from some startling mutation. Nothing so drastic is involved. There was no single cause, nor a linear process of causation. The process of adaptation is systemic and involves interaction between all aspects of the environment and the evolving species. Evolution is the slow unfolding of new forms. Looking back on the process we can see that arm-swinging is preadaptive to bipedalism; its advantages can have been realized only in fairly open or scattered woodland. Bipedalism is preadaptive to plains living; our hominid ancestors had some startling advantages for plains life, in terms of their sensory acuity, powers of prediction, and manipulative ability. The opportunistic nature of the evolutionary process allowed the exploitation of an ecological situation with great possibilities. Early hominids no longer used their eyes mostly for locomotion, as they had done in the forest; they could now use them largely for the more productive activities of food-finding, defense, and intraspecific communication.

One other adaptation, however, seems necessary as a postulate. Baboons have survived in open country to a great extent as a result of their highly evolved social structure. If early hominids were not fully preadapted in this respect, they almost certainly were partially preadapted. At any rate, it seems that plains life for such a defenseless creature as a primate necessitates the rapid evolution of a means of cooperative defense and a binding social structure. This social structure made possible the remarkable changes we see in the Pleistocene phase of human evolution.

11.4 The Australopithecus phase

THE EARLIEST CERTAIN hominid fossils with known archeological associations are those of *Australopithecus*. We have seen (3.9) that this genus ranges in time from 5.5 to 1.3 million years BP and that while the earliest known species are confined to Africa, the later forms also appear in Java, 8,000 miles away by land. The earliest securely dated occurrences are found at Omo in Southern Ethiopia; here both *A. africanus* and the robust *A. boisei* date from nearly 3.0 million years BP.

The most ancient evidence of tool use, and perhaps of hunting, comes from the site KBS at Koobi Fora, east of Lake Rudolf, excavated by Isaac (1971). Here there is an occupation level containing stone artifacts and introduced broken-up bones in a freshwater stream bed near a lake shore. Mingled with the limbs of a single hippopotamus were 31 artifacts and manuports. In all probability the site represents a home base briefly occupied. With a date in the region of 2.5 million years this is the oldest occupation level of any hominid, but older stone implements have been collected at Omo with a possible date of about 3 million years.

As we know from many sites at Olduvai, hunting was a fairly regular occupation of the early hominids after 2.0 million years BP. Game varied from small reptiles, frogs, birds, and rodents to middle-sized antelopes and pigs, to giraffe, buffalo, and even large pachyderms. Some of the remains may have been hunted, some scavenged. Some sites look like living sites (Fig. 7.20), some like butchery sites (Fig. 7.21), and some like workshop sites. The evidence of so much butchery, however, should not cause us to forget that hominids have almost certainly always eaten a preponderance of vegetable food.

Reliance on meat would have developed slowly. From a diet containing very little meat (such as that of a chimpanzee), the proportion would have increased, especially in the dry season, or when vegetable foods were scarce. That meat would ever have composed more than 50 per cent of the diet of early hominids seems extremely improbable; a far smaller percentage is a more likely figure. Today the Bushmen of the Central Kalahari desert survive in much more arid regions, and when game is absent they survive for long periods on vegetable foods. High meat diets are a recent adaptation to winter arctic desert conditions where there is no vegetation whatsoever. The dentition of the Hominids at all known stages is oriented toward chewing and grinding rather than cutting and slicing and carries clear evidence of a preponderantly vegetable diet.

We can therefore suppose that the early hominids came to exploit a range of fauna as a subsidiary food resource, with the adaptations of the bipedal stance and run and, later, the use of the cutting tool. There is no convincing evidence at this early stage of the use of weapons to kill game, but it is quite possible that sticks and bones would have been used as clubs (Wolberg 1970). Sticks are shown in agonistic display by apes (Hall 1963a) and would almost certainly have been used as clubs by hominids. Techniques of killing must have been developed that did not involve the use of sharp canine teeth.

There is little doubt that during this period of adaptation the social structure of the hominids underwent some modification. Even among living chimpanzees we see a tighter social structure in savanna woodland than in forest areas, and this pattern was undoubtedly followed by early hominids. The relationship between social structure and ecology has been clearly demonstrated among primates (Crook 1972). This relationship is important for our consideration of early hominid adaptations, because the social group sets the scene for social hunting.

Cooperative hunting cannot have arisen out of a vacuum, but must have come gradually from another related cultural adaptation. In the same way that a physical structure may prove preadaptive for different functions, so cultural structures may equally serve a novel purpose. From what we know of the behavior of baboons, it is not unreasonable to suppose that behavioral adaptations for cooperative defense gave rise to cooperative offense. There is, in fact, considerable evidence of cooperative hunting among baboons. One eyewitness has, during twenty years, repeatedly observed, in South Africa,

> apparently organized hunts which often result in the death of the intended victim. The baboons, usually led by a veteran of the troop, surround an unsuspecting three-parts grown Mountain Reedbuck, or Duiker, as the case may be, and on one occasion, a young Reedbuck doe was the victim. It would appear that on a given signal the baboons close in on their quarry, catch it and tear it asunder. . . . In nine cases out of ten, the game animal is devoured limb by limb, and after the affair is all over all that is to be found is the skull and leg bones [Oakley 1951, p. 78].

Such behavior may be unusual for baboons, but it does suggest that the difference between cooperative defense and cooperative offense is not very great. Many primates use signals for troop control; what is interesting here is that the baboons show the sort of discipline that we associate with dogs, when they act as cooperative hunters. It appears that primates are sufficiently versatile in their behavior to develop this particular pattern when it proves to be adaptive.

By the time of the deposition of Bed I, Olduvai Gorge, above the 1.8 million-year-old larva flow, a stable adaptation to woodland savanna was almost certainly achieved. Food remains on the living floors at many sites testify to successful hunting as well as scavenging. The move to the savanna was completed. We know, however, that the Olduvai sites were on a lake shore, and it seems certain that the hominids were dependent on both a constant supply of fresh water and the presence of trees (Fig. 11.3). Arid areas without fresh water but with saline lakes would not have been satisfactory sites for hominid occupation.

Early Pleistocene African sites usually lie along sandy stream channels. These ephemeral stream courses would carry strips of riverine bush as well as trees; shade and fruits would be available as well as essen-

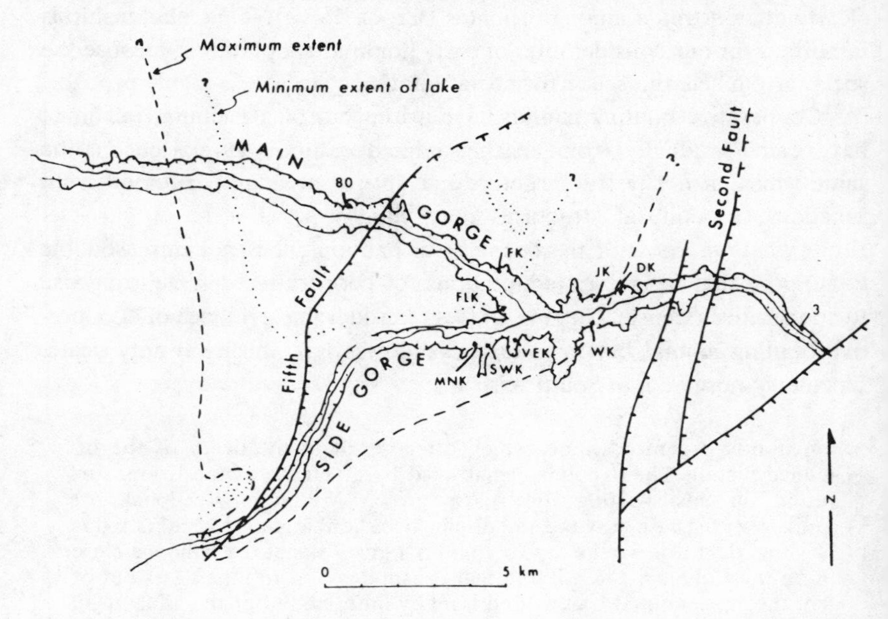

FIG. 11.3. Like most primates, early hominids required a regular supply of fresh water. All the hominid-bearing sites at Olduvai Gorge (which carry code letters on this plan) are situated in areas which once constituted the shore of an ancient lake, into which freshwater streams flowed. (From Campbell 1972; data from Hay 1970.)

tial drinking water. There is also less well documented evidence of settlement in montane forest. At present the data could be made to fit any dietary hypothesis, and it is the environmental diversity that should be stressed. We can certainly predict as well as recognize this characteristic among early Pleistocene hominids. An adaptability in this respect is a *sine qua non* of expansion into yet more varied environments including those of the north temperate regions.

Further adaptations to arid savanna regions must have appeared by Olduvai times. We know that the stone tool kit was by now quite extensive (M. Leakey 1971), and probably included tools for making implements of wood and bone. Large stones suitable for tool making were collected from point sources as much as eight miles from the lake shore. Animal products such as skins and ligaments would probably have been used and prepared with stone tools. One Olduvai site carries evidence that has been interpreted as indicating some sort of shelter from rain or wind (Fig. 7.20).

Modern social carnivores such as lion or hyena use all possible techniques to get meat: hunting, scavenging, and chasing other animals from their kill (Schaller and Lowther 1969). It therefore seems highly probable that *Australopithecus* similarly took advantage of any means available to get meat. This is also suggested by the animal bones on his occupation floors; there is great variety and a complete absence of regularity in his prey species.

Australopithecus africanus survived for millions of years, and the fossil evidence suggests that he originated in Africa and spread throughout the tropical regions of Eurasia through savanna and woodland. It is possible, however, that further discoveries will indicate his early appearance in Eurasia and his later entry into Africa. For the present there is a gap in the Eurasian fossil hominid record between the appearance of *Ramapithecus* in India (8–12 million years BP) and *Australopithecus* in Java (2 million years BP).

11.5 The first men

THERE IS NO DOUBT that throughout the Pliocene the evolving hominids survived as a result of their social structure as much as of their intelligence, for social life is one of the primates' most important adaptations. By the early Pleistocene, therefore, we would expect *Australopithecus* to have a stable and well-adapted social system.

It now seems possible that the expansion of the hominid populations into temperate zones was the most significant step in the evolution of *Homo*. It was more than a shift between adjacent tropical biomes such as occurred among earlier hominids; it involved a major climatic change and was accompanied by many important new adaptations. The most striking adaptations, however, were cultural and social rather than biological.

Adaptations to temperate climates in particular involved dependence not only on tools but on facilities, defined by Wagner (1960) as objects that restrict or prevent motion or energy exchanges (such as dams or insulation), so that anything that retains heat is included (tents, houses, or clothing). Containers of various sorts—skins for carrying food or water, pots or boats, fences or even cords—fall into this category. Temperate adaptation requires far more facilities than tropical adaptation; and their most important aspect is the extent to which they enabled man to become even marginally independent from certain limiting factors in the environment.

Unlike the tropical biomes, the temperate regions are subject to quite extensive seasonal fluctuations in temperature: there are usually two or three months of frosty weather in winter when plant growth ceases. The critical climatic factor limiting adaptation to temperate regions is not the mean annual temperature, but the seasonal variation—the lack of equability. One long harsh winter will stress temperate species to the utmost, and would have placed a premium upon those human groups who had the shelter, fire, and stored food necessary for survival. A hard winter was the most demanding experience hominids had faced since they spread into the open savanna country millions of years earlier. On the other hand, temperate biomes carry an abundant fauna and a richly diverse flora, and the temperate woodland is second only to the tropical rain forest in diversity. The rainfall is evenly distributed, and in the woodland regions lakes, permanent streams, and rivers are common.

Adaptations to this biome would have opened up very extensive food resources to early man. While geographical (mountain) barriers may well have delayed expansion into this environment, cultural adaptations to winter were undoubtedly the most important factor enabling man to move north.

From the period from two to one million years BP we have tropical fossil hominids from Java which are broadly similar to the *Australo-*

pithecus africanus fossils from East Africa. These populations could have entered Southeast Asia via a tropical woodland and savanna corridor at this time if not earlier. Perhaps the earliest archeological evidence of occupation of a temperate zone is that recorded at the Vallonet Cave in southern France (Alpes-Maritime), which is believed to be of Günz age (see Fig. 3.21). There is evidence of hearths in the Escale Cave in the nearby region of Bouches-du-Rhone; these deposits date from a period which is at least early Mindel. Neither of these sites, however, is so convincing as the inter-Mindel site at Vértesszöllös in Hungary, where there are hearths with burnt bones and numerous broken and split bones of a substantial mammalian fauna, particularly rodent bones but also those of bear, deer, rhino, lion, and dog. There are several hundred artifacts of chert and quartz, mainly choppers and chopping tools, flake tools, and side scrapers.

The great cave of Choukoutien, near Peking, is another important site of about the same age. It has very extensive deposits containing numerous hearths with food remains of 45 different species, including sheep, zebra, pigs, buffalo, and rhinoceros, together with deer, which form about 70 per cent of the total. The tool kit has much in common with that of Vérteszöllös and indeed with the "developed Oldowan" from Olduvai Gorge.

Dating from the late Mindel we have two other important European sites, and both show the use of fire. At Torralba and Ambrona in Spain (two contemporary sites, three kilometers apart) we have evidence of extensive butchery by bands of *Homo erectus* who appear to have been trapping elephants in a bog and butchering them on the spot. Bones of more than 20 elephants have been found in a small area together with remains of horses, cervids, aurochs, rhinoceros, and smaller animals; evidently extensive slaughter took place at this site from time to time. There is also evidence of shaped and polished tools of wood and bone and there is a rich Early Acheulian industry (see Fig. 11.4).

At Terra Amata, in present-day Nice, we have evidence of seasonal habitations on coastal dunes. There are ovoid arrangements of stones with regularly spaced post holes. Within the shelters these represent, the floors were covered with pebbles or animal hides (imprints are preserved). Hearths occur in holes or on stone slabs sheltered by low stone walls. Food residues include elephant, deer, boar, ibex, rhinoceros, small mammals, and marine shells and fish. The industry is of Early Acheulian type, and includes a few bone artifacts.

Although he probably entered this latitude during the preceding warm interglacial, on the basis of the present evidence man seems to have become well established in north temperate zones during the Mindel glaciation. By Mindel times (variously dated between 700,000 and 400,000 BP), highly efficient and productive hunting techniques were employed; and it appears that no animal was too large or too dangerous to be killed by hunting bands. The product of the hunt would have served to support

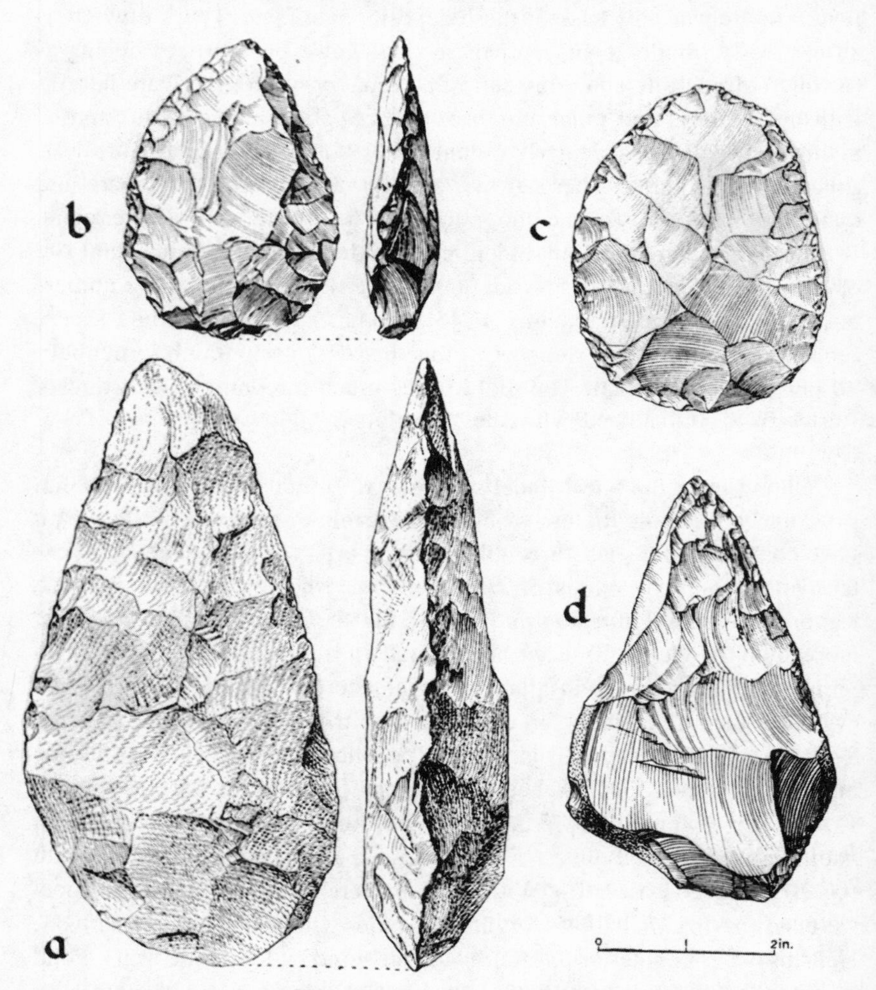

FIG. 11.4. Acheulian tools of the Middle Pleistocene, from Olorgesailie, Kenya (*A*), St. Acheul, France (*B*), Wady Sidr, Palestine (*C*), and Hoxne, England (*D*). (By permission of the trustees of the British Museum [Natural History].)

an increased population through an improved food supply and the use of other animal products, such as skins, that were needed in an advancing technology.

Because at most times of the year vegetable foods in the temperate regions would have been fairly plentiful, systematic hunting was primarily a means of supplementing a diverse vegetable diet. Mammal meat would have become a primary food resource only during late winter and early spring. Berries and seeds would have been eaten by this time, and the new succulent vegetation would not yet be grown. Even the game would have dispersed to find food. Because this limiting period of the temperate year would have kept hunting populations fairly sparse, both gene flow and the transmission of cultural traits would have been restricted. New developments in human evolution were in future to be tied to adaptations which increased man's extraction rate by the exploitation of additional food resources and by food storage techniques.

There is very limited evidence of biological adaptation to cold in modern man. Biologically suited to the tropics as he was, his survival through the cold winters of the temperate zone required extensive cultural adaptations unmatched by the evidence from Africa.

Whether in a grass hut shelter (such as those found at Terra Amata) or cave site, fire would have been maintained not only for warmth but as essential protection against giant carnivores, such as cave bears and brown bears, which were dangerous competitors for life space in cave sites. Archeological evidence strongly supports the notion that fire was used in temperate biomes for a very long period in human evolution. The earliest evidence of fire in Africa is dated about 55,000 BP as we have seen (7.12).

Cold winters also necessitated considerable development in social behavior. It seems inescapable that there would have been a fairly complete division of labor by this time: the men hunting and the women minding the babies and gathering vegetable foods, water (in skins), and fuel. Perhaps for the first time babies were put down and left in charge of siblings or aunts at the base camp. The division of labor and separation of the sexes must have increased the need to communicate abstract ideas by the development of language, and the vocabulary expanded. Perhaps the expression of the emotions (which language could replace) was first inhibited in a closely knit cave-dwelling band, and this emotional inhibition was to become increasingly important. It may prove to have been one of the most fundamental social developments which has shaped the psychology of modern man.

FIG. 11.5. Imaginative reconstruction of the life of *Homo erectus* from Peking. (By permission of the trustees of the British Museum [Natural History].)

From the skeletal evidence at Choukoutien we can deduce that more than 50 per cent of the population died before the age of 14, before they reached full reproductive age. This suggests that it would have been necessary to produce four children per family simply to maintain the population level, and it is probable that more than four children were born in each family. The impossibility for hunter-gatherers of carrying and nursing more than one child at a time indicates that a cultural adaptation such as infanticide was possibly quite common.

The cold made demands on man's ingenuity to devise protective facilities such as clothing and tents. It was surely an important factor in the evolution of human intelligence. All the necessary adaptations had their anatomical correlates in the brain, the skeleton, and the soft parts of the body. The people of Choukoutien show a great advance over *Australopithecus*. Their mean endocranial capacity is twice that of *Australopithecus* and falls into the range of modern man. At Choukoutien the cranial capacity varies between 915 and 1,225 cc, while at the Vértesszöllös site in Hungary (which is at least half a million years old), the cranial capacity has been estimated to be about 1,400 cc. This is over twice the mean for *Australopithecus* and above the mean for modern man (1,325 cc). The people at Choukoutien were also anatomically more advanced than *Australopithecus;* they were of greater stature and more sturdily built, with a well-developed bipedalism. Yet they still carried a heavily built masticatory apparatus that clearly distinguished them from modern man.

The cultural adaptations man developed to survive in temperate and arctic biomes were an extraordinary achievement, and northern winters were undoubtedly a factor of great importance in the evolution of *Homo sapiens.*

11.6 The rise of modern man

NONE OF THE SITES discussed so far carry evidence of permanent cold, but only of seasonal frost. The fauna and pollen data suggest a cool climate becoming either colder (as at Vértesszöllös) or warmer (as at Torralba). The climate at Terra Amata near the sea was certainly mild, and that at Choukoutien was of warm interglacial type.

Man at this time survived cold temperate winters, but there is no evidence that he had yet adapted to arctic conditions. Following the Mindel period, we have more northerly fossils of late Riss-Würm interglacial date (Swanscombe and Steinheim) which represent the expansion of human populations northward during warm temperate spells. In view of the extreme difficulty of survival in northern coniferous forest and arctic tundra, it is not to be expected that man would have entered these zones at a very early date. Today it is hard to see how early man could possibly have adapted to arctic conditions without domesticated animals such as reindeer or dogs. The presence of Neandertal man in the first Würm glaciation of western Europe is a surprising fact. It suggests an advanced use of both tools and facilities.

Though the present evidence suggests that Neandertal man did not survive throughout the first major advance of the Würm glaciation, it does clearly demonstrate that he could survive extreme cold and must have lived some thousands of years under arctic conditions. This Würm advance of the northern ice sheets brought a cold moist climate characterized by animals such as mammoth, woolly rhinoceros, reindeer, musk-ox, ibex, blue fox, and marmot. All these were hunted, together with the formidable cave bear.

FIG. 11.6. Imaginative reconstruction of the life of early *Homo sapiens* from Swanscombe. (By permission of the trustees of the British Museum [Natural History].)

Neandertal man was well established in southern and central Europe before the colder weather descended, and he survived the cold to a great extent by using caves and rock shelters. Judging by the extent of their cultural remains the Neandertal people adapted successfully to the climate and were able to exploit the huge herds of reindeer and other animals. In some areas such as the Dordogne, the local topography must have offset the very extreme conditions. The Dordogne river and its tributaries dissect deeply into a limestone plateau and offer a number of sheltered valleys. Possibly the vast herds of animals which must have undertaken regular seasonal migrations used these valleys as migratory routes. It seems possible that here, as in southwest Asia, these people came to rely on harvesting migratory animals. Many temperate animals, and more particularly arctic species, migrate regularly in the spring and autumn between coastal plain and mountain pasture. These people were thus not only able to harvest "earned" resources (which gain their food within the local habitat where they live), but to tap "unearned" resources—animals which pass through or spend some portion of their annual life cycle in one biome and yet gain most of their food (energy) in another biome.

To settle along the migration routes of herd mammals such as reindeer, musk-ox, or ibex, and intercept them between their summer and winter feeding grounds is a sophisticated adaptation that we can fairly safely attribute to late Neandertal man. It was a simple step to allow autumn-killed meat to dry and freeze for use during the winter, as many Eskimo do today. We can also deduce that the game was sufficient for their needs, for they must have relied to a great extent on meat during the winter.

An interesting clue is provided by the teeth of the man from La Ferrassie. They show a particular type of extreme wear also found today among the Eskimo and some other hunters caused by chewing animal skins to soften them for clothing. It is indeed highly probable that Neandertal man had exploited the whole range of animal products, and especially skins, to develop a well-differentiated material culture. He could well have made the kind of clothing that we find among the Eskimo, though the ready-made shelter of rocks and caves would probably have stood in place of the warm and intimate family igloo.

Probably the most difficult problem facing these arctic people was transport. Without dogs or sleds, they would be confined to a small area during the winter months, and their movements would have been limited to the valleys in which they lived. This restriction shows how successfully they had been able to exploit the local food resources of the region.

The successors of Neandertal man in western Europe, the Aurignacian and Magdelenian peoples, have left us a far more detailed picture of their adaptations than their predecessors. Migratory herds of reindeer were harvested in very large numbers, and often formed 85 to 90 per cent of the faunal assemblage. (Other animals hunted include mammoth, bison, and horse.) When climatic conditions became severe again, as they did toward the end of this period, the later Magdelenians (14,000–12,000 BP) began the systematic hunting of new kinds of unearned resources: migratory birds, aquatic mammals, and fish. The significance of these additions to their food supply can scarcely be overstated. Migratory fish and fowl appear in early spring, a time of maximum food shortage, and make possible the survival of a much larger population throughout the year. At the same time, it is probably important that fish oils (unlike terrestrial animal fats) contain vitamin D, and this may have been an essential vitamin source to people living in areas where insolation was low and clothing essential at all times. Evidence of rickets is present among Neandertal skeletal remains; as far as we know, these people were unable to catch fish and complement their own low vitamin D production. Shortage of this vitamin may indeed have been one of the factors that mitigated against their survival.

This extensive exploitation of migratory animals, which must have arisen slowly through the later stages of human evolution, was to have a profound effect on the evolution of man's social life. The most obvious result was that it allowed a more sedentary way of life to develop. Home bases could be occupied for longer periods of time during animal migrations and the ensuing winter, and there was no longer such a premium on the mobility of hunting bands. During the early phases of hunting and gathering, possessions and infants were limited by the need for mobility. A mother with a baby or a small child who must break camp frequently and transport her baby as well as her household gear will not welcome a second infant to care for and carry or a mass of material possessions. She will have few compunctions about taking any means necessary to limit family size. Today, hunter-gatherers practice infanticide, abortion, and other means of birth control to retain their essential mobility.

Sedentism changed all this. As soon as a band could remain more or less permanently in one place, an increase in possessions and in population densities was possible. The limitation on numbers was removed, and population could now expand to a level related to that dictated by the increased food supply. When we compare the sites of the earlier Magdelenian to those of the later period (since 14,000 BP), we find that the living sites are more numerous, larger, and more often situated low on river banks, frequently at places where the river narrows. Many of these sites have yielded evidence that they were inhabited throughout the year. Thus we find these permanent settlements associated with an increase in density and in group size. It is also clear that these developments may have required a much greater complexity of social structure, compared with the essentially egalitarian local bands that characterize most hunter-gatherer groups. It opened the way to the developments that characterize the Neolithic period.

In his adaptations as a hunter-gatherer, man remained part of a more or less stable natural ecosystem, as he adapted to the particular conditions of each biome and each region. Because food supplies were not the limiting factor in population growth, except perhaps at certain very critical periods, we do not find evidence of overkill or of any serious instability following the appearance of man in a region. As a hunter he was competing with carnivores for herbivores, but the ecosystem is characterized by a functional dynamism that allows it to equilibrate in the face of climatic and other minor changes, especially if it is diverse in species.

Farming is the protection of food plants and animals at the expense of wild forms of less nutritive value to man. It also involves domestication, which is the selective breeding of certain species for their tameness and their value as food. The overall effect of the practice is a reduction in the diversity of organisms in an area, which is balanced by an increase in the domestic species, and a larger proportion of the solar energy in the area is turned into human food, either as plants or meat. Pastoralism can be a surprisingly effective adaptation in this respect, especially when more than one animal species is herded in a single area. Where field agriculture results in whole areas of the ground being covered in one or two food species, the conversion rate of solar energy into human food is even higher. Although we cultivate only 10 per cent of the earth's land surface today, it has been estimated that the human population has increased from a potential maximum of about 10 million hunter-gatherers to its present size: some 3½ billion (many of whom suffer from malnutrition and starvation). But the ecological cost of the introduction of pastoralism and agriculture is high; it implies the destruction of the natural ecosystem and of the diversity of species which assure its stability. The Neolithic was the start of the destruction of man's natural environment, and as the rate of population increase grew, the rate of destruction increased.

In the past 5,000 years man has altered the ecosystem in many parts of the world and destroyed the natural balance. Pastoralism itself has been one of the most destructive forces; it is clear that wherever it has been carried out in semi-arid regions, whether in Australia, Asia, Africa, the Americas, or in limited areas in the Mediterranean regions of Europe, there has been degradation of the grasslands and the threat or reality of soil erosion. The local fauna has been destroyed and the existing ecosystem degraded beyond the point where it can naturally equilibrate. Where soils are eroded, the loss is irrevocable. The displacement of game and the destruction of their natural environment has done far more damage to natural life than all the hunting of the Pleistocene. In the same way agriculture and deforestation for timber have involved the destruction of vast areas of forest (areas of naturally high rainfall), and we have lost both the forest with its associated flora and the forest animals (which often have a very limited distribution). All these developments, though they may eventually prove to endanger his survival, have enabled man to increase his extraction rate from his environment and place himself increasingly at the top of the energy food chain of the biosphere.

11.7 *Modern man as a zoological species*

THE FINAL SCENES in this story of human evolution show the replacement of the large-jawed forms by the small-jawed people —ourselves. In the continuously evolving organic world, what we see around us today is, as it were, one frame in the continuing motion picture of human evolution; "mankind evolving" is the subject of Dobzhansky's important book (1962). We find man to be as we are today, living in the places we occupy, doing the things we are doing. Our genes and our environment together have created us as we find ourselves; our physical nature and our behavior are so determined. Our society and culture have brought about the situation in which we read this book now; it is all part of our complex process of adaptation to each other and to the planet on which we live.

But much has happened in these most recent phases of human evolution. Since the appearance of man indistinguishable from ourselves, which is recorded in the fossil record in France, in Kenya, in South Africa, and in southeastern Asia about 30,000–60,000 years ago, he has spread throughout the land masses of the Old and the New World and now occupies a vast range of environments. Stable populations are adapted for life in the arctic and in the desert. Man's adaptations are now both physical and cultural; variations in body form and body chemistry are adaptations to the environment, as are variations in culture.

We are not yet able to account for all such variations in terms of their functions, but we can understand a certain number, especially the most obvious ones, such as the ratio of surface area to weight (Fig. 11.7), which is correlated with temperature and relative humidity, or skin and eye color, which is correlated with solar radiation (see 8.9), or nose shape, which is correlated with absolute humidity (see 8.4). These kinds of bodily adaptations are clear enough to make recognizable the sorts of differences that zoologists would classify as subspecific, and, as a result, anthropologists recognize between six and nine different subspecies that have arisen as modern man's adaptations to different ecological situations (Garn 1961) (Table 11.1 and Fig. 11.8). Of course, their isolation is incomplete, and gene flow continues between them; otherwise

TABLE 11.1 SUBSPECIES OR GEOGRAPHICAL RACES OF MODERN MAN (AFTER GARN 1961) (SEE FIG. 11.8)

	Notes
1. Amerindian	
2. Polynesian	Polynesians, Melanesians and Micronesians
3. Melanesian	are combined by some authors
4. Australian	as a single group.
5. Asiatic	
6. Indian	Indians and Europeans are combined
7. European	by some authors as Caucasians.
8. African	
9. Micronesian	

they would be defined as distinct species. Man is a single widely dispersed polytypic species, showing adaptation to a wide range of environments.

One of the difficulties in understanding the adaptations that distinguish the different subspecies of modern man lies in his nomadic nature. Evolution is a slow process, but geographical expansion can be rapid, and, as a result, populations originally genetically adapted to climate A are now, as a result of the high adaptive value of culture, living in climate B. As we recognized earlier (see 10.1), organic adaptation in man therefore lags far behind environmental change. The problems introduced by migration and expansion in understanding racial differentiation and characterization are very great. To understand the relationship between the present pattern of *Homo sapiens* populations and their environment must involve, in practice, a consideration of their past history and cultural adaptations as well as their anatomical and physiological characters.

The recent phases in human evolution are striking at both an organic and a cultural level. As Washburn has pointed out in an important paper on the study of race (1963a), there were perhaps from three to five times as many Bushmen as there were Europeans only 15,000 years ago, when the ice sheets of the last glaciation reduced the habitable area of Europe to half that available to the Bushmen in eastern and southern Africa (Fig. 11.9). Since that time, differential population growth has changed the whole pattern of *Homo sapiens* on the face of the earth, so, as a result, although the majority of racial elements have

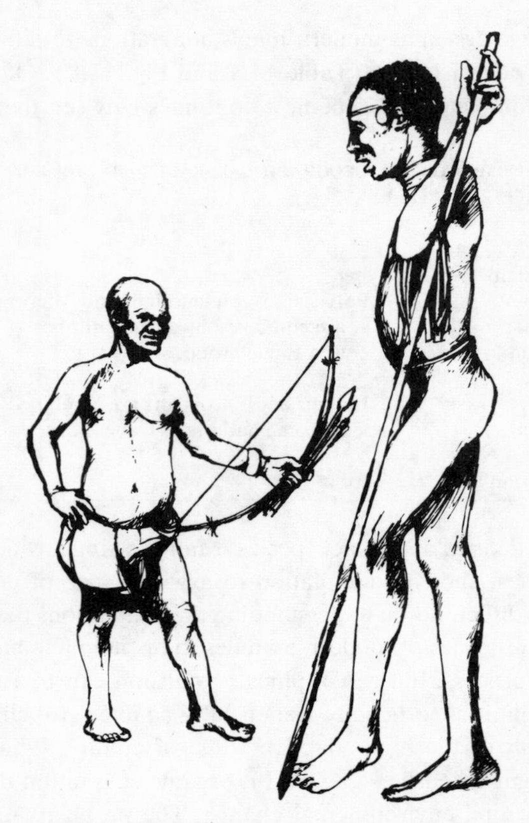

FIG. 11.7. Two neighboring tribes in East Africa show very different physical adaptations. On the left is one of the pygmies from the rain forests of the Congo; on the right, one of the Dinka, a tribe that lives in the desert regions of the upper Nile. They represent the tallest and shortest people in the world. (Courtesy British Museum [Natural History].)

probably survived, their proportions have changed out of all recognition.

As we all know, one after another population has undergone the so-called "cultural revolutions" that marked the beginnings of the ages of agriculture, of metallurgy, and of industrialization. Each revolution has made available greater resources and has resulted in population growth. Today we find industrialized countries supporting an immensely greater population per square mile in comparison with those at an earlier stage of cultural development. This increased population has been made possible by a whole range of adaptations, based on a greater food supply

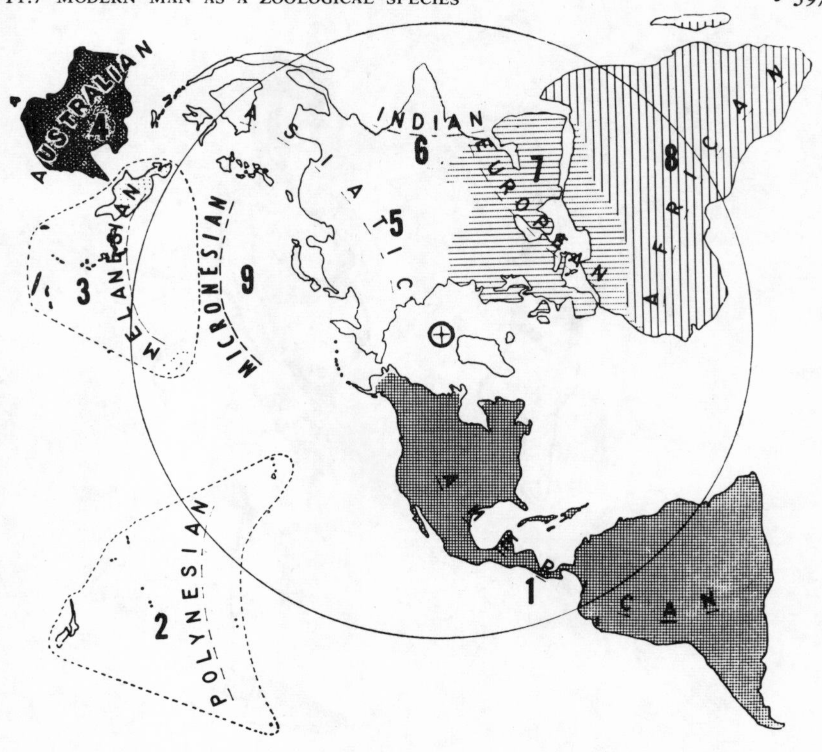

FIG. 11.8. Distribution of nine surviving geographical subspecies of *Homo sapiens*. Each is bounded by natural barriers in the form of sea, ice, mountains, or desert. (From Garn 1961.)

and, recently, advances in medical science. Life expectancy has improved beyond all recognition. In South Africa, a white woman can still expect nearly twenty-five more years of life than can a black woman (Washburn 1963a). Changes in life expectancy are readily subject to cultural adjustment; they can rapidly change the distribution pattern of *Homo sapiens* and will probably do so again as they have in the past.

As we find him today, *Homo sapiens* is in a state of rapid change, change that is due, not so much perhaps to an increase in rate of evolution (such a rate cannot at present be satisfactorily measured), as to an increase in rate of cultural change, which has brought with it greater mobility and rapid adaptation. These developments in culture place man in an unusual situation, which requires some further elucidation.

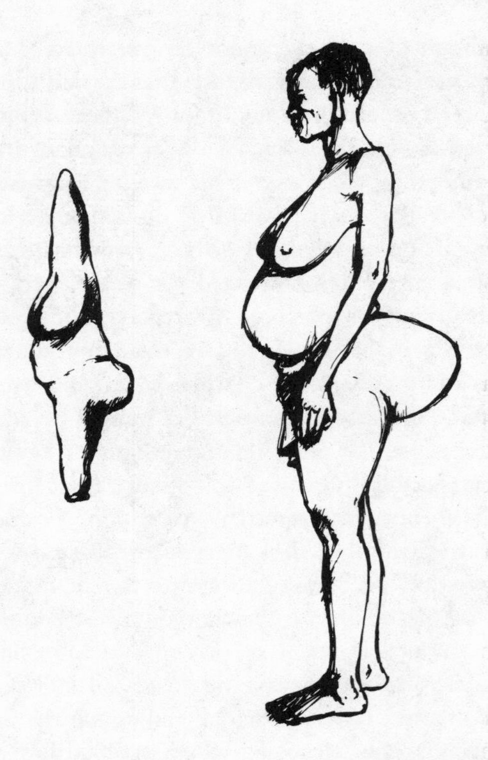

FIG. 11.9. On the right is shown a sketch of a Bushwoman from the Kalahari Desert in South Africa. She shows well the remarkable and characteristic steatopygia, which has been found in this extreme form among only two tribes (Bushmen and Hottentots). On the left is a drawing of a figurine (8 inches high) recovered from Savignano in Italy, which is possibly over 20,000 years old. Relics of this kind suggest that steatopygia was widespread in Europe in prehistoric times.

11.8 Tools and technology— man's material culture

FROM WHAT has been said in the preceding pages, it seems not unreasonable to trace the origins of man's material culture to two of his special mental attributes: his perception and his intelligence.

We have discussed the importance of the evolution of human perception at some length (see 10.2), and its significance as a correlate and determinant of culture cannot be overrated. Man came to see his environment in a unique manner, one that carried a greater informational content about its nature than any other. We now know that the real structure of our environment is beyond our perception, for it is an emptiness sparsely occupied by electric charges moving at great speed. There is no world corresponding to the world of our common experience. But the way the human organism came to perceive the average effect of these electrical phenomena has fashioned his world. His senses enabled him to form in his mind the percepts of objects, solid, colored, and textured; these objects lie in his mind, not in the environment, and what other animals make of these electric charges, which constitute the environment, we cannot tell. One special character of human perception was, however, that since it depended so much on manipulation it came in turn to facilitate manipulation; because it evolved to a great extent from the motor investigation of texture and form through exploration, it came in time to increase the motor investigation of the objective world as it was perceived.

The evolution of manipulation for the manufacture of tools, and eventually for the advance of technology, has come to depend, appropriately, on the development of perception. Man has used his tools to increase his sensory awareness and to reveal still further the nature of matter and organic life. By the beginning of the seventeenth century man had invented the telescope, which enabled him to see more clearly the form of the universe, and the microscope, which enabled him to perceive the minute structure of matter and organisms. Today, powerful telescopes and electron microscopes have vastly increased our visual perception, just as amplifiers, thermometers, and chemical analysis have effectively increased the range and sensitivity of our other senses.

Modern technology appears to be based on this expanding knowledge of our environment, which springs from our increased powers of perception. There is a limit to this increase, however, because we find that at the microcosmic level we are dealing with electric phenomena that can no longer be termed objects, since they do not have simultaneously a position and a velocity. It follows that, while there is much to be done in developing these aids to perception, the development of technology may in the future receive its impetus from the mechanization of other mental processes that characterize the brain, and, in particular, the processes of computation and prediction. The computer, another superorganic extension of the human brain, will in turn come to modify man's relationship with his environment.

The second factor that is complementary to man's highly evolved perception is, of course, his intelligence: his ability to make conscious deductions on the basis of his experience, his power to reason. We have mentioned the conscious nature of intelligent thought, and the hypothesis springs to mind that it was the adaptive value of reason that brought the evolution of self-awareness (see 10.8). Whether or not this was so, it seems that human perception and human reason, interacting under the competitive conditions of organic life, are the twin bases from which we may suppose the growth of human material culture to have developed.

We must also recall that human culture has one essential character: it embodies a complex means of communication. Just as sexual reproduction allows a more rapid spread of novel genetic information within the gene pool than would be possible by asexual heredity, so the aspects of cultural behavior that form systems of communication (speech and writing) allow cultural diffusion within an "idea pool" at a fast rate. Writing is, at the same time, culture's own memory, another attribute previously known only in organic life. Today, machines can sense, analyze, and record; memory of percepts and experience are a commonplace, stored in books or on videotape. Man has learned to project into a cultural form most of the activities of his own brain. Man's unique achievement was, therefore, in the first place, the extension of his motor function and, in the second place, the extension of his sensory and neural functions, all by means of a material culture.

The "brain-like" nature of culture, so obvious in its store of knowledge and ability to compute probabilities, is derived from the abilities not of one man but of a whole society. No one man, without education, could build a computer or write an encyclopedia. Culture is the long-accumulated experience and knowledge of the whole society. The basis of the idea of *involution* is that cooperation replaces evolutionary radiation and that by cooperation man is creating machines of a superhuman kind. Culture is in one sense the group mind of *Homo sapiens,* and it depends on cooperation.

The advantages of cooperation have of course resulted in the evolution of society not only among men but among many different kinds of animals. As we have seen, these societies have in turn evolved quite sophisticated means of communication, yet only among men have the material products of cooperation been preserved as an external record of experience. It is the development of human society's high-speed communication systems, its sensory and motor machines, and its memories and computors that differentiates it from any other. Soon, as a result of the development of "multiple access," we shall be able to dial our problems to the state computor by telephone. When we can do that, our function will no longer be to solve problems but to invent them. As Herrick has written, "the most significant characteristic of intelligence is the ability to invent problems. This capacity for imaginative or creative thinking marks the highest level of integration in the organic realm" (1956, p. 359).

11.9 The function of modern culture

WE HAVE TRIED to understand the adaptive nature of the three aspects of culture: mentifacts and socifacts (10.7, 10.8) as well as artifacts (11.8). Yet when we turn to consider the present state of man, it looks as though culture is failing as an adaptive mechanism —as though it may lead to destruction rather than preservation of the species. In this connection there are three features of modern culture that may throw some light on the complexity of our cultural adaptation and help us understand it.

In the first place, modern culture makes a very effective barrier between man and the external environment that is outside his cultural adaptations. The result is that under the shield of culture man is able to survive and breed with a genetic endowment that in the absence of culture would not allow him to survive (the survival of diabetics is an obvious case in point). The dangers of this situation have been repeatedly stressed and have been recently discussed by Dobzhansky (1962). They will not, therefore, be pursued here; instead, let us consider another aspect of this cultural barrier.

The present acceleration of cultural development may in practice be due not to a primary failure in adaptation but to an imprecision in cultural adaptation. Thus, a particular novel cultural trait that arises in response to a particular environmental condition may also alter a second environmental condition as a by-product. While such imprecision may also occur in organic adaptation, the speed and novelty of cultural change renders it more significant. For example, in north temperate regions, clothes were developed to cut down heat loss from the body, and, although they are most effective, they also have far-reaching effects on skin, humidity, solar radiation, abrasion, external parasites, and many characters of a sematic kind, especially epigamic features. The original equilibrium is adjusted only to result in a new imbalance; yet, while this imbalance may seem continually to increase, the power to adjust that is allowed us by modern culture can increase too. Human evolution is characterized by a great increase both in immediate environmental variability (through culture) and in cultural adjustment to it. Every development may be an attempt at adaptation, yet it may in turn bring with it a new factor requiring adaptation.

Second, it is clear that one of the central problems of the modern world arises from the fact that cultural adaptation has gone further in some social groups than in others and in any case is of a different nature in different places. Nations that have steadily evolved cultural adaptations find themselves today (in the absence of war) in a relatively stable environment. Nations with different and more limited cultural adaptations may as a result of modern communications and transport receive culture "secondhand" from culturally more "advanced" nations, which may amount to a gross change in the environment and in turn cause a biologically unstable period, requiring considerable cultural adjustment on the part of the native population. To give a simple example: in England the increase in the means of production that resulted in an increased food supply occurred at the same time as (if not before) the developments in medical science that raised the life expectancy from about forty to about sixty-five years of age. There was, in fact, enough food available for the vastly increasing population. Today, in many countries we see the results of medical science as an improvement in life expectancy, without its accompanying and necessary improvement in food supply; the result is intermittent famine. Culture contact, by its nature, introduces instability into the environment of human populations insofar as such populations absorb behavior patterns that for them are nonadaptive. They must then adapt to these patterns. In fact, it is a remarkable sign of the adaptive efficiency of culture that contact is not always disastrous. On occasions it has been fatal (the Tasmanians could not survive it), and it has often been nearly fatal (as for the North American Indians), but in most instances adaptation rapidly follows.

Third, we must recognize that cultural adaptations are not always directed toward the impact of the external environment. They also serve to protect society against the individual, for man's survival depends on the integrity of his society, and, as we have seen (10.7), cultural behavior is directed toward maintaining that integrity. The problem is particularly acute in the large social groups that characterize modern man. Possibly we may come to see that national enterprises, such as war-making between nations (unique to modern man) and, more recently, space exploration, are adaptations to maintain the bonding of the societies that undertake them. Modern advances in communication are certainly essential to the successful integration of the vast social conglomerates that are modern nation-states.

Today, western man is no longer occupied merely with his own survival. He is concerned with increasing his economic resources and the search for something he can rarely find, which he calls "happiness." When he is not "at work" (a concept which is by no means universal), he can ponder his own condition and look into his own mind with intelligent self-awareness; and he will find there something more intractable than the external environment. As Eiseley has written (1958), "ancestral man entered his own head, and he has been adapting ever since to what he finds there." There is no doubt that a great deal of human behavior that appears to be non-adaptive, often termed religious and frequently (though not necessarily) ritualized, arises in response to some form of projection into the environment of mental experience that is not rationally understood. In particular, the regular repression of emotion (which seems to be a species-specific character of *Homo sapiens*) can result in inappropriate symbolic projections that are associated with maladaptive (neurotic) behavior. With his increase in leisure, western man turns from his society's god to a personal god, as well as to his psychiatrist; he must come to terms with the phenomenon of his own consciousness.

We see then that human culture springs in some very general sense from the projection into the environment of human neural attributes and mental experience. Man extends his means of perception, his means of computation and of memory, into the outside world. He uses these extensions as adaptations to his environment. In understanding his needs, in using his reason, man comes finally to look into himself, and he must in turn adapt to what he finds in his own consciousness. Man has to live not only with his environment and with his fellows but also with himself, and his behavior is an adaptation to all three.

11.10 The dream animal

APART FROM our great ignorance of biological organization that is the structure of life, the supreme problem for the biologist lies in the evolution of consciousness and its relationship to matter. Because consciousness is often believed to be a specially human character and because of the difficulties of any attempt to study a phenomenon so subjective, this mysterious problem has received scant attention from biologists. Hinshelwood (1959, p. 445) is one of the few distinguished scientists who have drawn attention to it and discussed the form of the problem. As he said, "the question of the relation of the internal and external worlds cannot and should not be ignored by men of science."

Both he and Thorpe (1965) argue that human behavior is not machine-like, and it is clear that the mind-matter relation should not be totally ignored in any text on the evolution of man. The difficulty for the scientist arises from the fact that he is used to dealing with observable data, and, apart from one's own subjective experience, the existence of consciousness in others can only be inferred. But, as Hinshelwood points out, somewhat the same comment could apply to the existence of the atomic nucleus; in the absence of direct observation, we can regard its existence only on the basis of inference as in the highest degree probable. As was said in the Introduction to this book, all that science is really ever in a position to offer is a coherent body of evidence that has emerged from a large number of varied tests and observations.

The evidence for human consciousness arises ultimately from our social intercourse, and therefore it does not allow us to postulate with any certainty the existence of consciousness in animals; but evidence from those who have lived with animals intimately, whether with birds, dogs, lions, or primates, suggests strongly that some degree of consciousness exists, at least in the warm-blooded vertebrates (Thorpe 1965). As already suggested, it is probable that consciousness, like any other human character, evolved slowly, and it may indeed appear in its most simple form at much lower levels of life than has been supposed. If consciousness is a function of brain, and the two are certainly concomitant, then we can postulate its evolution to coincide at least with the evolution of the vertebrate central nervous system. This is, in fact, the simplest and therefore the most acceptable hypothesis, though at present it cannot be tested.

But man has something more than mere consciousness; he also has self-awareness, for, while the animal knows, only man knows that he knows. Here, we are on even more difficult ground, and indeed we have no evidence and no clue to the history of this heightened consciousness. For the present, we must stick to our interpretation of the fall of man (see 10.8). With that in mind, we may see with Dobzhansky (1962) and others that self-awareness is man's distinctive mental character, a character which, linked with imagination, has raised him to the status of lord of creation. Hallowell describes the adaptive value of self-awareness as follows:

> The attribute of self-awareness, which involves man's capacity to discriminate himself as an object in a world of objects other than himself, is as central to our understanding of the prerequisites of man's social and cultural mode of adjustment as it is for the psychodynamics of the individual. A human social order implies a mode of existence that has meaning for the individual at the level of self-awareness. A human social order, for example, is always a moral order. If the individual did not have the capacity for identifying the conduct that is his own and, through self-reflection, appraising it with reference to values and social sanctions, how would a moral order function in human terms? If I cannot assume moral responsibility for my conduct, how can guilt or shame arise? What conflict can there be between impulse and standards if I am unaware of values or sanctions? It is man's capacity for and development of self-awareness that makes such unconscious psychological mechanisms as repression, rationalization, and so on of adaptive importance for the individual. They would have no function otherwise. They allow the individual to function without full self-knowledge [1953, p. 614].

Hallowell also points out that the development of self-awareness is dependent on the projection of the internal experience in socially meaningful terms. Only speech, in fact, can reveal to an individual his self-objectification; speech is necessary for self-awareness and self-awareness is necessary for the moral order of human society. While some self-awareness may exist in apparently moral animals like dogs, its full development is surely confined to man.

With the evolution of modern man, therefore, a new loneliness comes into his life; his separate nature is revealed to him by the evolution of new mental processes. As Eiseley (1958, p. 125) has put it, "for the first time in four billion years a living creature has contemplated himself and heard with a sudden, unaccountable loneliness, the whisper of the wind in the night reeds." Man looked at his environment and knew himself to be no part of it; he was alone, and he was divided from the source of his being (see Frontispiece).

11.11 Human evolution

IN THE FIRST CHAPTER of this book evolution was described as a homeostatic adjustment in a living system in response to environmental change. Evolution is the product of organic homeostasis. We have seen throughout our account of man's adaptations to a changing environment that this homeostatic adjustment, whether at a physiological, anatomical, or social level, is made by shifts in the mean of variable characters. We have seen how the amount of stability found in the occurrence of bodily characters and behavior is maintained by the evolution of factors which, like exogamy, induce variability or, like ritual, instill stability into the dynamic living process.

We have seen how the primitive mammal—already a remarkable and immensely complex evolutionary product—became adapted to arboreal life. We have observed that the beginnings of erect posture occurred in the forest and have noted the concomitant changes in limb structure, in hand and foot. We have seen how primates were preadapted for plains life by their diet and posture either as quadrupeds or as bipeds, and we have tried to elucidate how they further adapted once they entered their new environment.

Perhaps the most essential humanizing adaptation was the increasing importance of society as a mechanism for survival; society and culture are what have saved us from extinction and made us men, just as we in turn create our society. In the words that Gordon Child chose as a title for his famous book, "man makes himself." And that is the curious feature of human evolution; man himself seems to be involved in the creative process, up to the present unconsciously, perhaps in the future consciously.

But what we have recounted is no more than a rough sketch of our subject. We know so little that we have been able only to glimpse a process of great duration and immense complexity. We might be forgiven for doubting that anything less than a miracle was needed to produce, by a process of adjustment from some strings of amino acids, a Mozart or an Einstein. But, although we have treated only part of this story, in only a very superficial way, we have not found any serious inadequacies in the hypothesis that man shows a particular kind of adaptation to his environment through his independence of it and is indeed the outcome of the process of natural selection.

Yet there are some problems to be acknowledged before we close.

One concerns the limits of science, and another concerns the limits of our understanding. There are some special difficulties that are encountered when we attempt to apply the scientific method to human problems (Campbell 1964), difficulties arising from the special relationship between the observer and the observed. We know that perception is far more greatly influenced by previous experience than by the nature of the object that is perceived. Posthypnotic suggestion can completely destroy the reliability of our perceptions, and, at a more everyday level, it can easily be demonstrated that people tend to perceive what they expect to perceive rather than what is before them. The frailty of human perception is notorious, and it is particularly subject to aberration when the subject and object already have some relationship, real or imagined. For that reason, our interpretation of the evidence relating to human evolution may be subject to prejudice of which we know nothing. The dispassionate intellect exists only as an ideal, and an approach to it is made increasingly difficult as the subject and object are more closely involved with each other. The difficulties are particularly intense when we attempt to study our own behavior and culture; they are well demonstrated by the different schools of psychoanalysis. A perfectly objective analysis of human behavior could not be made by any human being.

In this context it is also well to recall that we are in the power of the *zeitgeist,* the contemporary way of looking at the world and at man—another important kind of aberration that destroys the clear focus of our science. This book, like almost every other, is a product of the times, with an emphasis that is typically slanted. The distortions in human studies are inbuilt and unavoidable. It is clear, however, that any approach to the truth can result only from the accumulation and record of human experience of every kind.

Finally, we must recognize the limits of our understanding, if only to indicate the most exciting directions for future research. The evolution of life and the evolution of consciousness each presents a problem that biologists have not yet solved. These two problems probably have something in common, but, in view of the subject of this book, we shall confine ourselves to the second only. If, like Hinshelwood (1959), we accept the existence of consciousness—and few are likely to deny it today—we can account for it in two ways:

1. The potentiality for life and consciousness exists in every atom (of carbon, hydrogen, oxygen, nitrogen, etc.), and these properties are revealed to us as the atoms are combined into organic molecules of greater and greater complexity (see, for example, Rensch 1960). In other words,

the fundamental particles have a mental component that is apparent only when they constitute appropriate structures. Thorpe (1965, p. 26) maintains, however, that this suggestion will not stand because it means that an unbelievable degree of potential organization was in fact present in the randomly moving atoms of the gases of the primordial nebulae, a potentiality for both organic and mental synthesis of vast complexity.

2. An alternative explanation lies in the idea of emergence, the idea that at some stages in the process of evolution a completely new and, in principle, unpredictable quality may appear. Since physics and chemistry know nothing of mind in matter, we must confront this possibility. The only alternative, it seems, is to accept the first possibility—and allow biology to modify physics.

We cannot foresee the final solutions to these problems, though they may come somewhat nearer in our lifetime. But let us not suppose that the ultimate truth is at all simple. A simplification of any hypothesis or data is always a loss of truth, and we are guilty of such simplification at every turn. We would do well in closing to recall a comment by Whitehead (1920, p. 163):

> The aim of science is to seek the simplest explanation of complex facts. We are apt to fall into the error of thinking that the facts are simple because simplicity is the goal of our quest. The guiding motto in the life of every natural philosopher should be, seek simplicity and distrust it.

SUGGESTIONS FOR FURTHER READING

Probably the best modern popular account of man's evolution is the new Time-Life series entitled *The Emergence of Man* (New York: Time-Life, 1972–73).

The most stimulating and varied collection of review articles on man's evolution are being published as a series of volumes under the general editorship of S. L. Washburn, entitled *Perspectives on Human Evolution* (New York: Holt, Rinehart & Winston, 1968).

The most important biological adaptations which distinguish the modern races of man are described in J. S. Weiner, *The Natural History of Man* (New York: Universe Books, 1971).

For a symposium describing research on the neural correlates of consciousness and bringing the whole field of the mind-body problem into scientific focus see J. C. Eccles (ed.), *Brain and Conscious Experience* (Heidelberg: Springer-Verlag, 1966).

References

AMOROSO. E. C.
1952 Placentation. *In* A. S. PARKES (ed.), *Marshall's Physiology of Reproduction* 2: 127–311. 3d ed. London and New York, Longmans, Green & Co.

ANDREW, R. J.
1965 The origins of facial expressions. *Scient. Amer.* 213: 88–94.

ARAMBOURG, C., and HOFFSTETTER, R.
1963 *Le Gisement de Ternifine,* Vol. 1. Mem. Inst. Paléont. Hum., No. 32.

ASHTON, E. H., and ZUCKERMAN, S.
1951 Some cranial indices of *Plesianthropus* and other primates. *Amer. J. Phys. Anthrop.* 9: 283–96.

AVIS, V.
1962 Brachiation: the crucial issue for man's ancestry. *Southwest. J. Anthrop.* 18: 119–148.

BEACH, F. A.
1947 Evolutionary changes in the physiological control of mating behavior in Mammals. *Psych. Rev.* 54: 297–315.

BENEDEK, T.
1952 *Psychosexual Functions in Women.* New York, Ronald Press Co.

BIEGERT, J.
1963 The evaluation of characteristics of the skull, hands, and feet for primate taxonomy. *In* S. L. WASHBURN (ed.), *Classification and Human Evolution,* pp. 116–45. Viking Fund Publs. Anthrop., No. 37. Chicago, Aldine Publishing Co.

BINFORD, S.
1968 A structural comparison of disposal of the dead in the Mousterian and Upper Paleolithic. *Southwest. J. Anthrop.* 24: 139–154.

BIRCH. L. C.
1957 The meaning of competition. *Amer. Nat.* 91: 5–18.

BISHOP, W. W.
1964 More fossil primates and other Miocene mammals from northeast Uganda. *Nature, London* 203: 1327–31.

BLACK, D. R., CHARDIN, T. DE, YOUNG, C. C., and PEI, W. C.
1933 *Fossil Man in China: The Choukoutien Cave Deposits, with a Synopsis of Our Present Knowledge.* Mem. Geol. Surv. China, Ser. A, No. 11.

BONIN, G. VON
1963 *The Evolution of the Human Brain.* Chicago, University of Chicago Press.

BONIN, G. VON, and BAILEY, P.
1961 Pattern of the cerebral isocortex. *Primatologia.* Vol. 2, Part 2, chap. 10.

BROOM, R., and ROBINSON, J. T.
1952 *Swartkrans Ape-Man,* Paranthropus crassidens. Transvaal Mus. Mem., No. 6. Pretoria.

BROOM, R., ROBINSON, J. T., and SCHEPERS, G. W. H.
1950 *Sterkfontein Ape-Man,* Plesianthropus. Transvaal Mus. Mem., No. 4. Pretoria.

BROOM, R., and SCHEPERS, G.
1946 *The South African Fossil Ape-Men, the Australopithecinae,* Transvaal Mus. Mem., No. 2. Pretoria.

BURTON, M.
1962 *Systematic Dictionary of Mammals of the World*, London, Museum Press Ltd.

CAMPBELL, B. G.
1963 Quantitative taxonomy and human evolution. *In* S. L. WASHBURN (ed), *Classification and Human Evolution,* pp. 50–74. Viking Fund Publs. Anthrop. No. 37. Chicago, Aldine Publishing Co.
1964 Science and human evolution. *Nature, London* 203: 448–51.
1972 Conceptual progress in physical anthropology: Fossil man. *Ann. Rev. Anthrop.* 1: 27–54.

CANNON, W. B.
1932 *The Wisdom of the Body.* London, Kegan Paul.

CARPENTER, C. R.
1940 *A Field Study in Siam of the Behavior and Social Relations of the Gibbon.* Comp. Psychol. Monogr., No. 16, p. 5. Republished 1964 *in* C. R. CARPENTER *Naturalistic Behavior of Nonhuman Primates.* University Park, Pennsylvania, The Pennsylvania State University Press.

CHANCE, M. R. A.
1961 The nature and special features of the instinctive social bond of primates. *In* S. L. WASHBURN (ed.), *The Social Life of Early Man,* pp. 17–33. Viking Fund Publs. Anthrop., No. 31. Chicago, Aldine Publishing Co.

CLARK, J. D.
1970 *The Prehistory of Africa.* New York, Praeger.
CLARK, W. E. LE GROS
1947 Observations on the anatomy of the fossil Australopithecinae. *J. Anat.* (London) 81: 300–33.
1950 New palaeontological evidence bearing on the evolution of the Hominoidea. *Quart. J. Geol. Soc. London* 105: 225–64.
1955 The os innominatum of the recent Pongidae with special reference to that of the Australopithecinae. *Amer. J. Phys. Anthrop.* 13: 19–27.
1964 *Fossil Evidence for Human Evolution.* 2d ed. Chicago, University of Chicago Press.
1971 *The Antecedents of Man.* 3d ed. Chicago, Quadrangle Books.
CLARK, W. E. LE GROS, and LEAKEY, L. S. B.
1951 *The Miocene Hominoidea of East Africa.* London, British Museum (Nat. Hist.).
COLBERT, E. H.
1955 *Evolution of the Vertebrates.* New York, Wiley.
CONROY, G. C., and FLEAGLE, J. G.
1972 Locomotor behavior in living and fossil pongids. *Nature, London* 237: 103–104.
CROOK, J. H.
1972 Sexual selection, dimorphism, and social organization in the primates. *In* B. G. CAMPBELL (ed.), *Sexual Selection and the Descent of Man, 1871–1971,* pp. 231–281. Chicago, Aldine Publishing Co.
DART, R. A.
1949 Innominate fragments of *Australopithecus prometheus. Amer. J. Phys. Anthrop.* 7: 301–338.
1957 *The Osteodontokeratic Culture of* Australopithecus prometheus. Transvaal Mus. Mem., No. 10. Pretoria.
1958 A further adolescent australopithecine ilium from Makapansgat. *Amer. J. Phys. Anthrop.* 16: 473–479.
DARWIN, C.
1859 *On the Origin of Species by Means of Natural Selection.* London, Murray.
1871 *The Descent of Man, and Selection in Relation to Sex.* London, Murray.
DAVIS, P. R.
1964 Hominid fossils from Bed I Olduvai Gorge: a tibia and fibula. *Nature, London* 201: 967–968.
DAVIS, P. R., and NAPIER, J.
1963 A reconstruction of the skull of *Proconsul africanus* (R.S. 51). *Folia Primat.* 1: 20–28.
DAY, M. H.
1969 Femoral fragment of a robust australopithecine from Olduvai Gorge, Tanzania. *Nature, London* 221: 230.

1971 Postcranial remains of *Homo erectus* from Bed IV Olduvai Gorge. Tanzania. *Nature, London* 232: 383–387.

DAY, M. H., and NAPIER, J. R.

1964 Hominid fossils from Bed I Olduvai Gorge: fossil foot bones. *Nature, London* 201: 969–970.

DE BEER, G. R.

1954 *Embryos and Ancestors.* Rev. ed. Oxford, Oxford University Press.

DEVORE, I.

1963 A comparison of the ecology and behavior of monkeys and apes. *In* S. L. WASHBURN (ed.), *Classification and Human Evolution,* pp. 301–319. Viking Fund Publs. Anthrop., No. 37. Chicago, Aldine Publishing Co.

1965 *Primate Behavior: Field Studies of Monkeys and Apes.* New York, Holt, Rinehart and Winston.

DEVORE, I., and WASHBURN, S. L.

1963 Baboon ecology and human evolution. *In* F. CLARK HOWELL and F. BOURLIÈRE (eds.), *African Ecology and Human Evolution,* pp. 335–367. Viking Fund Publs. Anthrop., No. 36. Chicago, Aldine Publishing Co.

DOBZHANSKY, T.

1940 *Genetics and the Origin of Species.* New York, Columbia University Press.

1962 *Mankind Evolving.* New Haven, Conn., and London, Yale University Press.

1963 Cultural direction of human evolution: a summation. *Hum. Biol.* 35: 311–16.

DOBZHANSKY, T., and MONTAGU, M. F. A.

1947 Natural selection and the mental capacity of mankind. *Science,* 105: 587–90.

DUBOIS, E.

1894 Pithecanthropus erectus, *eine Menschenahnliche Uebergangsform aus Java.* Batavia.

1926 On the principal characters of the femur of *Pithecanthropus erectus. Proc. Kon. Acad. Wet. Amsterdam.* 29: 730–43.

1932 The distinct organization of *Pithecanthropus* of which the femur bears evidence now confirmed from other individuals of the same species. *Proc. Kon. Acad. Wet. Amsterdam.* 35: 716–22. (also 37: 139–45 and 38: 850–52.)

DUBRUL, E. L.

1958 *Evolution of the Speech Apparatus.* Springfield, Ill., Charles C. Thomas.

DUBRUL, E. L., and SICHER, H.

1954 *The Adaptive Chin.* Springfield, Ill., Charles C. Thomas.

ECKSTEIN, P., and ZUCKERMAN, S.

1956 The oestrous cycle in the Mammalia. *In* A. S. PARKS (ed.),

Marshall's Physiology of Reproduction, 1, Part 1, pp. 226–396. 3d ed. London and New York, Longmans Green & Co.

EISELEY, L.
1958 *The Immense Journey*. London, Victor Gollancz.

ENDO, B.
1966 Experimental studies on the mechanical significance of the form of the human facial skeleton. *J. Fac. Sci., Univ. Tokyo*. Sec. 5, 3: 1–106.

ERICKSON, G. E.
1963 Brachiation in New World monkeys and in anthropoid apes. *Symp. Zool. Soc. London* 10: 135–164.

ETKIN, W.
1963 Social behavioral factors in the emergence of man. *Hum. Biol.* 35: 299–310.

EVERNDEN, J. F., and CURTIS, G. H.
1965 The potassium-argon dating of late Cenozoic rocks in East Africa and Italy. *Curr. Anthrop.* 6: 343–85.

FORD, C. S., and BEACH, F. A.
1951 *Patterns of Sexual Behavior*. New York, Harper & Bros. Reprinted 1965; London, Methuen.

FREEDMAN, L. Z., and ROE, A.
1958 Evolution and human behavior. *In* A. ROE and G. G. SIMPSON (eds.), *Behavior and Evolution*, pp. 455–79. New Haven, Conn., Yale University Press.

FRISCH, K. VON
1950 *Bees, Their Vision, Chemical Senses and Language*. Ithaca, N.Y., Cornell University Press.

GARDNER, R. A., and GARDNER, B. T.
1971 Two-way communication with an infant chimpanzee. *In* A. SCHRIER and F. STOLLNITZ (eds.), *Behavior of Non-Human Primates*, Vol. 4, pp. 117–184. New York, Academic Press.

GARN, S. M.
1961 *Human Races*. Springfield, Ill., Charles C. Thomas.

GARN, S.M., and LEWIS, A. B.
1963 Phylogenetic and intraspecific variations in tooth sequence polymorphism. *Symp. Soc. Study Hum. Biol.* 5: 53–73.

GARN, S. M., LEWIS, A. B., and KEREWSKY, R. S.
1964 Relative molar size and fossil taxonomy. *Amer. Anthrop.* 66: 587–592.

GESCHWIND, N.
1964 The development of the brain and the evolution of language. *In* C. I. J. M. STUART (ed.), *Report of the 15th Annual R.T.M. on Linguistic and Language Studies*, pp. 155–169. Monogr. Ser. Languages and Linguistics, No. 17.
1972 Language and the Brain. *Scient. Amer.* 226(4): 76–83.

Goodall, J.
1968 The behaviour of free-living chimpanzees in the Gombe Stream Reserve. *Animal Behav. Monographs* 1: 161–311.
1971 *In the Shadow of Man.* Boston, Houghton-Mifflin.

Goodhart, C. B.
1960 The evolutionary significance of human hair patterns and skin colouring. *Adv. Sci.* 17: 53–59.

Gregory, W. K.
1928 Were the ancestors of man primitive brachiators? *Proc. Amer. Phil. Soc.* 67: 129–150.

Haldane, J. B. S.
1956 The argument from animals to men: an examination of its validity for anthropology. *J. Roy. Anthrop. Inst.* 86: 1–14.

Hall, K. R. L.
1963a Tool-using performances as indicators of behavioral adaptability. *Curr. Anthrop.* 4: 479–494.
1963b Variations in the ecology of the Chacma baboon, *Papio ursinus. Symp. Zool. Soc. London* 10: 1–28.
1963c Observational learning in monkeys and apes. *Brit. J. Psychol.* 54: 201–226.

Hallowell, A. I.
1953 Culture, personality, and society. *In* A. L. Kroeber (ed.), *Anthropology Today,* pp. 597–620. Chicago: University of Chicago Press.

Hamburg, D. A.
1963 Emotion in the perspective of human evolution. *In* P. Knapp (ed.), *Expression of the Emotions in Man,* pp. 300–317. New York, International University Press.

Hardy, M.
1934 Observations on the innervation of the macula sacculi in man. *Anat. Rec.* 59: 403–418.

Hay, R. L.
1970 Silicate reactions in three lithofacies of a semi-arid basin, Olduvai Gorge, Tanzania. *Mineral. Soc. Amer. Spec. Pap.* 3: 237–255.

Hayes, C.
1951 *The Ape in Our House.* New York, Harper.

Hediger, H.
1961 The evolution of territorial behavior. *In* S. L. Washburn (ed.), *The Social Life of Early Man,* pp. 34–37. Viking Fund Publs. Anthrop., No. 31. Chicago, Aldine Publishing Co.

Herrick, C. J.
1946 Progressive evolution. *Science* 104: 469.
1956 *The Evolution of Human Nature.* New York, Harper and Brothers.

Hill, J. P.
1932 The developmental history of the primates. *Phil. Trans. Roy. Soc.* B. 221: 45–178.

HINSHELWOOD, C.
1959 The internal and external worlds. *Proc. Roy. Soc.* A. 253: 442–49.

HOCKETT, C. F., and ASCHER, R.
1964 The human revolution. *Curr. Anthrop.* 5: 135–68.

HOWELL, F. CLARK
1960 European and northwest African Middle Pleistocene hominids. *Curr. Anthrop.* 1: 195–232.
1967 Recent advances in human evolutionary studies. *Quart. Rev. Biol.* 42: 471–513.

HUBEL, D. H.
1963 The visual cortex of the brain. *Sci. Amer.* 209: 54–62.

HUXLEY, J. S.
1958 Cultural process and evolution. *In* A. ROE and G. G. SIMPSON (eds.), *Behavior and Evolution,* pp. 437–54. New Haven, Conn., Yale University Press.

IMANISCHI, K.
1960 Social organization of sub-human primates in their natural habitat. *Curr. Anthrop.* 1: 393–407.

INGE, W. R.
1922 *Outspoken Essays.* 2d ser. London and New York, Longmans, Green & Co.

ISAAC, G. L., LEAKEY, R. E. F., and BEHRENSMEYER, A. K.
1971 Archeological traces of early hominid activities, east of Lake Rudolf, Kenya. *Science* 173: 1129–1134.

ITANI, J. S.
1965 Social organization of Japanese monkeys. *Animals* 5: 410–17.

JERISON, H. J.
1955 Brain to body ratios and the evolution of intelligence. *Science* 121: 447–49.

JONES, F. WOOD
1916 *Arboreal Man.* London, Arnold.

KAWAI, M.
1965 Japanese monkeys and the origin of culture. *Animals* 5: 450–55.

KEESING, F. M.
1958 *Cultural Anthropology.* New York, Rinehart.

KINSEY, A. C., and Staff of the Institute for Sex Research, Indiana University.
1953 *Sexual Behavior in the Human Female.* Philadelphia, W. B. Saunders Co.

KINSEY, W. G.
1971 Evolution of the human canine tooth. *Amer. Anthrop.* 73: 680–694.

KÖHLER, W.
1925 *The Mentality of Apes.* New York and London, Kegan Paul.

KOHTS, N.
1935 *Infant Ape and Human Child.* Moscow, Sci. Mem. Mus. Darwin.

KORTLANDT, A.

1965 Comment on the essential morphological basis for human culture. *Curr. Anthrop.* 6: 320–25.

KRANTZ, G. S.

1963 The functional significance of the mastoid process in man. *Amer. J. Phys. Anthrop.* 21: 591–93.

KROEBER, A. L., and KLUCKHOHN, C.

1952 *Culture: A Critical Review of Concepts and Definitions.* Pap. Peabody Mus. No. 47.

KUMMER, H.

1971 *Primate Societies.* Chicago, Aldine Publishing Co.

KURTÉN, B.

1972 *Not from the Apes.* New York, Pantheon.

KUYPERS, H. G. J. M.

1964 The descending pathways to the spinal cord, their anatomy and function. *In* J. C. ECCLES and J. P. SCHADE (eds.), *Progress in Brain Research,* Vol. 2, pp. 178–202. New York, Amsterdam Elsevier.

KUYPERS, H. G. J. M., SZWARCBART, M. K., MISHKIN, M., and ROSVOLD, H. E.

1965 Occipitotemporal corticocortical connection in the rhesus monkey. *Exp. Neurology* 11: 245–262.

LANCASTER, J. B.

1968 Primate communication systems and the emergence of human language. *In* P. C. JAY (ed.), *Primates: Studies in Adaptation and Variability,* pp. 439–457. New York, Holt, Rinehart and Winston.

LANCASTER, J. B., and LEE, R. B.

1965 The annual reproductive cycle in monkeys and apes. *In* I. DEVORE (ed.), *Primate Behavior,* pp. 486–513. New York, Holt, Rinehart and Winston.

LEACH, E.

1966 Ritualization in man in relation to conceptual and social development. *Phil. Trans. Roy. Soc.* B. 251: 403–408.

LEAKEY, L. S. B.

1959 A new fossil skull from Olduvai. *Nature, London* 184: 491–493.

1960 Recent discoveries at Olduvai Gorge. *Nature, London* 188: 1050–1052.

1965 *Olduvai Gorge, 1951–1961.* Cambridge, Cambridge University Press.

1968 Bone smashing by late Miocene Hominidae. *Nature, London* 220: 3–9.

LEAKEY, L. S. B., TOBIAS, P. V., and NAPIER, J. R.

1964 A new species of the genus *Homo* from Olduvai Gorge. *Nature, London* 202: 3–9.

LEAKEY, M. D.

1970. Early artifacts from the Koobi Fora Area. *Nature, London* 226: 228–230.

1971 *Olduvai Gorge, Vol. 3. Excavations in Beds I and II, 1960–1963.* Cambridge, Cambridge University Press.

LEE, R. B.
1968 What hunters do for a living, or, how to make out on scarce resources. *In* R. B. LEE and I. DEVORE (eds.), *Man the Hunter,* pp. 30–48. Chicago, Aldine Publishing Co.

LERNER, I. M.
1954 *Genetic Homeostasis.* New York, Wiley.

LEWIS, G. E.
1934 Preliminary notice of new man-like apes from India. *Amer. J. Sci.* 27: 161–179.

LEWIS, O. J.
1973 The Hominoid *Os Capitatum,* with special reference to the fossil bones from Sterkfontein and Olduvai Gorge. *J. Human Evol.* 2: 1–11.

LIEBERMAN, P., CRELIN, E. S., and KLATT, D. H.
1972 Phonetic ability and related anatomy of the newborn and adult human, Neanderthal Man, and the Chimpanzee. *Amer. Anthrop.* 74: 287–307.

LOWE, C. VAN RIET
1954 The cave hearths. *S. Afr. Arch. Bull.* 33: 25–29.

LUMLEY, H. DE, and LUMLEY, M. A. DE
1971 Découverte de restes humains anténéandertaliens datés du début du Riss à la Caune de l'Arago (Tautavel, Pyrénées-Orientales). *C. R. Acad. Sci. Paris* 272: 1739–1742.

MAGOUN, H. W.
1960 Concepts of brain function. *In* SOL TAX (ed.), *The Evolution of Man,* pp. 187–209. Chicago, University of Chicago Press.

MALTHUS, T. R.
1798 *An Essay on the Principle of Population.* London, Johnson.

MARK, V. H., and ERVIN, F. R.
1970 *Violence and the Brain.* New York, Harper and Row.

MARTIN, C. J.
1902 Thermal adjustment and respiratory exchange in monotremes and marsupials—a study in the development of homoeothermism. *Phil. Trans. Roy. Soc. London,* B. 195: 1–37.

MASON, A. S.
1960 *Health and Hormones.* London, Penguin Books.

MAYR, E.
1963 *Animal Species and Evolution.* Cambridge, Mass., Harvard University Press; London, Oxford University Press.

McCOWN, T., and KEITH, A.
1939 *The Stone Age of Mount Carmel,* Vol. 2. Oxford, Oxford University Press.

McHENRY, H.
1973 Early hominid humerus from East Rudolf, Kenya. *Science* 180: 379–741.

MEDNICK, L. W.
1955 The evolution of the human ilium. *Amer. J. Phys. Anthrop.* 13: 203–16.
METTLER, F. A.
1956 *Culture and the Structural Evolution of the Neural System.* New York, Amer. Mus. Nat. Hist. Also *in* M. F. ASHLEY MONTAGU (ed.), *Culture and the Evolution of Man,* pp. 155–201. New York, Oxford University Press, 1962.
MILLER, G. S.
1931 The primate basis of human sexual behavior. *Quart. Rev. Biol.* 6: 379–410.
MILLS, J. R. E.
1963 Occlusion and malocclusion of the teeth of primates. *Symp. Soc. Study Hum. Biol.* 5: 29–51.
MONTAGNA, W.
1965 The skin. *Scient. Amer.* 212: 56–66.
MONTAGU, M. F. ASHLEY
1962 Time, morphology and neotony in the evolution of man. *In* M. F. ASHLEY MONTAGU (ed.), *Culture and the Evolution of Man,* pp. 324–42. New York, Oxford University Press.
1971 *Touching: The Human Significance of the Skin.* New York, Columbia University Press.
MORTON, D. J.
1922 Evolution of the human foot. *Amer. J. Phys. Anthrop.* Part I, 5: 305–306; Part II, 7: 1–52.
1927 Human origin. *Amer. J. Phys. Anthrop.* 10: 173–203.
1964 *The Human Foot.* New York, Haffner Publishing Co.
MOSS, M. L., and YOUNG, R. W.
1960 A functional approach to craniology. *Amer. J. Phys. Anthrop.* 18: 281–92.
MUSGRAVE, J.
1971 How dextrous was Neandertal Man? *Nature, London* 233: 538–541.
NAPIER, J. R.
1960 Studies of the hands of living primates. *Proc. Zool. Soc. London* 134: 647–57.
1961 Prehensibility and opposability in the hands of primates. *Symp. Zool. Soc. London* 5: 115–32.
1962 The evolution of the hand. *Scient. Amer.* 207: 56–62.
1963a Locomotor functions of hominids. *In* S. L. WASHBURN (ed.), *Classification and Human Evolution,* pp. 178–89. Viking Fund Publs. Anthrop., No. 37. Chicago, Aldine Publishing Co.
1963b Brachiation and brachiators. *Symp. Zool. Soc. London* 10: 183–94.
1964 The evolution of bipedal walking in the hominids. *Arch. Biol.* (Liége) 75 (suppl.): 673–708.

NAPIER, J. R., and DAVIS, P. R.
1959 *The Forelimb Skeleton and Associated Remains of* Proconsul Africanus. London, British Museum (Nat. Hist.).
NAPIER, J. R., and NAPIER, P. H.
1967 *A Handbook of Living Primates.* London and New York, Academic Press.
NAPIER, J. R., and WALKER, A.
1967 Vertical clinging and leaping: a newly recognised category of locomotor behavior of primates. *Folia Primat.* 7: 204–219.
NOBACK, C. R., and MOSKOWITZ, N.
1963 The primate nervous system: functional and structural aspects in phylogeny. *In* J. BUETTNER-JANUSCH (ed.), *Evolutionary and Genetic Biology of Primates* 1: 131–177. New York, Academic Press.
OAKLEY, K. P.
1951 A definition of man. *Penguin Science News* 20: 69–81. Also *in* M. F. ASHLEY MONTAGU (ed.), *Culture and Evolution of Man*, pp. 3–12. New York, Oxford University Press, 1962.
1961 On man's use of fire, with comments on toolmaking and hunting. *In* S. L. WASHBURN (ed.), *The Social Life of Early Man,* pp. 176–93. Viking Fund Publs. Anthrop., No. 31. Chicago, Aldine Publishing Co.
1963 *Man the Toolmaker.* 5th ed. London, British Museum (Nat. Hist.).
1969 *Frameworks for dating fossil man.* 3d. ed. London, Weidenfeld and Nicholson; Chicago, Aldine Publishing Co.
OAKLEY, K. P., CAMPBELL, B. G., and MOLLISON, T. I.
1967–74 *Catalogue of Fossil Hominids, 3 Vols.* London, British Museum (Nat. Hist.).
ORNSTEIN, R. E.
1972 *The Psychology of Consciousness.* San Francisco, Freeman.
OVEY, C. D.
1964 *The Swanscombe Skull.* Occ. Pap. Roy. Anthrop. Inst. 20. London.
OXNARD, C. E.
1963 Locomotor adaptations in the primate forelimb. *Symp. Zool. Soc. London* 10: 165–82.
1968 A note on the Olduvai clavicular fragment. *Amer. J. Phys. Anthrop.* 29: 429–432.
1969 Evolution of the human shoulder: some possible pathways. *Amer. J. Phys. Anthrop.* 30: 319–322.
PATTERSON, B., and HOWELLS, W. W.
1967 Hominid humeral fragment from early Pleistocene of northwest Kenya. *Science* 156: 64–66.
PENFIELD, W., and RASMUSSEN, T. B.
1957 *The Cerebral Cortex of Man.* New York, Macmillan Co.
PENFIELD, W., and ROBERTS, L.
1959 *Speech and Brain Mechanisms.* Princeton, N.J., Princeton University Press.

PFEIFFER, J.
1955 *The Human Brain,* New York, Harper & Bros.
PILBEAM, D. R.
1972 *The Ascent of Man.* New York, Macmillan.
PILBEAM, D. R., and SIMONS, E. L.
1965 Some problems of hominid classification. *Amer. Sci.* 53: 237–59.
POCOCK, R. I.
1925 The external characters of the catarrhine monkeys and apes. *Proc. Zool. Soc. London* 1925: 1479–1579.
POLYAK, S.
1957 *The Vertebrate Visual System.* Chicago, University of Chicago Press.
PREMACK, D.
1971 On the assessment of language competence in the chimpanzee. *In* A. M. SCHRIER and F. STOLLNITZ (eds.), *Behavior of Nonhuman Primates,* Vol. 4, pp. 185–228. New York, Academic Press.
RASMUSSEN, T. B., and PENFIELD, W.
1947 Further studies of the sensory and motor cerebral cortex in man. *Fed. Proc. Amer. Soc. expt. Biol.* 6: 452–560.
RENSCH, B.
1956 Increase of learning capability with increase in brain size. *Amer. Nat.* 90: 81–95.
1959 Trends toward progress of brains and sense organs. *Cold Spring Harbor Symp. Quant. Biol.* 24: 291–303.
1960 *Evolution above the Species Level.* New York, Columbia University Press.
ROBINSON, J. T.
1953 Telanthropus and its phylogenetic significance. *Amer. J. Phys. Anthrop.* 11: 445–502.
1956 *The Dentition of the Australopithecinae.* Transvaal Mus. Mem., No. 9. Pretoria.
1963 Australopithecines: culture and phylogeny. *Amer. J. Phys. Anthrop.* 21: 595–605.
1965 Homo "habilis" and the australopithecines. *Nature, London* 205: 121–24.
1972 *Early Hominid Posture and Locomotion.* Chicago, University of Chicago Press.
ROBINSON, J. T., and MASON, R. J.
1957 Occurrence of stone artifacts with australopithecines at Sterkfontein. *Nature, London* 180: 521–24.
ROE, A.
1963 Psychological definitions of man. *In* S. L. WASHBURN (ed.), *Classification and Human Evolution,* pp. 320–31. Viking Fund Publs. Anthrop., No. 37. Chicago, Aldine Publishing Co.
ROMER, A.
1945 *Vertebrate Paleontology.* Chicago, University of Chicago Press.

1956 *The Osteology of Reptiles*. Chicago, University of Chicago Press.
1962 *The Vertebrate Body*. Philadelphia and London, W. B. Saunders Co.

ROSEN, S. I., and MCKERN, T. W.
1971 Several cranial indices and their relevance to fossil man. *Amer. J. Phys. Anthrop.* 35: 69–74.

SADE, D. S.
1964 Seasonal cycle in size of testes of free-ranging *Macaca mulatta*. *Folia Primat.* 2: 171–80.

SAHLINS, M. D.
1959 The social life of monkeys, apes and primitive man. *Hum. Biol.* 31: 54–73.

SCHALLER, G. B.
1963 *The Mountain Gorilla: Ecology and Behavior*. Chicago, University of Chicago Press.

SCHALLER, G. B., and LOWTHER, G. R.
1969 The relevance of carnivore behavior to the study of early hominids. *Southwest. J. Anthrop.* 25: 307–341.

SCHENKEL, R.
1947 Ausdrucks-studien an wölfen. *Behavior* 1: 81–129.

SCHLOSSER, M.
1911 Beiträge zur kenntnis der oligozänen land-säugetiere aus dem fayum: Ägypten. *Beitrage Paleont. Geol. Oesterreich-Ungarns* 24: 51–167 (Plate I).

SCHULTZ, A. H.
1924 Growth studies on primates bearing upon man's evolution. *Amer. J. Phys. Anthrop.* 7: 149–64.
1930 The skeleton of the trunk and limbs of higher primates. *Hum. Biol.* 2: 303–438.
1935 Eruption and decay of teeth in primates. *Amer. J. Phys. Anthrop.* 19: 489–581.
1936 Characters common to higher primates and characters specific for man. *Quart. Rev. Biol.* 11: 259–83, 425–55.
1937 Proportions of long bones in man and apes. *Hum. Biol.* 9: 281–328.
1941 Length of spinal regions in primates. *Amer. J. Phys. Anthrop.* 24: 1–22.
1942 Conditions for balancing the head in primates. *Amer. J. Phys. Anthrop.* 29: 483–97.
1948 The number of young at birth and the number of nipples in primates. *Amer. J. Phys. Anthrop.* 6: 1–23.
1950 Man and the catarrhine primates. *Cold Spring Harbor Symp. Quant. Biol.* 15: 35–53.
1953 Relative thickness of bones in primates. *Amer. J. Phys. Anthrop.* 11: 277–311.

1955 The position of the occipital condyles and of the face relative to the skull base in primates. *Amer. J. Phys. Anthrop.* 13: 97–120.

1956 Post-embryonic age changes. *Primatologia* 1: 887–964.

1960 Einige Beobachtungen und Masse am Skelett von *Oreopithecus. Z. Morph. Anthrop.* 50: 136–49.

1961 Some factors influencing the social life of primates in general and early man in particular. *In* S. L. WASHBURN (ed.), *The Social Life of Early Man,* pp. 58–90. Viking Fund Publs. Anthrop., No. 31. Chicago, Aldine Publishing Co.

1963a The foot skeleton in primates. *Symp. Zool. Soc. London* 10: 199–206.

1963b Age changes, sex differences, and variability as factors in the classification of primates. *In* S. L. WASHBURN (ed.), *Classification and Human Evolution,* pp. 85–115. Viking Fund Publs. Anthrop., No. 37. Chicago, Aldine Publishing Co.

SHOCK, N. W.

1951 Growth curves. *In* S. S. STEVENS (ed.), *Handbook of Experimental Psychology,* pp. 330–46. New York, John Wiley & Sons.

SIMONS, E. L.

1964 On the mandible of *Ramapithecus. Proc. Nat. Acad. Sci. U.S.A.* 51: 528–36.

1965 New fossil apes from Egypt and the initial differentiation of the Hominoidea. *Nature, London* 205: 135–39.

1968 A source for dental comparison of *Ramapithecus* with *Australopithecus* and *Homo. S. Afr. J. Sci.* 64: 92–112.

1969 Late Miocene hominid from Fort Ternan, Kenya. *Nature* 221: 448–451.

SIMONS, E. L., and PILBEAM, D. R.

1965 Preliminary revision of the Dryopithecinae (Pongidae, Anthropoidea). *Folia Primat.* 3: 81–152.

SIMPSON, G. G.

1935 The first mammals. *Quart. Rev. Biol.* 10: 154–180.

1945 The principles of classification and a classification of mammals. *Bull. Amer. Mus. Nat. Hist.* 85: 1–350.

1949 *The Meaning of Evolution.* New Haven, Conn., Yale University Press.

SIVERTSEN, E.

1941 On the biology of the harp seal. *Hvalradets Skrifter* 26: 1–166.

SPERRY, R. W.

1955 On the neural basis of the conditioned response. *Brit. J. An. Behaviour* 3: 41–44.

1966 Brain bisection and consciousness. *In* J. C. ECCLES (ed.), *Brain and Conscious Experience.* Heidelberg, Springer-Verlag.

SPUHLER, J. N.

1959 Somatic paths to culture. *Hum. Biol.* 31: 1–13.

STEWART, T. D.

1962 Neanderthal cervical vertebrae. *Bibliotheca primatol.* Fasc. 1: 130–54.

STRAUS, W. L.

1936 The thoracic and abdominal viscera of the primates. *Proc. Amer. Phil. Soc.* 76:1.

1949 The riddle of man's ancestry. *Quart. Rev. Biol.* 24: 200–23.

1963 The classification of Oreopithecus. *In* S. L. WASHBURN (ed.), *Classification and Human Evolution,* pp. 146–77. Viking Fund Publs. Anthrop., No. 37. Chicago, Aldine Publishing Co.

STRAUS, W. L., and CAVE, A. J. E.

1957 Pathology and posture of Neanderthal man. *Quart. Rev. Biol.* 32: 348–63.

TANNER, J. M.

1962 *Growth at Adolescence.* 2d ed. Oxford, Blackwell Scientific Publications.

TESTUT, L.

1928 *Traité d'Anatomie humaine.* 8th ed. Paris, Doin & Co.

THOMPSON, D. W.

1942 *On Growth and Form.* New ed. Cambridge, Cambridge University Press.

THORPE, W. H.

1965 *Science, Man and Morals.* London, Methuen & Co.

TOBIAS, P. V.

1963 Cranial capacity of *Zinjanthropus* and other australopithecines. *Nature, London* 197: 743–46.

1967 *Olduvai Gorge.* Vol 2, *The Cranium of Australopithecus (Zinjanthropus) boisei.* Cambridge, Cambridge University Press.

1971 *The Brain in Hominid Evolution.* New York, Columbia University Press.

TOBIAS, P. V., and VON KOENIGSWALD, G. H. R.

1964 A comparison between the Olduvai hominines and those of Java and some implications for hominid phylogeny. *Nature, London* 204: 515–518.

TOBIAS, P. V., *et al.*

1965 New discoveries in Tanganyika: their bearing on hominid evolution. *Curr. Anthrop.* 6: 391–411.

TRIVERS, R.

1971 The evolution of reciprocal altruism. *Quart. Rev. Biol.* 46: 35–57.

TUTTLE, R. H.

1967 Knuckle-walking and the evolution of hominoid hands. *Amer. J. Phys. Anthrop.* 26: 171–206.

1969 Knuckle-walking and the problem of human origins. *Science* 166: 953–961.

TYLOR, E. B.

1871 *Primitive Culture.* London, Murray.

UDRY, J. R., and MORRIS, N. M.
1968 Distribution of coitus in the menstrual cycle. *Nature, London* 220: 593–596.

VAN VALEN, L., and SLOAN, R. E.
1965 The earliest primates. *Science* 150: 743–745.

WAGNER, P.
1960 *The Human Use of the Earth*. Glencoe, Ill., Free Press.

WALKER, A. C.
1973 New *Australopithecus* femora from East Rudolf, Kenya. *J. Hum. Evol.* 2. In press.

WALLS, G. L.
1963 *The Vertebrate Eye and Its Adaptive Radiation*. New York, London, Haffner Publishing Co.

WASHBURN, S.L.
1950 The analysis of primate evolution with particular reference to the origin of man. *Cold Spring Harbor Symp. Quant. Biol.* 15: 57–78.
1957 Ischial callosities as sleeping adaptations. *Amer. J. Phys. Anthrop.* 15: 269–276.
1963a The study of race. *Amer. Anthrop.* 65: 521–531.
1963b Behavior and human evolution. *In* S. L. WASHBURN (ed.), *Classification and Human Evolution*, pp. 190–203. Viking Fund Publs. Anthrop., No. 37. Chicago, Aldine Publishing Co.
1968 *The Study of Human Evolution*. The Condon Lectures, Oregon State System of Higher Education.

WASHBURN, S. L., and DEVORE, I.
1961 Social behavior of baboons and early man. *In* S. L. WASHBURN (ed.), *Social Life of Early Man*, pp. 91–105. Viking Fund Publs. Anthrop., No. 31. Chicago, Aldine Publishing Co.

WASHBURN, S. L., and HAMBURG, D. A.
1965 The implications of primate research. *In* I. DEVORE (ed.), *Primate Behavior*, pp. 607–622. New York, Holt, Rinehart and Winston.

WEIDENREICH, F.
1938 Discovery of the femur and humerus of *Sinanthropus*. *Nature, London* 141: 614–617.
1939–41 The brain and its role in the phylogenetic transformation of the human skull. *Trans. Amer. Phil. Soc.* 31: 321–442.
1945 *Giant Early Man from Java and South China*. Anthrop. Pap. Amer. Mus. Nat. Hist., No. 40, Part 1.

WEINER, J. S.
1954 Nose shape and climate. *Amer. J. Phys. Anthrop.* 12: 615–618.

WHITE, L. A.
1959 The concept of culture. *Amer. Anthrop.* 61: 227–251.

WHITEHEAD, A. N.
1920 *The Concept of Nature*. Cambridge, Cambridge University Press.

WHYTE, L. L.
1965 *Internal Factors in Evolution*. London, Tavistock Publications.

WILLIAMS, B. J.
1973 *Evolution and Human Origins*. New York, Harper and Row.

WILSON, A. C., and SARICH, V. M.
1969 A molecular time scale for human evolution. *Proc. Nat. Acad. Sci.* 63: 1088–1093.

WOLBERG, D. L.
1970 The hypothesized osteodontokeratic culture of the Australopithecinae: a look at the evidence and opinions. *Curr. Anthrop.* 11: 23–37.

WOO, J. K.
1965 Preliminary report on a skull of *Sinanthropus lantianensis* of Lantian, Shensi. *Scientia sinica* 14: 1032–36.

WOOLLARD, H. H.
1927 The retina of primates. *Proc. Zool. Soc. London* 1927: 1–17.

WYNNE-EDWARDS, V. C.
1965 Self-regulating systems in populations of animals. *Science* 147: 1543–48.

ZIEGLER, A. C.
1964 Brachiating adaptations of chimpanzee upper limb musculature. *Amer. J. Phys. Anthrop.* 22: 15–32.

ZUCKERMAN, S.
1930 The menstrual cycle of the primates. *Proc. Zool. Soc. London* 1930: 691–754.

Glossary

Italicized words appearing in the definitions are cross-references to other entries in the Glossary.

ABDUCTOR: Term applied to muscles that move a part of the body away from the middle line of the body. (Compare ADDUCTOR.)

ACETABULUM: A cup-shaped depression on the external surface of the *innominate bone* into which the head of the *femur* fits. (See Fig. 5.5.)

ACHILLES TENDON: Tendon of the heel that joins the *gastrocnemius* and *soleus* muscles to the *calcaneus* or heel bone. (See Fig. 5.15.)

ACROMIOCLAVICULAR JOINT: *Articulation* of *acromion* and *clavicle*. (See Fig. 6.2.)

ACROMION: The outer end of the spine of the *scapula* that projects over the *glenoid fossa:* it *articulates* with the *clavicle* at the *acromioclavicular joint*. (See Fig. 6.2.)

ADAPTATION (adj., ADAPTIVE): A *character* or set of *characters* of a *population* selected by the *environment* that by its existence improves the chance of survival of the population. Adaptations also allow exploitation of the *environment*.

ADDUCTOR: Term applied to muscles that move a part of the body toward the middle line of the body. (Compare ABDUCTOR.)

ADDUCTOR MAGNUS: A muscle, part of which extends the thigh in man and part of which adducts it; it arises on the lower borders of the *pubis* and *ischium*, including the *ischial tuberosity,* and is inserted in the shaft and *distal* end of the *femur*.

ADRENALIN: A *hormone* produced by the adrenal *glands* under conditions of stress or fear.

ALLANTOIS: A *fetal* membrane that in egg-laying *vertebrates* lies close to the shell and forms an *organ* of *respiration;* in *mammals* it contributes to the formation of the *placenta*.

AMINO-ACID: An organic compound based on a carbon chain or ring with amine (NH_2) and carboxyl (COOH) groups of atoms. Amino-acids are all synthesized by green plants, and some are synthesized by animals; in the case of man some eight out of about twenty must be obtained "ready made" and cannot be synthesized. *Proteins* are synthesized from amino-acids.

AMNION: The membrane that forms the fluid-filled sac in which the embryo grows.

AMNIOTE EGG: An egg containing an *embryo* provided with an *amnion.*

AMYGDALA: A component of the *limbic system* believed to be responsible for the expression of rage, and the level of activity of the *hypothalamus.* (See Fig. 8.12.)

ANATOMY: The science of the *morphology* and structure of *organisms.*

ANESTRUS (adj., ANESTROUS): A period of sexual quiescence between the mating seasons of *mammals;* a condition not regularly found in man except following *parturition.*

ANTERIOR: The front of, or front part of, any object or *organism.*

ANTERIOR INFERIOR ILIAC SPINE: A small protuberance of bone on the *anterior* edge of the *ilium* to which is attached the *rectus femoris* muscle, which flexes the thigh. (See Fig. 5.8.)

ANTERIOR SUPERIOR ILIAC SPINE: A small protuberance of bone on the *anterior* edge of the *ilium* to which are attached the *oblique abdominal muscles.* (See Fig. 5.5.)

ANTHROPOCENTRIC: A view of nature that places man in the center and examines the world from man's point of view.

ANTHROPOIDEA (adj., ANTHROPOID): A *suborder* of the *order Primates* containing the monkeys, apes, and men.

ANTIBODIES: *Proteins* in the blood of *mammals* that are formed in response to *antigens* and protect the body from their toxic action.

ANTIGEN: A substance which is foreign to the body of an animal and usually toxic to it. It is capable of stimulating the formation of an *antibody.*

APOCRINE: The term applied to *glands* in the skin that *secrete* the odorous components of sweat.

ARBOREAL: Refers to trees; tree-living.

ARCHIPALLIUM: The older part of the *cortex* of the *cerebral hemispheres,* distinguished from the *neopallium,* which evolved more recently. (See Fig. 8.5.)

ARCHOSAURIA: The dinosaurs, the great ruling reptiles (like *Tyrranosaurus*) which dominated the earth during *Cretaceous* and *Jurassic* times before the great *radiation* of the *mammals.* (See Fig. 4.5.)

AREOLA (pl., AREOLAE): A ring of pigmented skin surrounding the nipple.

ARTICULATION (adj., ARTICULAR): A joint or connection between two bones allowing movement.

ARTIFACTS: Objects formed by the purposeful activity of animals, especially man.

ARTIODACTYLS: The even-toed *ungulates*—an order of *mammals* including pigs, sheep, and cattle.

ASSOCIATION CORTEX: That part of the outer layer of the *cerebral hemispheres* believed to function as a result of the connection there of two or more other areas of the *cortex* concerned with sensory or motor activity.

ASTRAGALUS: See TALUS.

AUDITORY CORTEX: That part of the outer layer of the *cerebral hemispheres* believed to be responsible for the reception and onward transmission of auditory input to the brain.

AUSTRALOPITHECUS: A genus of fossil *Hominidae* first discovered in South Africa in 1924 and so named by Raymond Dart in 1925. It contains two species, *A. africanus* and *A. boisei.*

AXILLARY: Relative to the armpit.

AXILOSPINAL ANGLE: The angle between the *spine* of the *scapula* and the *lateral* margin of that bone.

AXONS: Nerve fibers.

BACULUM: A small rod-shaped bone which serves to stiffen the *penis* of many nonhuman *primates.*

BASAL GANGLIA: Part of the brain which forms a component of the *limbic system.*

BASAL METABOLIC RATE: The rate of heat production in an individual at the lowest level of body activity in the waking state.

BEHAVIOR: The totality of motor activity of an animal.

BICEPS: Refers here to the muscle that flexes the forearm in relation to the upper arm. (See Fig. 6.10.) Originating on the *scapula,* it is *inserted* upon the *bicipital tuberosity* of the *radius.*

BICIPITAL TUBEROSITY: A small protuberance upon the shaft of the *radius* where the *biceps* muscle is *inserted.* (See Fig. 6.10.)

BICUSPID: Having two *cusps.*

BILOPHODONT: Having the *cusps* of the tooth arranged to form two ridges with a valley between, characteristic of *Cercopithecoidea* among the *primates.* (See Fig. 7.13.)

BIOLOGICAL EFFICIENCY: The efficiency of a species in converting food materials into *biomass.*

BIOLOGY: The branch of science that deals with living *organisms.*

BIOMASS: The total mass of living substance that constitutes a particular species.

BIOSPECIES: A group of naturally or potentially interbreeding *populations* of animals or plants between which *gene flow* can occur, but which in nature is reproductively isolated from other such groups or populations.

BIPEDALISM: The act of locomotion on two feet.

BP: literally, years Before Present; in practice the ages of archaeological sites termed BP are calculated as years before 1950 AD.

BRACHIATION: The act of locomotion by means of the arms.

BRACHYCEPHALIC: Having a head that is relatively short and broad. (Compare DOLICHOCEPHALIC.)

BRAINSTEM: That part of the base of the brain which leads to the *spinal cord.*

BROCA'S AREA: An area on the *cerebral cortex* that Broca discovered to be in some way responsible for speech. (See Fig. 10.8.)

CALCANEUS: The heel bone. (See Fig. 5.15.)

CALOTTE: The bones of the roof of the skull.

CALVARIA (pl., CALVARIAE): The *neurocranium* without the facial bones.

CANTILEVER: A beam supported from one end.

CARNIVORA: An order of flesh-eating *mammals,* including lions, cats, and dogs.

CARNIVORE: An animal living exclusively or almost exclusively on other animals. (Compare HERBIVORE.)

CARPAL BONES: The small bones of the wrist. (See Fig. 6.15.)

CARTILAGE: A skeletal tissue of *vertebrates* distinct from bone, consisting of a resilient translucent matrix containing fibres.

CATARRHINA: An alternative term for the *Cercopithecoidea.*

CEBOIDEA (adj., CEBOID): The *superfamily* of *primates* found in Central and South America, containing 16 *genera* and known as New World monkeys. (See Table 3.2.)

CECUM: A short and blind-ended branch of the intestine, found in certain *herbivorous* mammals, that has evolved to facilitate the digestion of cellulose.

CENTER OF GRAVITY: A point, the support of which allows a body to remain balanced in any position.

CENTRAL NERVOUS SYSTEM (*CNS*): That part of the nervous system of animals which includes the brain and *spinal cord.*

CERCOPITHECOIDEA (adj., CERCOPITHECOID): A *superfamily* of *primates* found in the Old World, containing 14 *genera* and known as Old World monkeys. (See Table 3.2.)

CEREBELLUM: The *posterior* part of the brain, lying, in man, beneath the *occipital* lobe of the *cerebral hemispheres* and attached to the *brainstem.*

CEREBRAL CORTEX: The outer layer of the *cerebral hemispheres* or *cerebrum.* (see Fig. 8.8.)

CEREBRAL HEMISPHERES: The two parts of the *cerebrum* connected by the *corpus callosum.* (See Fig. 8.11.)

CEREBRUM: The principal and most *anterior* portion of the human brain, divided into two hemispheres. (See Fig. 8.8.)

CERVICAL REGION: The neck of the vertebral column.

CETACEANS: The *mammals* that have evolved *adaptations* to a marine *environment,* including whales, porpoises, and dolphins.

CHANNEL: Used here to denote the precise mode of communication between individual *organisms,* e.g., vocal-auditory, or chemical-olfactory.

CHARACTER: Refers here to a particular attribute of an animal.

CHEMORECEPTOR: A sensory *organ* of the nervous system that is stimulated by certain chemical substances (e.g., an *organ* of taste or smell.)

CHIRIDIA: Complex *organs* that terminate the limbs, e.g., paws, hands, or feet.

CHROMOSOME (adj., CHROMOSOMAL): A thread-shaped structure, numbers of which occur in the nucleus of every plant and animal cell. Chromosomes consist largely of *DNA* and carry the *genetic* code of the individual or *gamete.*

CHRONOLOGY: The science of the study of time, age, and sequence.

CHRONOSPECIES: A *biospecies* with a temporal dimension; that is, a lineage of interbreeding *populations* during a certain defined period of time; a fossil species.

CINGULUM (CINGULATE GYRUS): In Neurology, a part of the *cerebral cortex* which lies low down on the *medial* surface of the hemispheres; a component of the *limbic system.* (See Fig. 8.12.)

CLAN: A unilateral (one-line) descent group of large size.

CLASS: In biology, this is a taxonomic rank. (See TAXONOMY.)

CLAVICLE: The collarbone, which forms part of the *pectoral girdle* and connects the *scapula* and forelimb to the *sternum.* (See Fig. 6.2.)

CLITORIS: A small cylindrical *organ* situated at the anterior end of the *labia* in female *primates.* Its stimulation effects erection and erotic response under appropriate conditions. (See Fig. 9.2.)

COCCYX: A small bone at the end of the vertebral column consisting of a few much reduced *vertebrae,* present only in primates without tails. (See Fig. 5.5.)

COCHLEA: A cone-shaped cavity in the ear region of the skull consisting of a spinal canal containing the *organ of Corti.*

COITUS: Sexual intercourse.

COMPARATIVE ANATOMY: The study and comparison of the form and structure of different animals.

COMPOUND GIRDER: A beam or girder constructed of smaller units with spaces between them.

CONCEPT: A mental abstraction from experience involving classification of that experience.

CONCEPTUAL MEMORY: That part of the memory which records *concepts* as an analysis and classification of experience.

CONDITIONING: A kind of learning from which the resulting behavior pattern becomes fully automatic and involves no conscious thought processes.

CONDYLE: A kind of joint in the skeleton in which paired but separate bearing surfaces allow a limited hingelike movement, e.g., *occipital condyles, mandibular condyles.*

CONE: A sensory *organ* of the nervous system stimulated by light and able to distinguish differences in wave frequency; it is found in the *retina* of animals able to see color.

CONTACT RECEPTOR: A sensory *organ* of the nervous system stimulated by contact with objects, including *organs* sensitive to touch and taste.

COPULATION: Sexual union between male and female individuals; *coitus.*

CORACOID PROCESS: A beak-like projection of the *scapula* upon which is *inserted* the *biceps* muscle. (See Figs. 6.6 and 6.10.)

CORONOID PROCESS: A small prominence on the *mandibular ramus* in which the *temporalis* muscle is inserted. (See Fig. 7.2.)

CORPUS CALLOSUM: A mass of nerve fibers running across the *cerebrum* between the two *cerebral hemispheres.* (See Figs. 8.9 and 8.11.)

CORPUS LUTEUM (pl., CORPORA LUTEA): A small yellowish ductless *gland* that develops in the *ovary* immediately after *ovulation.* Its most important product is the *hormone progesterone*. If pregnancy does not occur, the *gland* rapidly degenerates.

CORTEX (adj., CORTICAL): The outer layer or mantle of an *organ*—used here to describe the important outer layer of the *cerebrum*.

CORTICOSPINAL FIBERS: Nerve fibers connecting the *cerebral cortex* and the *spinal cord*.

CRANIUM (adj., CRANIAL): The skull, excluding the mandible.

CRETACEOUS: A *period* of the earth's history believed to have lasted from about 135–63 million years ago. (See Table 2.1.)

CRURAL JOINT: The ankle joint of the hind leg. (See Fig. 5.15.)

CULTURE: The totality of *behavior* patterns of a social group of animals that are passed between generations by learning; socially determined *behavior* learned by observation, imitation, and instruction.

CUSP: A protuberance on the grinding surface of a tooth.

DECIDUATE: Here applied to a *placenta;* the term indicates that when the *placenta* is shed, it brings with it the outer layers of the *endometrium* with which it is intimately associated.

DELTOID: A muscle of the upper arm, with its *origin* in the *clavicle* and *scapula* and its *insertion* in the shaft of the *humerus*. (See Fig. 6.3.)

DEME: A local interbreeding population; a group of individuals so situated that any two of them have equal probability of mating with each other and producing offspring (Mayr 1963); equivalent to *population*.

DENTAL ARCADE: The curved line formed by the teeth in the *maxilla* and *mandible;* the dental arch.

DENTITION: The teeth as a whole.

DEOXYRIBONUCLEIC ACID (DNA): A chemical substance that constitutes *chromosomes* and carries the genetic code, the *genotype,* of each individual within every cell.

DEVELOPMENTAL HOMEOSTASIS: The capacity of the developmental pathways to produce a normal *phenotype* in spite of developmental or environmental disturbances (Mayr 1963).

DIASTEMA (pl., DIASTEMATA): The gap formed in each jaw to receive the canine tooth of the opposite jaw, found in animals with canines much larger than their other teeth.

DIGASTRIC MUSCLES: Muscles that lower the jaw and stabilize the *hyoid bone*. Their *origin* lies near the *mastoid process* at the base of the skull, and they are *inserted* into the lower border of the *symphyseal* region of the *mandible*. (See Fig. 10.7.)

DIGITIGRADE: A type of locomotion in which the ventral surfaces of the *digits* only carry the weight of the animal, and the tarsal and carpal bones are held clear of the ground, as in the dog or cat. (See PLANTIGRADE.)

DIGITS: Fingers or toes.

DIPHYODONTISM: The property of producing two sets of teeth during an individual's life—the milk teeth or deciduous teeth, and the permanent teeth.

DIPLOË: A sponge-like bony structure separating and joining the two layers of compact bone (the *inner* and *outer table*) of the *cranium*.

DISCRIMINATION: The act of perceiving slight differences.

DISTAL: The part of the body farthest from the middle line, or from the trunk; referring to the extremities of the body. (Compare PROXIMAL.)

DISTANCE RECEPTOR: A sensory organ of the nervous system stimulated by physical disturbances emitted by objects at a distance (e.g., light, sound, scent).

DNA: See DEOXYRIBONUCLEIC ACID.

DOLICHOCEPHALIC: Having a head that is relatively long and narrow. (Compare BRACHYCEPHALIC.)

DORSAL: Relating to the back; the opposite of *ventral.*

DRYOPITHECINAE: A *subfamily* of the *Pongidae* containing fossil apes of different *genera,* including *Dryopithecus* and *Gigantopithecus.*

DRYOPITHECUS: A fossil *genus* of the *subfamily Dryopithecinae* (of the *family Pongidae*) containing many *species.*

ECCRINE: The name of *glands* of the skin that secrete sweat.

ECOLOGICAL NICHE: See NICHE.

ECOLOGY: The science of the mutual relations of different *organisms* and their *environments.*

EFFECTOR: An organ that receives nerve impulses and reacts by movement or *secretion* (e.g., a muscle or a *gland*).

EMBRYO (adj., EMBRYONIC): Term applied to organisms during early stages of *ontogeny* while dependent on maternal food supplies; within the egg or during the early stages of gestation. (See also FETUS.)

ENDOCAST: A cast of the interior of the *cranial* cavity.

ENDOCRINE: Describes a system and its components which control bodily activity by chemical, as distinct from nervous, means. The chemicals produced are termed *hormones,* and the *glands* that produce them are termed "ductless glands."

ENDOGAMY: Inbreeding; the selection of a mate from within a small group by another member of that group. (Compare EXOGAMY.)

ENDOMETRIUM: The sponge-like *vascular* lining of the *mammalian uterus,* into which the *placenta* penetrates.

ENTROPY: The phenomenon of randomness, which, according to the second law of thermodynamics, in any closed system can only increase, never decrease.

ENVIRONMENT: The total surroundings of an individual.

EPIDERMIS: The outer layers of the skin.

EPIGAMIC: Relating to sexual reproduction or to *copulation,* and serving to attract or stimulate individuals of the opposite sex during courtship.

EPIGLOTTIS: A small plate of *cartilage* at the root of the tongue that folds back over the aperture of the *larynx,* covering it during the act of swallowing. (See Fig. 10.6.)

EPITHELIOCHORIAL PLACENTA: A *placenta* in which the maternal and *fetal* bloodstreams are separated by both maternal and *fetal tissues.* (Compare HEMOCHORIAL PLACENTA, and see Fig. 9.1.)

EQUABILITY: A relative absence of fluctuation or *variation;* refers here to climates with relatively small seasonal variation in temperature.

ERA: A span of time; geological term referring to certain major subdivisions in the earth's *chronology.* (See Table 2.1.)

ERECTOR SPINAE: Muscles of the back that help maintain the erect trunk in man. They find their *origin* on the *sacrum* and *ilium* and are *inserted* into the ribs and *vertebrae.*

ESOPHAGUS: Part of the digestive tract between the *pharynx* and the stomach.

ESTROGEN: A group of *hormones,* mainly produced in the *ovary,* which induce *estrus* and *ovulation* in a mature female individual. During growth, the increasing production of estrogen brings about the development of *secondary sexual characters* and *menarche.*

ESTROUS CYCLE: The series of *uterine, ovarian,* and other changes that occur in mammals and are responsible for *copulation,* pregnancy, etc.

ESTRUS: The stage of the *estrous cycle* around *ovulation,* when the female animal is receptive to and encourages males in *copulation.*

EVOLUTION: Cumulative change in the characteristics of populations of *organisms,* occurring in the course of successive generations related by descent.

EXOGAMY: Outbreeding; the selection of a mate from outside a small group of individuals. (Compare ENDOGAMY.)

EXPERIENTIAL MEMORY: That part of the memory which records experience in temporal sequence.

EXTENSOR: A muscle the contraction of which tends to straighten the bones about a joint. (Compare FLEXOR.)

EXTERNAL PTERYGOIDS: See LATERAL PTERYGOIDS.

EXTRAPYRAMIDAL SYSTEM: Descending *nerve tracts* that are interrupted by *synapses* and form relatively indirect links between the *brainstem* and the muscles. (See Fig. 3.10.)

FACILITY: an *artifact* designed and used to restrict or prevent motion or energy exchanges (such as dams or insulation); the simplest include all heat retaining structures and containers of liquids or solids (Wagner 1960).

FALLOPIAN TUBES: Two tubes leading from the two *ovaries* to the *uterus.* The eggs are shed into the open end of these tubes and so pass into the *uterus.*

FAMILY: Usually refers to a group of individuals related by blood and/or marriage including at least one adult male, one adult female, and one or more young, though it cannot be strictly defined. The term is also a taxonomic rank. (See TAXONOMY.)

FEMUR (pl., FEMORA): The *proximal* bone of the leg, the thighbone. The head of the *femur* fits into the *acetabulum.* (See Fig. 5.13.)

FENESTRA OVALIS: A small opening, sealed by a *membrane,* through which vibration is transmitted from the *middle ear* to the *cochlea.*

FETUS (adj., FETAL): The unborn young of a *viviparous* animal after it has taken form in the *uterus.*

FIBULA (pl., FIBULAE): One of the two *distal* bones of the leg. (See also TIBIA.)

FITNESS: In an evolutionary context, this term refers to the possibility of survival of a *population* over a long period of time in a changing environment.

FLEXOR: A muscle the action of which is to flex a joint. (Compare EXTENSOR.)

FOLLICLE, OVARIAN: or Graafian follicle—a *vesicular* body in the *ovary* containing the egg.

FOLLICLE, HAIR: Pit within mammalian skin surrounding and supporting the hair root.

FOLLICLE-STIMULATING HORMONE (FSH): A hormone produced by the *anterior pituitary gland* that stimulates the *ovarian follicles*, inducing their maturation and the liberation of *estrogen*.

FOSSA: Refers here to certain recognized hollows in bones, such as the fossa iliaca shown in Fig. 5.5.

FOSSIL MAN: Descriptive of fossil bones, skulls, etc., of great age, belonging to early man and other early *hominids*. Some of the most important specimens are listed in the Appendix to Chapter 3.

FOVEA CENTRALIS: A small pit in the surface of the *retina* where the *photoreceptor* cells are not overlaid by nerve fibers and blood capillaries, an area of the *retinal* surface permitting optimum optical *discrimination* (also called "yellow spot").

FRONTAL BONE: The bone of the skull that rises above and behind the *orbits* and forms the *anterior* part of the *neurocranium*.

FRONTAL LOBES: The *anterior* lobes of the *cerebral hemispheres*. (See Fig. 8.8.)

FSH: See FOLLICLE-STIMULATING HORMONE

FUNCTION: In *biology*, the activity of a biological mechanism.

FUNCTIONAL COMPLEX: A group of *anatomical* and *physiological* characters that jointly bring about a particular necessary and integrated activity in an animal.

GAMETE: The germ cells or sex cells that are produced by male and female sexually reproducing *organisms*; gametes of each sex must fuse together in fertilization to initiate the growth of a new individual.

GANGLION: An aggregation of nerve cells.

GASTROCNEMIUS: A muscle of the calf of the leg that has its *origin* in the *distal* end of the *femur* and its insertion by the *achilles tendon* into the *calcaneus*. (See Fig. 5.15.)

GENE (adj., GENETIC): A unit of the material of inheritance.

GENE FLOW: The passage of genes through a population or between *populations* over a period of time which is the result of sexual reproduction. *Endogamy* will restrict gene flow between populations, *exogamy* will increase such gene flow.

GENE POOL: The totality of *genes* of a given population existing at a given time (Mayr (1963).

GENETIC DRIFT: Random changes in the *gene pool* not due to *selection*, or immigration and characteristic of small *populations*.

GENETIC HOMEOSTASIS: The property of a population of balancing its *genetic* composition so as to resist sudden changes (Mayr 1963).

GENIAL TUBERCLE: A very small protuberance of bone on the inner surface of the *symphyseal* region of the *mandible* in which are *inserted* the *geniohyoid muscles* that anchor the tongue to the *mandible*.

GENIOHYOID: A muscle that lies above the *mylohyoid* on the floor of the mouth, and like it raises the *hyoid bone* or lowers the jaw, and carries the tongue. It has its *origin* on the *genial tubercles* of the *mandible* and is *inserted* on the *hyoid bone*. (See Fig. 10.6.)

GENOTYPE: The totality of *genetic* factors that make up the *genetic* constitution of an individual.

GENUS (pl., GENERA): An important taxonomic rank including a group of species which have more in common with each other than they have with other similar groups. (See TAXONOMY.)

GESTATION: The period of pregnancy during which the *embryo* (and *fetus*) develops and grows. It begins with *implantation* and ends with *parturition*.

GLAND: An *organ* producing a *secretion*, either onto the surface (as sweat glands) or into a cavity (as digestive glands) or into the bloodstream (the *endocrine* glands).

GLANS PENIS: The conical expansion that forms the end of the *penis;* it is very sensitive to mechanical stimulation.

GLENOID CAVITY: A socket formed by two *articular* depressions on the edge of the *scapula* into which the head of the *humerus* fits. (See Fig. 6.2.)

GLUTEUS MAXIMUS: A muscle that effects *abduction* of the thigh in nonhuman primates but is an *extensor* in man; it has its *origin* at the *posterior* end of the *iliac* crest and is *inserted* upon the upper part of the shaft of the *femur*. (See Fig. 5.8.)

GLUTEUS MEDIUS: A muscle that effects *abduction* of the thigh; it has its *origin* on the outer surface of the *ilium* and is *inserted* upon the *greater trochanter* of the *femur*. (See Fig. 5.11.)

GLUTEUS MINIMUS: A muscle that effects *abduction* of the thigh; it has its *origin* on the outer surface of the *ilium* and is *inserted* upon the *greater trochanter* of the *femur*. (See Fig. 5.11.)

GONADS: *Organs* that act both as *endocrine* glands and in the generation of *gametes;* in the male the *testes*, in the female the *ovaries*.

GORILLA: A living *species* of ape confined to central Africa—*Pan gorilla*. There are two subspecies, the lowland gorilla of West Africa and the mountain gorilla of the east-central region, on the Zaire-Uganda frontier.

GREATER TROCHANTER: A large bony prominence near the upper end of the *femur* in which are *inserted* the *gluteus medius* and *gluteus minimus* muscles. (See Fig. 5.13.)

HALLUX: The great toe; the first *digit* of the *pes*.

HAND-AXES: Stone artifacts of a particular shape and with certain functions first produced in the Middle *Pleistocene*. (See Fig. 11.4.)

HEMOCHORIAL PLACENTA: A *placenta* in which the maternal and *fetal* bloodstreams are separated only by *fetal* tissues; the maternal tissues have disintegrated. (Compare EPITHELIOCHORIAL PLACENTA, and see Fig. 9.1.)

HERBIVORE: An animal feeding exclusively on vegetable matter. (Compare CARNIVORE.)

HETERODONT: Having teeth of varying shapes: a characteristic, typical of *mammals,* that facilitates *mastication.* (Compare HOMODONT.)

HETEROSIS: The *selective* superiority of individuals with dissimilar paired *genes.*

HIPPOCAMPUS: A small structure which lies on the floor of the lateral ventricle of the brain and is a component of the *limbic system.* (See Fig. 8.12.)

HOMEOSTASIS (adj., HOMEOSTATIC): The maintenance of a dynamic equilibrium in living processes; the self-regulating property of *organic* systems.

HOMINIDAE (adj., HOMINID): The *family* of *primates* including man and fossil *species* related to him. Three *genera* are now recognized: *Ramapithecus, Australopithecus,* and *Homo.* (See Table 3.3.)

HOMINOIDEA (adj., HOMINOID): A *superfamily* of the *primates* containing the *families* of apes and men.

HOMO: A *genus* of the family *Hominidae* containing two *species: Homo sapiens,* which includes man, and *Homo erectus,* which is known only from fossils. Its earliest members probably date from about 1.3 million years ago.

HOMO ERECTUS: A *species* of the *genus Homo* known only from fossils and probably extant from about 1.3 million to about 300,000 years ago.

HOMO HABILIS: A name referring to a group of fossils discovered at Olduvai, Tanzania. The *species* is not included in Table 3.3 because the author believes that these specimens are better classified either as *Australopithecus africanus* or as *Homo erectus.*

HOMO SAPIENS: The most recent *species* of the genus *Homo,* which includes modern man. It is believed to date from about 300,000 years ago.

HOMODONT: Having teeth of similar shape; a characteristic of lower vertebrates. (Compare HETERODONT.)

HOMOIOTHERMY: The property, characteristic of birds and *mammals,* of maintaining the body at a constant temperature by means of a complex *homeostatic* mechanism.

HORMONE: A chemical substance formed in one part of the body by a ductless *gland* and carried by the blood to another part, which it stimulates to functional activity. The chemical messenger of the *endocrine* system.

HUMERUS (pl., HUMERI): The bone of the upper arm, *articulating* with the *scapula* above and the *radius* and *ulna* at its *distal* end. (See Fig. 6.10.)

HYLOBATINAE: A *subfamily* of the *Pongidae* including the smaller *genera* of apes that are today found in Asia. There are two living *genera,* the gibbon (*Hylobates*) and the siamang (*Symphalangus*).

HYOID: A small bone to which the tongue is anchored at its base. (See Figs. 10.6 and 10.7.)

HYPERTROPHY: Increase in size of a part or *organ.*

HYPOCONULID: The fifth *cusp* found on the lower molar teeth of most *Hominoidea.*

HYPOTHALAMUS: A small mass of nervous tissue at the base of the brain; an important control center in nervous and *endocrine* function. (See Fig. 8.11.)

HYPOTHESIS: A supposition advanced as a basis for reasoning or argument or as a guide to experimental investigation; a tentative theory.

ILIAC PILLAR: A thickening of the *ilium* that helps to resist the compression of the bone produced between the *acetabulum* and the *origin* of the muscles that flex the thigh.

ILIO-PSOAS: A muscle that flexes the *femur* on the trunk; it has its *origin* at the *lumbar* vertebrae and the internal surface of the *ilium* (iliac fossa) and is *inserted* upon the lesser trochanter of the *femur.*

ILIUM (adj., ILIAC): The hipbone, a large flat bone that forms part of the *pelvis.* (See Fig. 5.5.)

IMPLANTATION: The entry of the fertile egg of *mammals* into the *endometrial* lining of the *uterus;* it is the beginning of pregnancy.

INCEST: *Copulation* with an individual within a forbidden group of close kin.

INDEX OF CEPHALIZATION: An index that relates the weight of the brain to the weight of the body. Index $= \frac{\text{Brain weight}}{\text{Body weight}^{0.66}}$

INFERIOR: Lower, in relation to another structure. (Compare SUPERIOR.)

INNATE BEHAVIOR: Behavior which arises spontaneously in individuals without learning. Sometimes referred to as unlearned, such behavior is difficult to detect in higher *primates,* and can be identified by rearing an animal away from its kind but in an otherwise natural environment.

INNER EAR: That part of the ear which contains the *mechanoreceptors* for sound, balance, and movement.

INNER TABLE: The layer of compact bone that forms the inner surface of the *neurocranium.*

INNOMINATE BONE: The structure of bone formed by the fusion of the *ilium, ischium,* and *pubis* in the adult and constituting the *lateral* part of the *pelvis.* The two innominate bones articulate with each other *anteriorly* and with the *sacrum posteriorly.* Also called *os coxae.*

INSECTIVORA: An order of *mammals* that contains small insect-eating creatures now living (such as the shrew in Fig. 3.1) as well as certain very primitive forms.

INSERTION: Refers to the more movable of the two points of attachment of a muscle. (Compare ORIGIN.)

INSTINCT: Sometimes defined as an unlearned behavior pattern, sometimes as a conditioned behavior pattern, and sometimes as a psychological drive. The word has therefore little value in science.

INSTITUTION: An established form of *cultural* group *behavior*.

INTERMEMBRAL INDEX: An index relating the different lengths of the hindlimbs and forelimbs. Index $= \frac{\text{length of } humerus + radius}{\text{length of } femur + tibia} \times 100$.

INTERNAL ENVIRONMENT: The *environment* of the body cells, consisting primarily of a complex fluid maintained constant in composition and temperature by a wide range of *homeostatic* mechanisms.

INTERNAL FERTILIZATION: The fertilization of the egg within the body of the female rather than in water; an important adaptation of *terrestrial vertebrates* to dry land.

INTERNAL PTERYGOIDS: See MEDIAL PTERYGOIDS.

INTERNUNCIAL NEURON: A nerve cell that relays impulses from the descending nerve fibers to the *motor neuron,* which effects muscular contraction.

INTEROSSEUS MEMBRANE: Refers here to the *ligamentous membrane* conecting the *ulna* to the *radius* and transmitting forces of tension between them.

INTERSPINOUS LIGAMENTS: *Ligaments* connecting the *spinous processes* of the *vertebrae* and transmitting tension between them.

INTERVERTEBRAL DISC: A disc interposed between the bodies of adjacent *vertebrae,* consisting of an outer fibrous part and an inner gelatinous mass.

INTROMISSION: A term usually applied to the insertion of the *penis* within the *vagina.*

INVOLUTION: The opposite of *evolution*; the loss of organic *variety*; in contrast to evolutionary radiation, organic life becomes limited and *specialized.*

ISCHIAL CALLOSITIES: A thickening of the outer layer of the skin overlying the *ischial tuberosities,* found in Old World monkeys. (See Fig. 9.2.)

ISCHIAL SPINE: A small bony protuberance on the *posterior* margin of the *ischium* in which *originate* the muscles that form the floor of the abdomen. (See Fig. 5.5.)

ISCHIAL TUBEROSITY: A rough bony projection at the lower end of the body of the *ischium.* (See Fig. 5.8.)

ISCHIUM (adj., ISCHIAL): One of the three bones of the *innominate bone.* In adult *mammals* it is joined to the *ilium* and *pubis.* (See Fig. 5.5.)

ISOIMMUNIZATION: The development of an *antibody* as a result of *antigens* introduced from another individual; for example, *antigens* may pass from the *fetal* to the maternal blood streams during pregnancy and cause isoimmunization of the mother.

LABIA: Refers here to the two pairs of lips, the *labia majora* and *labia minora,* which protect the *clitoris, urethra,* and *vagina.*

LACTATION: The production of milk by *mammals* from the *mammae* or breasts following *parturition.*

LARYNX: The organ of voice production in man; the upper part of the respiratory tract between the *pharynx* and the *trachea.* (See Fig. 10.6.)

LATERAL: On the side; to the side of the midline. (Compare MEDIAL.)

LATERAL GENICULATE BODY: A swelling at the angle of each optic nerve tract, at the base of the brain, to the side of the midline. (See Fig. 3.12.)

LATERAL PTERYGOIDS: Also known as external pterygoids, these paired muscles have their *origin* in the *sphenoid bone* at the base of the skull and are *inserted* into the *mandible*. Contracted together, they move the *mandible* forward. (See Fig. 7.4.)

LATISSIMUS DORSI: These broad muscles of the back have their *origin* in the *thoracic* and *lumbar vertebrae, sacrum,* and *iliac crest,* and are *inserted* into the *humerus*. They *adduct,* rotate, and extend the forelimb. (See Fig. 6.3.)

LAW: A politically formulated rule supported by sanctions.

LEMNISCAL SYSTEM: A system of *nerve tracts* making a relatively direct link between the *peripheral* sensory nerve cells and the *cerebral cortex*. (See Fig. 3.11.)

LESSER TROCHANTER: A bony prominence near the upper end of the *femur,* in which are *inserted* some of the muscles that flex the thigh.

LH: See LUTEINIZING HORMONE.

LIGAMENT (adj., LIGAMENTOUS): A strong band of fibres connecting two bones at a joint, to restrict movement and prevent dislocation.

LIMB GIRDLE: A complex structure of bone that connects the limbs and trunk and transmits forces of compression and tension between them. (See PECTORAL GIRDLE and PELVIC GIRDLE.)

LIMBIC SYSTEM: A group of interconnected structures in the brain which lie below the *cerebral cortex* and which jointly generate motivation and mediate emotional responses, such as rage and fear. (See Fig. 8.12.)

LINGUAL: Adjective referring to the tongue.

LOAD ARM: That part of a beam which lies between the pivot and the point at which the load is applied. (See Fig. 7.6.)

LOAD LINE: Line of action of forces through a structure.

LUMBAR REGION: The third region of the spine, consisting of *vertebrae* to which ribs are not *articulated;* the lower back, between *thorax* and *pelvic girdle*.

LUNATE SULCUS: A furrow on the surface of the *cerebral hemispheres,* readily identifiable in *primates,* which forms a boundary to the *visual cortex;* sometimes termed *simian sulcus*. (See Fig. 8.5.)

LUTEINIZING HORMONE (LH): A hormone *secreted* by the *anterior pituitary gland;* its prime function is to stimulate development of the *corpus luteum*.

MALAR BONE: One of the bones of the face lying below the *orbit*. (See Fig. 8.1.)

MAMMAE: Milk-secreting *glands* of *mammals;* breasts.

MAMMALIA: A class of *vertebrate* animals the majority of which are characterized by *homoiothermy, mastication, viviparity,* and the secretion of milk for the nourishment of their young. They are divided into three *subclasses, Placentalia, Marsupialia,* and *Monotremata*.

MANDIBLE (adj., MANDIBULAR): The bone of the lower jaw that bears the lower *dentition*. (See Fig. 8.1.)

MANDIBULAR CONDYLES: Bony eminences on each side of the *mandible* that fit into a groove in the base of the skull and about which the jaw pivots. (See Fig. 7.2.)

MANDIBULAR CORPUS: The horizontal part of the *mandible* that carries the teeth. (See Fig. 7.2.)

MANDIBULAR RAMUS: That part of the *mandible* which carries no teeth and to which the masticatory muscles are attached; in higher *primates* it is formed at right angles to the *mandibular corpus*.

MANDIBULAR TORUS: A thickening in the *symphyseal* region of the *mandibular corpus* on the inner side, which helps to strengthen it. (See Fig. 7.9.)

MANUBRIUM: The upper segment of the *sternum*. (See Fig. 6.2.)

MANUS: The hand; in animals, the forepaw or front foot.

MARSUPIALIA: A *subclass* of *mammals* (including the opossum and kangaroo) living in America and Australia in which a *placenta* of a peculiar kind is developed, with the young born in a very undeveloped state. After birth they continue growth in a pouch containing the milk glands.

MASSETER: Paired muscles of *mastication;* each *originates* upon the *zygomatic arch* and is *inserted* upon the outer surface of the *ramus* and *coronoid process* of the *mandible*. (See Figs. 7.2 and 7.3.)

MASTICATION: The breakdown of foodstuffs by the dentition, involving cutting, chewing, grinding, tearing, etc.

MASTOID PROCESS: A bony prominence at each side of the base of the skull behind the ear. It carries the insertion of the *sternomastoid muscle*. (See Fig. 4.16.)

MATERIAL CULTURE: The totality of *artifacts* produced by a population of social animals, or men.

MAXILLA: The bone of the face constituting the upper jaw. (See Fig. 8.1.)

MECHANORECEPTOR: Sensory *organ* of the nervous system that is stimulated by pressure or movement; e.g., organs of touch, hearing, balance, etc.

MEDIAL: At the middle line—the line that divides a bilaterally symmetrical organism into two halves that are mirror images.

MEDIAL PTERYGOIDS: Also known as internal pterygoids, these paired muscles have their *origin* in the base of the skull (at the *sphenoid* and *maxilla* bones) and are *inserted* in the inner surface of the *mandibular ramus;* they close the jaw. (See Fig. 7.3.)

MEDULLA (adj., MEDULLAR): The center of a structure or *organ*.

MEGANTHROPUS: A *genus* of the family *Hominidae,* not now believed to be distinct from the *genus Australopithecus*. The name was given to a *mandibular* fragment discovered in Java in 1941.

MELANIN: A dark brown or black pigment.

MELANOCYTES: Special cells, the function of which is to produce *melanin;* they are normally present in the skin of *primates*.

MEMBRANE (adj., MEMBRANOUS): Fine sheet-like tissue lining parts of the body.

MENARCHE: The establishment during growth of the *menstrual cycle* in higher *primates,* as shown by *menstruation.*

MENSTRUAL CYCLE: A form of the *estrous cycle* found in man and some *primates,* characterized by a regular discharge of blood from the *uterus.*

MENSTRUATION: The regular discharge of blood from the *uterus* characteristic of some *primates;* it occurs more or less monthly and is an event in the *estrous cycle.*

MENTIFACTS: Assumptions, ideas, values, intentions (Huxley 1958).

MESOZOIC: A geological *era,* including the *Cretaceous, Jurassic,* and *Triassic epochs.* (See Table 2.1.)

METACARPUS (adj., METACARPAL): Small bones of the *manus* between the *carpal* bones and fingers, common to all *primates.* (See Fig. 6.15.)

METATARSUS (adj., METATARSAL): Small bones of the *pes* between the *tarsal* and toe bones common to all *primates.* (See Fig. 5.15)

MID-BRAIN: Part of the brain lying below the *cerebral hemispheres,* which is a component of the *limbic system.* (See Fig. 8.12).

MIDDLE EAR: Small cavity in the bones of the skull through which sound vibrations are transmitted by small bones, between the eardrum and the *fenestra ovalis.*

MIOCENE: A geological *period* of the *Tertiary era,* believed to have lasted from about 25 million to 12 million years ago.

MOBILITY: The ability to be moved by an outside force. (Compare MOTILITY)

MOMENT OF BENDING: A moment (rotational force) tending to cause a beam to bend; a product of the magnitude of the force upon the beam and the perpendicular distance between the pivot and the line of action of the force.

MONOTREMATA: A *subclass* of *mammals* confined to Australia and New Guinea, clearly distinguished from other *mammals,* since they lay eggs and possess many other reptilian features.

MORPHOLOGICAL STATUS: An estimate, based on *morphology,* of the position of an *organism* in its relation to other *organisms.*

MORPHOLOGY: The science of the form and configuration of animals.

MORTALITY RATE: The death rate, usually calculated as the number of deaths per annum per 1000 of the population.

MOTILITY: The ability to move actively, peculiar to animals. (Compare MOBILITY.)

MOTOR CORTEX: A part of the *cerebral cortex* that has been discovered to function as a transmitter and receiver of nerve impulses associated with muscular contraction.

MOTOR GYRUS: A prominent rounded elevation on the surface of the *cerebral hemispheres* that includes the *motor cortex.*

MOTOR NEURON: A *neuron* whose *axon* connects to a muscle fibre and which transmits impulses from the *central nervous system* to effect muscular contraction.

MUCOPERIOSTEUM: The fine *membranous* lining of bone characteristic of the roof of the mouth and the interior of the nose, which, in the latter, carries the *olfactory* receptors.

MUSCULATURE: The arrangement of muscles on the skeleton of an animal.

MUTAGENIC: Having the property of inducing *mutation*.

MUTATION: Refers in *biology* to a sudden and relatively permanent change in a particular *gene* or in *chromosomal* structure.

MUZZLE: That part of the face containing the nasal cavity, the *turbinal bones*, and *olfactory* organ; the protruding nose, typical of dogs.

MYLOHYOID: A muscle that forms the floor of the cavity of the mouth, and either raises the *hyoid bone* or lowers the jaw. It arises on the inner margin of the *mandible* and is *inserted* on the *hyoid bone*. (See Fig. 10.6.)

MYOTOMES: Muscles lying to the side of the spine that, on contraction, cause *lateral* curvature of the spine; characteristic of lower *vertebrates*, fish, etc.

NATURAL SELECTION: The principal mechanism of *evolutionary* change described by Darwin in 1859; the mechanism whereby those individuals best *adapted* to the environment contribute more offspring to succeeding generations than do the remainder, so that as their characteristics are inherited, the composition of the population is changed.

NEANDERTAL: A valley in western Germany famous for the discovery in a limestone cave in 1856 of the first fossil man, clearly distinct from modern man, to be identified as such.

NEGATIVE FEEDBACK: A concept that refers to the mechanism of *homeostasis*, whereby changes in one direction effect adjustment in an opposite direction.

NEOPALLIUM: The *phylogenetically* more recent part of the *cerebral hemispheres* that has evolved since the older *archipallium* and overlaid it in man. (See Fig. 8.5.)

NEOTENY: An *evolutionary* change in which young states of development of early forms persist into the maturity of later forms by the evolution of earlier sexual maturation or the retarded development of individual structures.

NERVE PATHWAYS: See NERVE TRACTS.

NERVE TRACTS: Bundles of nerve fibers connecting different points of the nervous system.

NEUROCRANIUM: That part of the skull which encloses the brain; the brain box, excluding the jaws and facial bones.

NEURONS: Nerve cells; special cells the function of which is to transmit electrical impulses along very long extremities—the nerve fibers or *axons*.

NICHE: That part of the *environment* occupied by a *species* or *subspecies* with particular reference to food and other natural resources upon which the species depends for survival.

NOMEN (pl., NOMINA): Refers here to the latin names given to organic *species* according to the International Code of Zoological Nomenclature.

NUBILITY: The stage in the growth of a girl at which eggs are regularly produced; the beginning of the reproductive span of life.

NUCHAL: Adjective referring to the neck region at the back of the skull; the nuchal area on the *occipital* bone is where the nuchal muscles are inserted. (See Fig. 4.15.)

NUCHAL CREST: A bony crest that develops on the skull of some *primates* at the boundary of the *nuchal* area and serves to increase the actual area of attachment of the *nuchal* muscles upon the *occipital* bone.

OBLIQUE ABDOMINAL MUSCLES: Muscles forming the sidewalls of the abdominal cavity; they have their origin in the lower ribs and their insertion in the crest of the ilium.

OCCIPITAL: Adjective referring to the back of the head, and in particular to the bone that forms the back of the skull. (See Fig. 8.1.)

OCCIPITAL CONDYLES: A pair of rounded *articular* surfaces on the *occipital* bone at the base of the skull that form the joint between the skull and the first *cervical vertebra.* (See Fig. 7.10.)

OCCIPITAL LOBE: That part of the *cerebral hemispheres* which lies at the back of the head. (See Fig. 8.8.)

OCCLUSAL PLANE: The plane of *occlusion* of the teeth; the plane on which the teeth meet when the jaw is closed. (See Fig. 7.5.)

OCCLUSION: The way in which the dentitions of the upper and lower jaws come together when the jaws are closed.

OLFACTORY APPARATUS: The totality of structures that contribute to the sense of smell.

OLFACTORY BULBS: The two small bulb-like extremities of the olfactory region of the brain that receive the olfactory nerves from the nose. (See Fig. 8.5.)

OLIGOCENE: A geological *period* of the *Tertiary era,* believed to have lasted from about 36 to 25 million years ago.

ONTOGENY: The course of development and growth during the life of an individual. (Compare PHYLOGENY.)

OPTIC CHIASMA: The structure formed beneath the forebrain by nerve fibers from the right eye crossing to the left side of the brain and vice versa. (See Fig. 3.12.)

ORBIT: A cavity in the skull surrounded by a ring of bone that contains and protects the eyeball.

ORDER: A taxonomic rank; see TAXONOMY.

OREOPITHECUS: A *genus* of the *Hominoidea* from the *Pliocene period,* not clearly related to either *Hominidae* or *Pongidae*.

ORGAN: Any part of an animal that forms a structural or functional unit.

ORGAN OF CORTI: The auditory *receptor.*

ORGANISM: An individual living thing.

ORGASM: The culmination of *copulation,* characterized by the pleasurable release of nervous tension, by muscular contraction, and ejaculation of semen by the male.

ORIENTED: Past participle of verb "to orient," which describes the direction in which an object lies.

ORIGIN: Refers here to the less movable of the two points of attachment of a muscle. (Compare INSERTION.)

ORTHOGENESIS: *Evolution* of lineages supposedly following a predetermined pathway not subject to natural selection.

OS CALCIS: See CALCANEUS.

OS COXAE: See INNOMINATE BONE.

OS PENIS: See BACULUM.

OSTEOLOGY: The science of the skeleton and its structure; the study of bones.

OUTER EAR: That part of the ear which is visible externally and consists of a flap of skin upon *cartilage*.

OUTER TABLE: The layer of compact bone that forms the outer surface of the *neurocranium*.

OVARIAN FOLLICLE: See FOLLICLE, OVARIAN.

OVARY: The *gonad* of female animals that produces eggs and, in *mammals, hormones*.

OVIDUCT: The duct that carries the eggs from the *ovary* to the exterior; in *mammals* it consists of three parts, the *fallopian tube*, the *uterus*, and the *vagina*.

OVULATION: The discharge of a ripe egg from the *ovarian follicle* into the opening of the *fallopian tubes*.

PALMAR: Relating to the palms of the hands and feet; the *volar* areas of the *chiridia*.

PAN: A well-known *genus* of the family *Pongidae* found in equatorial Africa comprising the chimpanzee and gorilla.

PAPILLA: A small projection on the skin.

PAPIO: The baboon—a well-known *genus* of the Old World monkeys, the *Cercopithecoidea;* the baboon is a common *quadrupedal*, ground-living form found throughout much of Africa.

PARABOLIC GIRDER: A girder or beam of parabolic form used in the construction of *cantilevers* of certain types.

PARIETAL BONE: The bone that forms the side of the *neurocranium;* the parietal bones meet at the midline on top of the skull and are elsewhere fused, in adults, with the *occipital, temporal,* and *frontal bones.* (See Fig. 8.1.)

PARIETAL LOBE: Those parts of the *cerebral hemispheres* which lie approximately under the *parietal bones* of the skull, between the *frontal* and *occipital lobes*. (See Fig. 8.8.)

PARTURITION: Giving birth, childbirth.

PATELLA: The kneecap; a small disc of bone lying over the knee joint that transmits the tension developed by the *quadriceps* muscle to the *tibia.*

PECTORAL GIRDLE: The structure of bones that suspends the body between the forelimbs in quadrupeds and that suspends the arms from the body in man; in *primates,* it consists of two bones on each side, the *clavicle* and the *scapula.*

PECTORALIS MAJOR: A muscle that has its *origin* in the *clavicle, sternum,* and ribs and its *insertion* in the *humerus*. It *adducts* and rotates the arm. (See Fig. 6.2.)

PEDOMORPHOSIS: See NEOTENY.

PELVIC CANAL: The central opening of the bony ring of the *pelvis* through which the young must pass at birth to reach the external world.

PELVIC GIRDLE: The structure of bones that transmits the weight of the body to the hindlimbs. Unlike the *pectoral girdle*, it is fused to the vertebral column, where the *sacral vertebrae* themselves are fused together. (See PELVIS.)

PELVIS (pl., PELVES; adj., PELVIC): A bony structure consisting of the two *innominate bones* fused together *anteriorly* and to the *sacrum posteriorly*, forming a basin-like ring of bone in man. (See also PELVIC GIRDLE.)

PENIS: The male sex organ evolved as an organ of internal fertilization containing erectile tissues and, in most primates, a small bone, the *baculum*.

PENTADACTYLY: The possession of five *digits* on *manus* and *pes*.

PERCEPT: The mental product of perception, a mental construct quite distinct from the thing perceived.

PERIOD: Refers here to subdivisions in the earth's *chronology*. (See Table 2.1.)

PERIPHERAL: The opposite of central; applied to the surface of an *organ* or the body of an animal.

PERISSODACTYLS: The odd-toed *ungulates*, an *order* of *mammals* including the rhinoceros and horse.

PES: The foot; in animals the paw or hind foot.

PHALANGEAL FORMULA: A simple formula indicating the order of length of the *digits*.

PHALANGES: See PHALANX.

PHALANX (pl., PHALANGES): The small bones of the *digits peripheral* to the *metacarpals* and *metatarsals;* the finger and toe bones.

PHARYNX: The throat, the shared continuation of the *esophagus* and *trachea*. (See Fig. 10.6.)

PHENOTYPE: The sum of the characters manifest in an *organism*, to be contrasted with the *genotype;* the phenotype is formed by the interaction of the fertilized egg and its *environment*, in the process called growth.

PHEROMONE: A chemical substance produced by an animal, either by a scent *gland*, or as a waste product, which acts as a signal in communication.

PHONATION: The act of making speech sounds.

PHOTORECEPTOR: The sensory organ of the nervous system that is stimulated by electromagnetic waves of certain frequencies (380–760 $m\mu$), the nerve impulses from which are interpreted in the brain as light.

PHYLOGENY (adj., PHYLOGENETIC): The *evolutionary* lineage of *organisms;* their *evolutionary* history.

PHYSIOLOGY: The science of organic function, of the processes of *organisms* that constitute their life.

PIRIFORM LOBE: A lobe of the *archipallium* associated with the analysis of *olfactory input*. (See Fig. 8.5, *A*.)

PITHECANTHROPUS: A generic name previously given to certain hominid fossils now usually classified as *Homo erectus*.

PITOCIN: A *hormone* produced by the *posterior pituitary gland,* responsible for muscular contractions of the *uterus.*

PITUITARY BODY: A compound *gland* lying beneath the base of the brain close to the *hypothalamus;* it has been described as the "master" gland of the *endocrine system.* (See Fig. 8.11.)

PLACENTA: An *organ* peculiar to *mammals* consisting of *embryonic* tissues evolved to absorb nourishment from the wall of the *uterus* and there to discharge waste products; it is connected to the *fetus* by the umbilical cord and produces *hormones* that keep the *uterus,* and indeed the mother, adapted to the state of pregnancy.

PLACENTALIA: A *subclass* of *mammals* with world-wide distribution, in which a *placenta* is formed from the *fetal allantois* for the nourishment of the *fetus.* This most widespread *subclass* includes *primates, ungulates,* and many other *orders.* (See Fig. 2.1.)

PLANTAR LIGAMENTS: The *ligaments* of the sole of the foot that maintain the arched form of the foot bones.

PLANTIGRADE: A type of locomotion in which the whole *ventral* surface of the foot comes into contact with the ground. (See DIGITIGRADE.)

PLANUM TEMPORALE: An area on the upper surface of each *temporal lobe* which lies in the *Sylvian sulcus* and shows differential development on the two sides in many individuals. (See Fig. 8.9.)

PLASTICITY: Refers here to the plasticity of the *phenotype* during *ontogeny;* its variable response to differences in *environment* in spite of constancy in the *genotype*—a more obvious character of plants than of animals.

PLATYRRHINA: An alternative term for the *Ceboidea,* or New World monkeys.

PLEISTOCENE: A geological *period* of the *Quaternary era,* believed to have lasted from about 2 million to 10,000 years ago.

PLIOCENE: A geological *period* of the *Tertiary era,* believed to have lasted from about 12 million to 2 million years ago.

POLLEX: The first *digit* of the *manus,* the thumb.

POLYANDRY: That form of polygamy in which one woman is formally permitted to cohabit with more than one man.

POLYGYNY: That form of polygamy in which one man has several wives. (See POLYANDRY.)

PONGIDAE (adj., PONGID): The *family* of *primates* including living apes such as gorilla, chimpanzee, and orang-utan, and many fossil forms. (See Table 3.3.)

PONGO: The orang-utan, a *genus* of the *family Pongidae* confined to Sumatra and Borneo.

POPULATION: A local or breeding population; a group of individuals so situated that any two of them have an equal probability of mating with each other, which generally find their mates within the group, but which are also able to mate with members of neighboring populations; a *deme.*

POSTERIOR: Behind or after; back of any organism.

POWER ARM: That part of a beam which lies between the pivot and the point at which (muscular) forces are exerted. (See Fig. 7.6.)

PRECISION GRIP: A grip, characteristic of the human hand, in which the tip of the thumb can be opposed to the tips of the other fingers to give a precise yet firm grip. (See Fig. 6.14.)

PREFRONTAL AREA: The most *anterior* part of the *cerebral cortex,* so named because it lies to the front of the frontal area on the *frontal lobes.* (See Fig. 8.8.)

PRIMATES: An *order* of the *class Mammalia,* characterized by *arboreal adaptations* and including man himself.

PROGESTERONE: A *hormone* that is *secreted* mainly by the *corpus luteum* and that prepares the *endometrium* for *implantation* and brings about many of the changes associated with pregnancy.

PROGNATHISM: With jaws projecting beyond the rest of the face.

PROLACTIN: A *hormone secreted* by the *anterior pituitary,* responsible for the onset of milk production after *parturition.*

PRONATION: Position, or rotation toward the position, of the forelimb so that the *manus* is twisted through 90 degrees relative to the elbow, the *radius* and *ulna* being crossed. (Compare SUPINATION.)

PROPLIOPITHECUS: A *genus* of fossil ape from the *Oligocene epoch* in Egypt.

PROPRIOCEPTORS: A sensory *organ* of the nervous system that is found in muscles as well as other parts of the body and detects stretch or contraction.

PROSIMII (adj., PROSIMIAN): A *suborder* of the *order Primates* containing various Old World *genera,* including the lemurs, lorises, and tarsiers.

PROTEIN: A very complex organic compound containing chains of *amino-acid* molecules; proteins occur in infinite variety and are the basis of most living substance.

PROXIMAL: Part of the body nearest to the trunk or to the middle line. (Compare DISTAL.)

PSYCHOLOGY: The science of conscious life, of mental and emotional processes.

PTERYGOID MUSCLES: Muscles controlling movement of the lower jaw. (See LATERAL PTERYGOIDS and MEDIAL PTERYGOIDS.)

PTYALIN: A digestive chemical present in the saliva of some *mammals,* including man, which brings about the breakdown of starch into sugar.

PUBIC SYMPHYSIS: The *anterior* area of union of the two *innominate bones* —the two pubic bones. Fusion at this point is completed in man only at full maturity. (See Fig. 5.3.)

PUBIS: A bone that, with the *ilium* and *ischium,* forms the *innominate bone;* it is the most *anterior* of the three. (See Fig. 5.5.)

PULVINAR: The most *posterior* part of the *thalamus;* it is significantly larger in higher *primates* than in lower forms.

PYRAMIDAL SYSTEM: Descending *nerve tracts* that form a relatively direct link between the cerebrum and the muscles. (Compare EXTRAPYRAMIDAL SYSTEM.)

QUADRUPEDALISM: The act of locomotion upon four feet.

QUATERNARY: The most recent geological era, believed to date from about 2 million years ago and containing the *Pleistocene* period, as well as recent time.

RACE: A group of *populations* of a *species* which are distinct in at least a few *characters* from other races of the same *species*. Very similar in meaning to the terms variety and subspecies.

RADIATION: Refers here to *evolutionary* radiation of a variety of *species* from a single *species,* all bearing a proportion of characters in common.

RADIUS (pl., RADII): The *lateral* and shorter of the two bones of the forearm. (See Fig. 6.10.)

RAMAPITHECUS: A *genus* of fossil *Hominoidea* from the *Pliocene* and/or *Miocene periods* in northern India and Kenya, classified in the *family Hominidae*.

RAMUS (adj., RAMAL): Literally a branch; here commonly applied to the vertical part of the *mandible* upon which is *inserted* the musculature and which bears no teeth.

RECEPTORS: A term applied to *organs* of the nervous system with a sensory function.

RECTUS ABDOMINIS: A muscle of the abdominal wall *originating* on the *pubic* bone and *inserted* upon the *sternum* and lower ribs, which maintains tension between *pelvis* and *thorax*. (See Fig. 4.18.)

RECTUS FEMORIS: One of four *extensors* of the knee joint, this muscle has its *origin* in the *anterior inferior iliac spine* and upper margin of the *acetabulum,* and its *insertion* through the *patella* to the *tibia*.

REFERENT: An object referred to; an external object that is perceived.

REFLEXES: An involuntary reaction on the part of an *organism* to a particular stimulus; reflexes may be *innate* or conditioned by learning.

REPTILES: A class of *vertebrates,* usually *terrestrial,* dominant in the *Mesozoic era*. They evolved from amphibians, and in turn *mammals* evolved from them. Examples include alligators, lizards, and snakes.

RESPIRATION: Refers here to the inhalation of air for the absorption of oxygen into the bloodstream, and its exhalation, together with carbon dioxide, by the lungs.

RETICULAR SYSTEM: An indirect and *phylogenetically* old system of ascending *nerve tracts* forming a relay of *neurons* between the sensory *neurons* and the *thalamus* and *cortex*. (See Fig. 3.11.)

RETINA (adj., RETINAL): Part of the eye containing nerve *receptors* (*rods* and *cones*) sensitive to light. (See Fig. 3.12.) These receptors form a dense layer on the inner surface of the eyeball.

RHINAL SULCUS: A *sulcus* on each *cerebral hemisphere* that separates the *archipallium* and *neopallium*. (See Fig. 8.5.)

RHINARIUM: The moistened, hairless, tactile-sensitive skin that surrounds the nostrils of many mammals, seen typically in the dog.

RHINENCEPHALON: The *olfactory* brain; that part of the brain concerned with olfaction, the sense of smell.

RHODESIAN MAN: A fossil skull and other bones discovered in a mine at Broken Hill, Zambia, in 1921.

RITES DE PASSAGE: Rituals connected with important stages in the development of individuals as members of society.

RODENT: An *order* of *mammals* characterized by teeth evolved for gnawing; it is the largest *order,* with 350 *genera,* including rats, mice, squirrels, and porcupines.

RODS: A sensory cell of the nervous system stimulated by light, especially of very small intensities. (See CONES and RETINA.)

SACCULUS: A sensory *organ* that, with the *utricle,* is sensitive to the direction of gravity and changes in that direction; part of the *inner ear.* (See Fig. 8.7.)

SACRUM (adj., SACRAL): A single curved bone that is part of the *pelvis,* formed in man by the fusion of five sacral *vertebrae.* (See Fig. 5.5.)

SADDLE JOINT: A joint with a saddle-shaped *articular* surface allowing movement in two planes as well as rotation.

SAGITTAL: The sagittal plane of an *organism* is that through the midline dividing the body into two symmetrical halves.

SAGITTAL CREST: A crest that develops along the *sagittal* line on top of the skull in certain *primates* and serves to increase the area of *origin* of the *temporal muscles.* (See Fig. 8.2.)

SALTATION: Jumping; a mode of progression found among certain *prosimians,* usually with the backbone erect.

SANCTION: A mechanism whereby society as a whole enforces *behavior* patterns upon individuals, for example by punishment.

SCAPULA (pl., SCAPULAE): A bone of the *pectoral girdle,* the shoulder blade. (See Figs. 6.2 and 6.3.)

SEBACEOUS GLANDS: Oil-producing *glands* of the skin.

SECONDARY SEXUAL CHARACTER: A *character* peculiar to males or females but without a function directly related to reproduction.

SECRETION (vb., SECRETE): An activity, specialized in *gland*-cells, involving the passage of a substance produced by a cell from within it to the surrounding tissues.

SECTORIAL: Cutting; referring to teeth that have evolved a cutting edge.

SELECTION (adj., SELECTIVE): As used here, this term refers to *natural selection.*

SEMATIC: Acting as a signal to other animals.

SEMICIRCULAR CANALS: Sensory *organs* of the *inner ear* evolved to detect the direction and acceleration of movement. (See Fig. 8.7.)

SENSORY CORTEX: See SOMATIC SENSORY CORTEX, VISUAL CORTEX, AUDITORY CORTEX.

SERRATUS ANTERIOR: This muscle has its *origin* in the lower ribs and its *insertion* in the *scapula;* it rotates the *scapula* and pulls it forward. (See Fig. 6.3.)

SEXUAL DIMORPHISM: The characteristic differences between the sexes of a single *species.*

SIMIAN SHELF: A small bony shelf on the inner surface of the *symphyseal* region of the *mandible,* serving to strengthen it. (See Fig. 7.9.)

SIMIAN SULCUS: See LUNATE SULCUS.

SINANTHROPUS: A generic name previously given to certain hominid fossils now usually classified as *Homo erectus.*

SINUS: A hollow in a bone.

SOCIFACTS: Acts of cultural *behavior.*

SOFT PALATE: The back of the palate not directly supported by bone, which separates the back of the mouth from the nasal cavity.

SOLEUS: A muscle of the calf of the leg having its *origin* in the *proximal* parts of the *tibia* and *fibula* and its *insertion* by the *achilles tendon* into the *calcaneus.* (See Fig. 5.15.)

SOMATIC SENSORY CORTEX: That part of the *cortex* in which are located *neurons* which receive and transmit the *somesthetic* input to the brain. (See Fig. 8.8.)

SOMESTHETIC: Refers to sense *receptors* of the skin (e.g., receptors of touch, temperature, and pain).

SOREX: A shrew, a *genus* of small animals of the *order Insectivora* found in both the Old and the New World. (See Fig. 3.1.)

SPECIALIZATION: In the context of *evolutionary* studies, a *character* evolved for a particular and limited function—the opposite of a generalized *character.*

SPECIATION (vb., SPECIATE): The division of a *biospecies* lineage over a period of time as a result of geographical isolation. Speciation results in the establishment of two *biospecies.*

SPECIES: A group of *populations* of organisms between which gene flow can occur, and which is reproductively isolated from other such groups. (See BIOSPECIES, CHRONOSPECIES.)

SPERM: (pl., SPERMATOZOA): The male sex cell or *gamete.*

SPERMATOGENESIS: The generation of *spermatozoa* that occurs in the *testes.*

SPHENOID BONE: A butterfly-shaped bone situated at the anterior part of the base of the skull, of which it forms the floor.

SPINAL CORD: The extension of the *central nervous system* within the vertebral column, and consisting of neurons and nerve tracts in the form of a hollow tube.

SPINOUS PROCESS: Connotes the bony projection on the *dorsal* side of most *vertebrae.*

STATUS: Social position, or position in relation to related objects.

STERNOCLAVICULAR JOINT: *Articulation* of *sternum* and *clavicle.* (See Fig. 6.2.)

STERNOMASTOID: Paired muscles which effect rotation of the head and have their origin in the *sternum* and *clavicle,* their insertion in the *mastoid process* and *occipital bone.* (See Fig. 4.16.)

STERNUM: The bone that *articulates* with the ribs on the *ventral* side of the *thorax.* (See Fig. 6.2.)

SUBFAMILY: A taxonomic rank. (See TAXONOMY.)

SUBORDER: A taxonomic rank. (SEE TAXONOMY.)

SULCUS: A fissure; one of the grooves or furrows on the surface of the brain, especially on the *cerebral hemispheres.*

SUPERFAMILY: A taxonomic rank. (See TAXONOMY.)

SUPERIOR: Higher, in relation to another structure. (Compare INFERIOR.)

SUPINATION: Position of the forelimb such that the *manus* lies in the same plane as the elbow joint so that the *radius* and *ulna* lie parallel. (Compare PRONATION.)

SUPRAORBITAL TORUS: A rounded thickening of the *frontal* bone along the upper edge of the *orbits,* evolved to carry some of the forces developed by a powerful masticatory apparatus.

SUPRASPINOUS LIGAMENTS: *Ligaments* connecting the *spinous processes* of the *vertebrae.*

SWEAT GLANDS: *Glands,* situated in the skin of *mammals,* that secrete the components of sweat. (See APOCRINE and ECCRINE.)

SYLVIAN SULCUS: A deep infolding of the *cortex* of the *cerebral hemispheres* which separates the *temporal lobe* from the *frontal* and *parietal lobes.* (See Figs. 8.9, 8.12.)

SYMBOL: An object, activity, or concept representing and standing as a substitute for something else.

SYMPHYSIS (adj., SYMPHYSEAL): A union of two bones (e.g., *pubic symphysis*).

SYNAPSE (pl., SYNAPSES): The point of contact of an *axon* and another *neuron* between which nerve impulses can pass; the connecting point between two nerve cells.

TALUS: The ankle bone (*astragalus*), which *articulates* with the *tibia* and *fibula* to form the ankle joint. (See Fig. 5.15.)

TARSUS (adj., TARSAL): The root of the *pes,* consisting of seven bones, including the *talus* and *calcaneus.* (See Fig. 5.15.)

TAXONOMY: The science of the classification of plants or animals which involves placing them in classes according to their relationships and ordering these classes in hierarchies. The conventional ranks used in this book are shown in Table 1.1 and 3.1.

TEMPORAL BONE: A bone found on each side of the skull between the *occipital, parietal,* and *frontal* bones. (See Fig. 8.1.)

TEMPORAL LINE: A slight depression, visible on the skull, which delineates the edge of the area of *origin* of the *temporalis* muscle. (See Fig. 8.1.)

TEMPORAL LOBE: A lobe of the *cerebral hemispheres* lying low down on each side of the brain. (See Fig. 8.8.)

TEMPORAL MUSCLE: See TEMPORALIS.

TEMPORALIS: The largest of the muscles that close the jaws; it has its *origin* on the sides and roof of the skull and its *insertion* in the *coronoid process* of the *mandible.* (See Fig. 7.2 and 7.3.)

TERRESTRIAL: Refers to the earth; living on the ground.

TERTIARY: A geological *era* believed to have lasted from about 63 million to 2 million years ago characterized by the *radiation* of the *Mammalia.*

TESTIS (pl., TESTES): The *gonad* of male animals that produces *spermatozoa,* and, in *mammals,* the male sex *hormones.*

THALAMUS: A large egg-shaped mass of tissue within the brain that serves as a relay for sensory stimuli to the *cortex.* (See Fig. 8.12.)

THORAX (adj., THORACIC): The chest; the upper part of the trunk between neck and abdomen containing a basket-like structure of ribs and *sternum.* (See Fig. 5.4.)

TIBIA (pl., TIBIAE): A bone of the calf that carries all the weight transmitted down the hind legs in higher *primates.* (See Fig. 5.15.)

TOOL: An *artifact* that may be considered as an extension of the manipulative *organs.*

TRACHEA: The windpipe or breathing tube running from the *epiglottis,* through the neck, to the point where it divides into two branches that lead to the lungs. (See Fig. 10.6.)

TRAPEZIUS: A muscle of the neck and shoulder with its *origin* in the *nuchal* area of the skull, the *cervical vertebrae* and *thoracic vertebrae,* and its *insertion* in the *scapula.* (See Fig. 6.3.)

TRICEPS BRACHII: A muscle that extends the forearm; it has its *origin* in the *scapula* and head of the *humerus* and its *insertion* on the *ulna* at the elbow. (See Fig. 6.10.)

TROCHANTER: Bony protuberance at the *proximal* end of the *femur.* (See GREATER and LESSER TROCHANTER.)

TROCHLEA: The grooved and rounded surface of the lower end of the *humerus* that *articulates* as a hinge joint with the *ulna.* (See Fig. 6.8.)

TURBINAL BONES: The fine bones of the nose that carry the *olfactory mucoperiosteum.*

ULNA (pl., ULNAE): The *medial* and larger of the two bones of the forearm. (See Fig. 6.10.)

UNGULATES: A group of *mammals* including two *orders* of plains-living *species,* the *Perissodactyla* and *Artiodactyla.*

URETHRA: Tube leading from the urinary bladder of mammals to the exterior; in female humans it opens within the vulva, in males by the *penis.*

UTERUS: The womb, a hollow muscular organ evolved from the *oviduct* in which the fertilized egg develops into a *fetus.* (See Fig. 2.4.)

UTRICLE: A *membranous* fluid-filled sac lying in the *inner ear,* which functions with the *sacculus* to detect the direction of gravity and movement. (See Fig. 8.7.)

VAGINA: The lowest part of the *oviduct* in *mammals* which joins the *uterus* to the exterior. Evolved to receive the *penis* in *copulation.*

VARIATION: Naturally occurring differences between individuals of a single *species* that are due to differences in *genotype* and *environment.*

VASCULAR: Containing blood vessels.

VENTRAL: Relating to the belly or abdomen; opposite of *dorsal.*

VERSATILITY: Applied here to the flexibility of *behavior* that results from evolved learning and perception.

VERTEBRA: Complex structure of bone, a number of which form the backbone or vertebral column in *vertebrates;* beside constituting the main structural components of the body, the vertebrae also protect the *spinal cord.*

VERTEBRATES: The most important *subphylum* (major group) of the animal kingdom; containing all animals with backbones, including fish, reptiles, birds, and *mammals.*

VESICULAR: Sack-like; a small cellular structure like a vessel or sack, often containing fluid.

VESTIGIAL CHARACTER: A rudimentary structure in an animal corresponding to a fully formed structure in an earlier or related form, sometimes assumed to have lost its function.

VIBRISSA (pl., VIBRISSAE): A long hair, such as a whisker, with sensory nerve endings at its base, evolved as a highly sensitive detector of mechanical stimulus.

VISUAL CORTEX: That part of the *cerebral cortex* involved in the reception and analysis of nerve impulses from the eyes.

VIVIPARITY (adj., VIVIPAROUS): The ability to give birth to living young, as contrasted with laying eggs.

VOLAR PADS: The hairless pads with special skin found on the friction surfaces of the *manus* and *pes* in *primates* as well as on the tails of some New World monkeys.

WERNICKE'S AREA: Part of the surface of the *cerebral cortex,* on the upper surface of the *temporal lobe* and adjoining the angular gyrus (See Fig. 10.8.) It plays an important part in the symbolization involved in human speech and is found only in man.

WORD MEMORY: That part of the memory which records words and the motor patterns for speech and writing.

ZINJANTHROPUS: A *genus* of the family *Hominidae,* not now believed to be distinct from the *genus Australopithecus.* The name was given to a skull discovered in Tanzania in 1959.

ZYGOMATIC ARCH: An arch of bone from the cheek to the ear region, called the cheekbone, carrying the *origin* of the *masseter* muscle. (See Fig. 7.2.)

Index

a

j

k

l

n

o

p

HUMAN EVOLUTION, SECOND EDITION
BY BERNARD G. CAMPBELL

Publisher / Alexander J. Morin
Production Editor / Georganne E. Marsh
Production Manager / Mitzi Carole Trout

Designed by Aldine Staff
Composed by Typoservice Corporation,
Indianapolis, Indiana
Printed by Printing Headquarters, Inc.,
Arlington Heights, Illinois
Bound by Brock and Rankin, Chicago